Landmarks in Mapping:
50 Years of The Cartographic Journal

Edited by
KENNETH FIELD
and
ALEXANDER J. KENT

The British
Cartographic
Society

Promoting the
Art and Science
of Map Making

Landmarks in Mapping:
50 Years of The Cartographic Journal

Edited by
KENNETH FIELD
and
ALEXANDER J. KENT

LONDON AND NEW YORK

First published in paperback 2024

First published 2014 by Maney Publishing

First issued in hardback 2019

Published 2024 by Routledge
4 Park Square, Milton Park, Abingdon, Oxon OX14 4RN

and by Routledge
605 Third Avenue, New York, NY 10158

Routledge is an imprint of the Taylor & Francis Group, an informa business

Publisher's Note
The publisher has gone to great lengths to ensure the quality of this reprint but points out that some imperfections in the original copies may be apparent.

Statements in the volume reflect the views of the authors, and not necessarily those of the Society, editors or publisher.

ISBN: 978-1-909662-38-4 (hbk)
ISBN: 978-1-03-292025-2 (pbk)
ISBN: 978-1-351-19123-4 (ebk)

DOI: 10.4324/9781351191234

Typeset by The Charlesworth Group

CONTENTS

Introduction

KENNETH FIELD

Editor, The Cartographic Journal / Esri Inc

ALEXANDER J. KENT

Assistant Editor, The Cartographic Journal / Canterbury Christ Church University

Founded by the British Cartographic Society (BCS) and first published in June 1964, *The Cartographic Journal* (TCJ) was the first general-distribution journal of cartography to appear in English. Today, it is an established international journal of record and comment containing authoritative peer-reviewed articles on all aspects of cartography: the science and technology of presenting, communicating and analysing spatial relationships by means of maps and other geographical representations of the Earth's surface. This includes coverage of related and emerging technologies where appropriate, for example, remote sensing, geographical information systems (GIS), the Geoweb, global positioning systems and mobile mapping. This volume of classic papers and invited reflections is being specially produced to celebrate 50 years of the Journal's publication and the enormous contribution that scholars, cartographers and map-makers have made to the publication in earning its place as the leading international journal of cartography. By way of an introduction, we tell a little of the story of the Journal's first 50 years as we mark its anniversary and look forward to many successful publishing years to come.

The creation of the British Cartographic Society was first discussed at a Cartographic Symposium in Edinburgh in 1962 and the Society was formally established the following year in Leicester. The first BCS Council was elected in September 1963 and included many leading British cartographers and geographers of the time, notably, G. R. Crone, Brigadier R. A. Gardiner, I. A. G. Kinniburgh, D. H. Maling, and W. D. C. Williams. The Council selected an editorial committee, appointed J. S. Keates as Editor, and in November 1963 bestowed the title 'The Cartographic Journal'. The first issue was published the following June and carried a foreword by the new Society's Chairman, Brigadier D. E. O. Thackwell (1964), who announced the Journal's intention 'to give news of all general cartographic activities, to include reports of symposia, meetings, and of their discussions, to publish articles of cartographic interest, and to review maps, atlases and books on cartography'. The Journal continues to uphold these aims for members of the Society, as well as for its wider international readership and proclaims its interest as being that of 'The World of Mapping' — as reflected in the original cover illustration designed by Mary Spence (a former President of the Society and awarded an MBE for services to Cartography). Since 2007 (Volume 44, No. 1) on the front cover of the Journal, the BCS logo has appeared alongside that of the International Cartographic Association (ICA) to recognize its official affiliation as an ICA publication. The relationship with the ICA has seen the Journal publish special selections of leading refereed papers presented at the International Cartographic Conferences in 2011 (Paris) and 2013 (Dresden). A special report on the history of the Journal was published in ICA News to mark the fiftieth anniversary of the ICA and the forthcoming Volume Six of the History of Cartography Project (edited by Mark Monmonier) will include an entry on the contribution of the Journal to twentieth-century cartography.

The very first Issue of the Journal included an eclectic mix of papers exploring the experimental portrayal of relief (Carmichael, 1964), globe construction (Fullard, 1964), a review of cartography as a university discipline (Maling, 1964) and a discussion of cartographic activity in Great Britain during the period 1961–1964 (Cartography Sub-Committee of the British National Committee for Geography, 1964) — a national report, first presented to the Second General Assembly of the ICA that continues to be updated and presented to each General Assembly. *The Cartographic Journal* was published bi-annually from 1964 to 2003 when the National Report to the ICA became a third, themed, 'Special Issue' (Volume 40, No. 2). Special Issues have since reported proceedings of the International Geographical Congress in Glasgow (Volume 41, No.2), explored map curation (Volume 42, No. 3), discussed contemporary research at Ordnance Survey (Volume 44, No. 3), examined the re-emergence of use and user issues (Volume 45, No. 2), proposed geovisual analytics as a development of geovisualisation (Volume 45, No. 3), brought together niche advances in cinematic cartography (Volume 46, No. 1) and art and cartography (Volume 46, No. 4), as well as cartographies of fictional worlds (Volume 48, No. 4). The most recent Special Issue (Volume 49, No. 4) explored aspects of cognition, behaviour and representation. The 50-year celebrations of the British Cartographic Society were marked in the Journal with the publication of a unique Issue (Volume 50, No.2) that included 17 articles from key players in the cartographic world reflecting on what cartography means to them personally and professionally. The Issue is destined to become a landmark as we reflect on the state of cartography in general and on some of the profound changes that have taken place in academia, industry and popular mapping over the last 50 years.

At present, the Journal is published quarterly with at least one Special Issue per Volume. The final Issue of each year contains details of the awards presented by the BCS at their annual Symposium as well as the Society's annual report and financial statement as a Society record. Obituaries of notable cartographers, cartographic scholars or those who have given outstanding service to BCS or to the Journal are also reported (including that of J. B. Harley in 1991 within Volume 28, No. 2). Additionally, the Journal contains regular sections identifying recent cartographic literature, reviewing cartographic-related texts and reporting new map accessions to the library of the Royal Geographical Society, London, which acts as an authoritative listing of newly published international maps. Reflecting the high quality of its content, the Journal has also been characterized by its excellent standards of production (especially for maps and illustrations), and long before it became a full-colour publication in 1999. *The Cartographic Journal* was the first journal in cartography to be rated by the Institute for Scientific Information and the entire back catalogue is available digitally.

The role of the British Cartographic Society in the ongoing production of the Journal has been to appoint an Editor, who is often an academic from a British university, with a special interest and involvement in the subject. The Editor works with a group of five international regional editors who have a remit to commission papers in the areas of Africa, Europe, North America, the Pacific Rim, and East and South East Asia. The Editor is also supported by an Editorial Board and Assistant Editors who often hold specific responsibilities (e.g. for book reviews, recent literature and map cabinet sections). Since 1975 the BCS has also presented an annual award, sponsored by Lovell Johns[1], to the author of the most outstanding article published in *The Cartographic Journal* as judged by the Editorial Board. The Editor reports to the BCS Council and a BCS Publications Committee. The current Editor, Dr Kenneth Field, has to date overseen eight Volumes and 29 of the 117 Issues in the first 50 Volumes of the Journal and, along with Alexander Kent, is delighted to be charged with bringing together this celebratory book.

The diverse nature of cartography has been well represented throughout the Journal's history with articles on every aspect of the subject. Far more than providing a barometer of change in cartography, together they offer a fascinating insight into the influence of maps and mapping on society and the evolution and impact of technology, particularly with respect to digital media, dynamic mapping and the upsurge in GIS. In 1967, a paper was published in the Journal on the SYMAP Programme for Computer Mapping (Robertson, 1967), while some 40 years later the paper downloaded most often from the online archive was on the importance of maps in GIScience (Kraak, 2006). The Journal contains numerous papers that explore specific areas of work such as computational approaches to new map projections, alternative generalization algorithms and mapping for users with specific needs, as well as papers that explore customized content, international map products and the changing landscapes of commercial and academic cartography. The broad scope of work published remains a key feature of the Journal in both its refereed and non-refereed sections that encourage high-quality academic research and general interest papers.

Many eminent cartographers and cartographic scholars have chosen *The Cartographic Journal* as a vehicle for disseminating their work, establishing the publication's distinguished reputation and contributing to the development of many careers as well as to cartography as a discipline. Noting a few of these illustrates the breadth of the field, the diverse content of the Journal, its international scope and the key themes that have characterized its publication. A brief survey might include, for example, cartographic education (Maling, 1964; Keates, 1974; MacEachren, 1986), relief portrayal (Carmichael, 1964; Jenny and Hurni, 2006), map curation (Wallis, 1965; Bond, 1991), land-use mapping (Coleman, 1965; Balchin, 1981), photogrammetry and remote sensing (Merriam, 1965; Fuller *et al.*, 2002), the history of cartography (Woodward, 1966; 1971; Harley, 1967; 1971; Robinson and Wallis, 1967; Robinson, 1971; Thrower, 1972; Stone and Gemmell, 1977; Wallis, 1981; Smith, 1986; Hodson, 2001, 2002; Fleet, 2007), extra-terrestrial cartography (Maling, 1965), automated cartography (Tobler, 1965; Rhind, 1971; Keates, 1974), colour in map design (Coleman, 1965; Cuff, 1973; Brewer, 1996; Harrower and Brewer, 2003), map projections (Tobler, 1966; Snyder, 1984; Loxton, 1985; Dent, 1987; Deakin, 1990; Van Wijk, 2008), perception and human factors in map design (Wood, 1968, 1972; Bartholomew

---

1. Named after Colonel Henry Johns who founded the company in 1965 as a traditional cartographic service, primarily for meeting the mapping needs of map publishers.

and Kinniburgh, 1973; Monmonier, 1977, 1979; Eastman, 1985), cartographic communication (Olson, 1975; Robinson and Petchenik, 1975), methods of visualizing data (Muller, 1976; Visvalingham, 1981; Slocum, 1984; MacEachren, 1991), topographic map design (Castner and Wheate, 1979; Forrest *et al.*, 1996; Kent and Vujakovic, 2009), art and aesthetics (Karssen, 1980; Krygier, 1995; Kent, 2005), computer-assisted cartography (Moellering, 1980; Drummond, 1984; Whitehead and Hershey, 1991), international cartography (Ormeling, 1982), automated generalization and expert systems (Li, 1988; Forrest, 1993; Hardy, 1998), the role of national mapping organizations (Rhind, 1991; Lawrence, 2002, 2004), cartographic design for census mapping (Dorling, 1993), lettering (Fairbairn, 1993), three-dimensional map design (Kraak, 1993), web cartography (Green, 1997; Midtbø and Nordvik, 2007), journalistic cartography (Vujakovic, 1999), animated cartography (Harrower, 2007), tactile mapping (Rowell and Morley, 2003), accessibility mapping (Beale *et al.*, 2006), community mapping (Perkins, 2007), cinematic cartography (Caquard and Taylor, 2009) and application of integrated media (Cartwright, 2009). Articles such as these have been complemented by especially insightful reflective or prophetic contributions, such as those by Sorrell (1979), Collinson (1981), Mumford (1981), Wilkinson and Fisher (1987), Hobbs (1989), (Visvalingham, 1989), Taylor (1994), Board (1993), Fisher (1998), Vujakovic (2002), Wood (2003), Field (2005), Kraak (2006), Cassettari (2007) and Lawrence (2007).

Clearly, then, this book has not been easy to compile for one very simple reason: there are simply too many 'classic' papers to include without it turning into something encyclopaedic. Our intention is to bring a flavour of the quality and breadth of the Journal into one volume that spans its history. As a reference, it highlights some of the very best work and, perhaps, allows readers to discover or re-discover a paper from the annals. As we constantly strive for new insights we cannot ignore the vast repository of material that has gone before as this has shaped cartography today and will continue to do so. A number of different models for including papers were proposed, with the Editorial Board sharing lists based on various rationales, such as the number of citations, the number of downloads, and personal preference. As with any list, when placed under real scrutiny, as many flaws emerged as good reasons for the decisions made, so we eventually settled on a democratic process. We compiled a new list comprising the papers that had won the Henry Johns Award, papers that began as slow burners and became critical to the development of the discipline (often which no-one could have foreseen), and those papers which members of the Editorial Board deemed worthy of inclusion. The task of whittling this long list down to a representative few must be akin to a long-lived band exploring their back catalogue and trying to create a two-hour set that will please everyone. A line needed to be drawn somewhere and, ultimately, papers only made the final cut if, in light of the above criteria, the reviewer made a solid justification for its inclusion.

We therefore include 16 classic papers from the archive that are reprinted here. Of course, some will be well known to the cartographic community, while others may not be as obvious. Each is reviewed by a member of the Editorial Board to place the work in some context and perhaps offer some wider thoughts. We believe the selection meets our aims for the book and provides a fitting celebration — and a lasting record — of the best of the first 50 years of *The Cartographic Journal*.

January 2014

REFERENCES

Balchin, W. G. V. (1981). 'Land use mapping in the 1980s', The Cartographic Journal, 18, pp. 44–45.

Bartholomew, J. C. and Kinniburgh, I. (1973). 'The factor of awareness', The Cartographic Journal, 10, pp. 59–62.

Beale, L., Field, K., Briggs, D., Picton, P. and Matthews, H. (2006). 'Mapping for wheelchair users: route navigation in urban spaces', The Cartographic Journal, 43, pp. 68–81.

Board, C. (1993). 'Neglected aspects of map design', The Cartographic Journal, 30, pp. 119–122.

Bond, B. (1991). 'Map and chart collections in crisis: change or decay?', The Cartographic Journal, 28, pp. 217–220.

Brewer, C. (1996). 'Guidelines for selecting colours for diverging schemes on maps', The Cartographic Journal, 33, pp. 79–86.

Caquard, S. and Taylor, D. R. F. (2009). 'What is cinematic cartography?', The Cartographic Journal, 46, pp. 5–8.

Carmichael, L. D. (1964). 'Experiments in relief portrayal', The Cartographic Journal, 1, pp. 11–17.

Cartography Sub-Committee of the British National Committee for Geography, (1964). 'Cartographic activity in Great Britain 1961–64', The Cartographic Journal, 1, pp. 10

Cartwright, W. (2009). 'Applying the theatre metaphor to integrated media for depicting geography', The Cartographic Journal, 46, pp. 24–35.

Cassettari, S. (2007). 'More mapping, less cartography: tackling the challenge', The Cartographic Journal, 44, pp. 6–12.

Castner, H. W. and Wheate, R. (1979). 'Re-assessing the role played by shaded relief in topographic scale maps', The Cartographic Journal, 16, pp. 77–85.

Coleman, A. (1965). 'Some technical and economic limitations of cartographic colour representation on land use maps', The Cartographic Journal, 2, pp. 90–94.

Collinson, A. (1981). 'Is cartography in the doldrums? — a personal view', The Cartographic Journal, 18, pp. 58–59.

Cuff, D. (1973). 'Colour on temperature maps', The Cartographic Journal, 10, pp. 17–21.

Deakin, R. E. (1990). 'The "Tilted Camera" perspective projection of the Earth', The Cartographic Journal, 27, pp. 7–14.

Dent, B. (1987). 'Continental shapes on world projections: the design of a poly-centred oblique orthographic world projection', The Cartographic Journal, 24, pp. 117–124.

Dorling, D. (1993). 'Map design for census mapping', The Cartographic Journal, 30, pp. 167–183.

Drummond, J. (1984). 'Polygon handling at the experimental cartography unit', The Cartographic Journal, 21, pp. 3–12.

Eastman, J. R. (1985). 'Cognitive models and cartographic design research', The Cartographic Journal, 22, pp. 95–101.

Fairbairn, D. (1993). 'On the nature of cartographic text', The Cartographic Journal, 30, pp. 104–111.

Field, K. S. (2005). 'Maps still matter — don't they?', The Cartographic Journal, 42, pp. 81–82.

Fisher, P. F. (1998). 'Is GIS hidebound by the legacy of cartography?', The Cartographic Journal, 35, pp. 5–9.

Fisher, P., Dykes, J. and Wood, J. (1993). 'Map design and visualisation', The Cartographic Journal, 30, pp. 136–142.

Fleet, C. (2007). 'Lewis Petit and his plans of Scottish fortifications and towns, 1714–16', The Cartographic Journal, 44, pp. 329–341.

Forrest, D. (1993). 'Expert systems and cartographic design.' The Cartographic Journal, 30, pp. 143–148.

Forrest, D., Pearson, A. and Collier, P. (1996). 'The representation of topographic information on maps — a new series', The Cartographic Journal, 33, p. 57.

Fullard, H. (1964). 'The construction of globes', The Cartographic Journal, 1, pp. 22–23.

Fuller, R. M., Smith, G. M., Sanderson, J. M., Hill, R. A. and Thomson, A. G. (2002). 'The UK land cover map 2000: construction of a parcel-based vector map from satellite images', The Cartographic Journal, 39, pp. 15–25.

Green, D. R. (1997). 'Cartography and the Internet', The Cartographic Journal, 34, pp. 23–27.

Hardy, P. F. 1998. 'Map production from an active object database, using dynamic representation and automated generalisation', The Cartographic Journal, 35, pp. 181–189.

Harley, J. B. (1967). 'Uncultivated fields in the history of British cartography', The Cartographic Journal, 4, pp. 7–11.

Harley, J. B. (1971). 'Place-names on the early ordnance survey maps of England and Wales', The Cartographic Journal, 8, pp. 91–104.

Harrower, M. (2007). 'Unclassed animated choropleth maps', The Cartographic Journal, 44, pp. 313–320.

Harrower, M. and Brewer, C. (2003). 'ColorBrewer.org: an online tool for selecting colour schemes for maps', The Cartographic Journal, 40, pp. 27–37.

Hobbs, D. (1989). 'Reminiscences of mapping before the computer', The Cartographic Journal, 26, pp. 44–45.

Hodson, Y. (2001). 'MacLeod, MI4, and the directorate of military survey 1919–1943', The Cartographic Journal, 38, pp. 155–175.

Hodson, Y. (2002). 'Ordnance survey and the definitive map of public rights of way of England and Wales', The Cartographic Journal, 39, pp. 101–124.

Jenny, B. and Hurni, L. (2006). 'Swiss-style colour relief shading modulated by elevation and by exposure to illumination', The Cartographic Journal, 43, pp. 198–207.

Karssen, A. J. (1980). 'The artistic elements in map design', The Cartographic Journal, 17, pp. 124–127.

Keates, J. S. (1974). 'Automation and education in cartography', The Cartographic Journal, 11, pp. 53–55.

Kent, A. J. (2005). 'Aesthetics: a lost cause in cartographic theory', The Cartographic Journal, 42, pp. 182–188.

Kent, A. J. and Vujakovic, P. (2009). 'Stylistic diversity in european state 1:50 000 topographic maps', The Cartographic Journal, 46, pp. 179–213.

Kraak, M.-J. (1993). 'Three-dimensional map design', The Cartographic Journal, 30, pp. 188–194.

Kraak, M.-J. (2006). 'Why maps matter in GIScience', The Cartographic Journal, 43, pp. 82–89.

Krygier, J. B. (1995). 'Cartography as an art and a science?', The Cartographic Journal, 32, pp. 3–10.

Lawrence, V. (2002). 'Mapping out a digital future for ordnance survey', The Cartographic Journal, 39, pp. 77–80.

Lawrence, V. (2004). 'The role of national mapping organizations', The Cartographic Journal, 41, pp. 117–122.

Lawrence, V. (2007). 'Perspectives on change in geographic information', The Cartographic Journal, 44, pp. 195–201.

Li, Z. (1988). 'An algorithm for compressing digital contour data', The Cartographic Journal, 25, pp. 143–146.

Loxton, J. (1985). 'The Peters phenomenon', The Cartographic Journal, 22, pp. 95–101.

MacEachren, A. (1986). 'Map use and map making education: attention to sources of geographic information', The Cartographic Journal, 23, pp. 115–122.

MacEachren, A. (1991). 'The role of maps in spatial knowledge acquisition', The Cartographic Journal, 28, pp. 152–162.

Maling, D. H. (1964). 'Cartography as an university discipline', The Cartographic Journal, 1, pp. 33–41.

Maling, D. H. (1965). 'Suitable projections for maps of the visible surface of the moon', The Cartographic Journal, 2, pp. 95–99.

Midtbø, T. and Nordvik, T. (2007). 'Effects of animations in zooming and panning operations on web maps: a web-based experiment', The Cartographic Journal, 44, pp. 292–303.

Merriam, M. (1965). 'The conversion of aerial photography to symbolised maps', The Cartographic Journal, 2, pp. 9–14.

Moellering, H. (1980). 'Strategies of real-time cartography', The Cartographic Journal, 17, pp. 12–15.

Monmonier, M. (1977). 'Regression-based scaling to facilitate the cross-correlation of graduated circle maps', The Cartographic Journal, 14, pp. 89–98.

Monmonier, M. (1979). 'Modelling the effect of reproduction noise on continuous-tone area symbols', The Cartographic Journal, 16, pp. 86–96.

Muller, J. C. (1976). 'Objective and subjective comparison in choroplethic mapping', The Cartographic Journal, 13, pp. 156–166.

Mumford, I. (1981). 'Is the British Cartographic Society in the doldrums?', The Cartographic Journal, 18, pp. 128–129.

Olson, J. (1975). 'Experience and the improvement of cartographic communication', The Cartographic Journal, 12, pp. 94–108.

Ormeling, F. (1982). 'International cartography', The Cartographic Journal, 19, pp. 85.

Perkins, C. (2007). 'Community mapping', The Cartographic Journal, 44, pp. 127–137.

Rhind, D. W. (1971). 'Automated contouring-an empirical evaluation of some differing techniques', The Cartographic Journal, 8, pp. 145–158.

Rhind, D. W. (1991). 'The role of the ordnance survey of Great Britain', The Cartographic Journal, 28, pp. 188–199.

Robertson, J. C. (1967). 'The SYMAP programme for computer mapping', The Cartographic Journal, 4, pp. 108–113

Robinson, A. H. (1971). 'The genealogy of the isopleth', The Cartographic Journal, 8, pp. 49–53.

Robinson, A. H. and Petchenik, B. B. (1975). 'The map as a communication system', The Cartographic Journal, 12, pp. 7–15.

Robinson, A. H. and Wallis, H. (1967). 'Humboldt's map of isothermal lines: a milestone in thematic cartography', The Cartographic Journal, 4, pp. 119–123.

Rowell, J. and Morley, S. (2003). 'Tactile mapping', The Cartographic Journal, 40, pp. 219–220.

Slocum, T. (1984). 'A cluster analysis model for predicting visual clusters', The Cartographic Journal, 21, pp. 103–111.

Smith, D. (1986). 'Jansson versus Blaeu', The Cartographic Journal, 23, pp. 106–114.

Snyder, J. (1984). 'Minimum-error map projections bounded by polygons', The Cartographic Journal, 21, pp. 112–120.

Sorrell, P. (1979). 'Cartographic outlook? a postscript', The Cartographic Journal, 16, pp. 56–58.

Stone, J. C. and Gemmell, A. M. D. (1977). 'An experiment in the comparative analysis of distortion on historical maps', The Cartographic Journal, 14, pp. 7–11.

Taylor, D. R. F. (1994). 'Cartography for knowledge, action and development: retrospective and prospective', The Cartographic Journal, 31, pp. 52–55.

Thackwell, D. E. O. (1964). 'Foreword', The Cartographic Journal, 1, p. 2

Thrower, N. (1972). 'Cadastral survey and county atlases of the United States', The Cartographic Journal, 9, pp. 43–51.

Tobler, W. R. (1965). 'Automation in the preparation of thematic maps', The Cartographic Journal, 2, pp. 32–38

Tobler, W. R. (1966). 'Notes on two projections', The Cartographic Journal, 3, pp. 87–89

Van Wijk, J. (2008). 'Unfolding the earth: myriahedral projections', The Cartographic Journal, 45, pp. 32–42.

Visvalingham, M. (1981). 'The signed chi-score measure for the classification and mapping of polychotomous data', The Cartographic Journal, 18, pp. 32–43.

Visvalingham, M. (1989). 'Cartography, GIS and maps in perspective', The Cartographic Journal, 26, pp. 26–32.

Vujakovic, P. (1999). 'A new map is unrolling before us: cartography in news media representations of post-cold war Europe', The Cartographic Journal, 36, pp. 43–57.

Vujakovic, P. (2002). 'From north-south to west wing: why the "Peters Phenomenon" will simply not go away', The Cartographic Journal, 39, pp. 177–179

Wallis, H. (1965). 'Report on the library classification of books and maps', The Cartographic Journal, 2, pp. 14–15.

Wallis, H. (1981). 'The history of land use mapping', The Cartographic Journal, 18, pp. 45–48.

Whitehead, D. C. and Hershey, R. R. (1991). 'Desktop mapping on the apple macintosh', The Cartographic Journal, 27, pp. 113–118.

Wilkinson, G. G. and Fisher, P. F. (1987). 'Recent development and future trends in geo-information systems', The Cartographic Journal, 24, pp. 64–70.

Wood, M. (1968). 'Visual perception and map design', The Cartographic Journal, 5, pp. 54–64.

Wood, M. (1972). 'Human factors in cartographic communication', The Cartographic Journal, 9, pp. 123–132.

Wood, M. (2003). 'Some personal reflections on change ... the past and future of cartography', The Cartographic Journal, 40, pp. 111–115.

Woodward, D. (1966). 'A note on the history of scribing', The Cartographic Journal, 3, pp. 58.

Woodward, D. (1971). 'A centre for the history of cartography', The Cartographic Journal, 8, pp. 48.

Automation in the Preparation of Thematic Maps[*]

W. R. TOBLER

Originally published in *The Cartographic Journal* (1965) 2, pp. 32–38.

Automation is of particular importance in the construction of thematic maps, especially those derived from extensive statistical data. A staff member of the Department of Geography, University of Michigan, examines the fundamental problems involved, and shows how standard computer facilities are being used.

The types of maps to which we wish to refer can be characterized in part by the following properties, as given by MacKay;[1] (a) they are generally at a small scale; (b) they are commonly prepared from start to finish by one individual working on a research project and not by a large mapping agency; (c) often they are never printed or distributed at all as they are prepared solely to aid the researcher and exist only in manuscript form, like a scientist's notebook; (d) when reproduced they usually appear along with an explanatory text in books, atlases, and periodicals designed for both specialized and general readers; only a small proportion are available for direct distribution or sale as separate map sheets to the public. To these general criteria one can add that (e) the information shown is often obtained from physical, social or economic surveys which differ considerably in method and in instrumentation from land surveying and from each other, and (f) that the information is often not directly visible in the field or is even not directly detectable by any means whatever at the location involved, but is instead derived from observational data by a series of more or less complicated manipulations suggested by the current theoretical knowledge in the subject matter area concerned.

These six properties serve, pragmatically, to distinguish thematic cartography from, say, the production of a topographic map series. Automation, of course, is being investigated in relation to topographic mapping. The major efforts in this area seem to be directed toward the rapid preparation of topographic maps directly from aerial photographs. Reviews of this work have recently appeared in the literature, and will not be repeated here.

CHARACTERISTICS OF THEMATIC MAPS

The criteria given above suggest some of the problems involved in the automated preparation of thematic maps For example, the fact that such maps are at a small

---

[*] *Paper given at the Technical Symposium of the International Cartographic Association, Edinburgh, 1964.*

1. MacKay, J. R. (1962) 'Some Cartographic Problems in the Field of Special (Thematic) Maps', **Canadian Cartography**, 1, pp.42–47.

8

scale implies that they cover large areas, often the entire world. As a consequence, the information to be shown generally comes from a heterogeneity of agencies and countries. In particular, the conditions, assumptions, and reporting format vary from one region to another. This heterogeneity of input data frustrates attempts at automation. The fact that large areas are covered also complicates the problem of choosing a map projection since the choice is no longer restricted to two or three projections but may require selection from among twenty to thirty projections. Also, a small scale map usually requires the greatest amount of generalization; the problem of generalization may not differ in kind for small scale maps, but such maps seem to require more of it. In a like manner, the fact that thematic maps are often an individual product reduces the economies of scale advantageous to automation. That many such maps are most useful in manuscript form suggests that they need not be at all attractive; the barest level of legibility suffices. On the other hand, such maps are, with increasing frequency, published in books, atlases, and journals. This requires that they conform to contemporary standards. Complete automation here would imply production of the final drawing, in colour or as a set of colour separations. The types of manipulations applied to subject matter information before it is mapped also differ considerably from one field to another. The distinct procedures used to collect map-able information in different fields again complicate attempts at automation. Clearly, then, the diversity of types of maps which can be included under the heading of thematic cartography is somewhat inimical to automation.

Another approach is to look briefly at the general intellectual process involved in producing a map. This will be a greatly simplified outline of cartography but can be usefully examined for areas of potential automation.

AUTOMATION AND CARTOGRAPHIC PROCESSES

First, some phenomenon is recognized as being of importance. This is perhaps the most fundamental and complex step in the entire procedure. An attempt is then made to measure (that is, to classify, to enumerate, to rank, or to scale) this phenomenon and to associate it with a definite terrestrial location, or set of locations. Alternatively the metric and/or location is inferred or deduced from related observational information. The terrestrial locations are then related to positions on a piece of paper by a map projection. The next step, and a difficult one, is to assign symbols to the phenomenon. These symbols should bear an unambiguous correspondence to the phenomenon to be depicted. They should be sufficiently few in number so as to result in a comprehensible map, and they should be visually representative of the character of phenomenon involved (in the sense that a translation from the visual to the mental construct is possible, correct, and reversible). Lastly, the symbols are placed in their appropriate position on the paper.

How many of these steps have been automated? Direct *in situ* measurement is sometimes possible as evidenced by automatic weather instruments and remote sensing techniques. Navigational devices are available for the instantaneous and continuous determination of location, with fair accuracy. The deductive or inferential steps can be automated to the same extent that they can be presented as a series of logical operations. The calculation, but as yet not the choice, of map projections can be completely automated. The step involving the choice of a symbol, which is really a

9

collection of individual steps, is difficult. Certain elementary operations, such as selecting a particular symbol on the basis of unambiguous criteria from a pre-designated collection of symbols, is fairly simple. Generalization of symbols, as must accompany a reduction in scale, remains difficult. The process of placing the symbols on paper is easily automated, if all of the previous steps were well performed, and is quite automated today.

AUTOMATION AND DIFFERENT TYPES OF THEMATIC MAP

A third introductory point of view which can be employed to obtain a perspective on automation in relation to cartography is to contemplate some of the uses to which maps are put. The most humdrum but also most practical of these is the use of maps as graphic locational inventories or records. It is now common knowledge that computer tapes can store vast amounts of information. The ease (speed, flexibility, and so on) with which information can be extracted from an inventory will govern whether it should be stored on a map or on a computer tape. Maps turn out to be a fairly efficient method of storing information for direct human use. But there are situations in which it is more convenient to have the information on magnetic tape. Because of this there have been developed a number of devices which will extract information from maps and record it on punched cards or tape. There are many problems associated with computer-related storage of inventories of geographical information. Geographical specification and the great volume are two of the most obvious difficulties. Nevertheless, a number of cities in the United States are experimenting with metropolitan data 'banks'. It can be anticipated that increasing amounts of geographical information of an inventory nature will be available in punched card or magnetic tape form. Fortunately it is usually possible to obtain a map of this information rather quickly through the use of a computer plotter combination.

A second highly important and very common use of maps is as nomographs; Mercator's projection and the ubiquitous road map are both employed in this fashion. Here the questions asked are of a kind which can be answered by direct computation; where is the nearest filling station, what is the quickest route; how many acres are in residential land; what is the loxodromic direction between two ports; what is the slope; and so on. Graphic solutions as obtained from maps are often sufficiently precise for the purpose at hand and are usually less tedious than the mathematical solutions. This is directly comparable to the two alternative methods employed to find the intersection of two straight lines; in the analytical method one calculates the intersection and in the graphical method one draws the lines and records their intersection point. The choice of methods depends on one's training, on the computational facility available, and on the speed and accuracy with which the result is required. Computers seem likely to replace a large share of the nomographic functions of maps, but the simple (though less precise) graphic methods will always be useful. Providing the computer with the geographical information in a convenient form may be somewhat of a problem, but once this is done the results of the computations may be of use to the cartographer. For example, the United States Bureau of the Census now makes available information by latitude and longitude co-ordinates. From this one could, if one wished, calculate the spatial rate of change of the population density (the gradient) and allow the mechanical devices to draw a map of the results, without ever

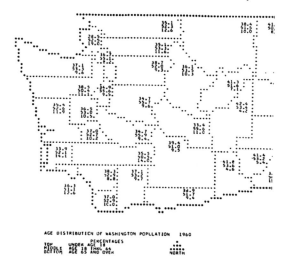

FIG. 1.   Simple computer printing of information in a geographical format. Part of outline map
printed along with census information. *Courtesy of B. Beyers, University of
Washington, Seattle.*

producing a map of the actual densities (manual procedures would require production
of a density map and then application of Wentworth's method to obtain the gradient
map[2]). It can be stated with considerable certainty that many novel, interesting, and
useful types of thematic maps can be expected in the future. This is one benefit brought
by the increased computational ability of digital computers.

Another class of problems for which maps are employed are such that they cannot
be solved by the mathematical methods available today. An extreme example is
pattern recognition. Automated photo interpretation, in spite of intensive efforts, is
an example of pattern recognition which has not met with much success. The geographic scientist employs pattern recognition when he uses the map as an hypothesis
generating device. A geographer contemplates a map and detects (or thinks he detects)
a pattern. He then formulates an explanation for this postulated regularity. The theoretical construct is subsequently tested by the rigorous, accepted and well-known
strategies of scientific decision making, in order to decide whether the explanation
can be accepted at some level of confidence. The test may involve the comparison of
a theoretical map with maps based on actual empirical observations. If the theoretical
construct is formulated in mathematical terms the production of 'expected' patterns
on a map by the computer is particularly simple. Hägerstrand's Monte Carlo simulation study of the migration of ideas is perhaps the best known example.

A final use of maps is for propaganda purposes; that is, to illustrate a point. This
may be for laymen or to give provisional credence to a purported statement. With
this type of map there is always a danger of misrepresentation since emphasis can be

2. Wentworth, C. K. (1930) 'Average Slopes of Land Surfaces', **Am. J. of Science**, series 5, 20, pp. 184–194.

varied greatly for anyone set of data by the details of symbol selection. Such maps are typically prepared in an artistic style and in several colours. Computers are not very good at this, although coloured maps can be produced by automatic devices.

THE USES OF AUTOMATION

Turning now to a brief discussion of specific equipment and techniques, a common element in all maps is that the symbols are positioned by relative location. Some people call this drafting, but it can also be regarded as a matter of arithmetic, since relative location can be defined as a co-ordinate count from some origin. Computers can do this counting at a rate of approximately 250,000 counts per second. All that is necessary is to construct a device which can position symbols according to the appropriate count. The simplest and most widely available such device is the typewriter. Positions across the page give the count in one direction and positions down the page give the count in the other direction. A simple dictation statement can be prepared, which, if followed rigorously, would enable one to produce a simple map on a typewriter. Nobody dictates maps to secretaries-it would be a rather slow business, but common computer-output printers yield maps which look as though they came from a typewriter, although the actual printer may operate quite differently. Printer speeds vary considerably; the one at the University of Michigan prints six hundred lines per minute: the resulting maps cost somewhat less than 50 cents per page. These maps are very crude and have several disadvantages, including an inability to print between the lines and a restricted choice of symbols. They are considerably more useful than their appearance suggests. Programming the computer for such maps can be along two lines. In the primitive case one can give the computer a set of instructions which essentially amount to a lengthy dictation statement, pre-specifying each printing location. This takes about two hours working with an outline map and an overlay of the computer printing positions. A more flexible procedure is to attach locational co-ordinates to the information and to programme the computer to prepare its own 'dictation statement'. This has the advantage that it is not tied to any particular map area, projection, or scale. Such map-printing computer programmes are now fairly common and are being used by city and highway planners, ecologists, geographers, geologists, meteorologists, and sociologists throughout the United States, particularly in research applications at the universities.

The next class of machines are the co-ordinate plotters, specifically designed for the highly accurate positioning of symbols. In application, these do not differ appreciably from the computer printers, but have several advantages. These include finer resolution, more appropriate symbols, larger size, the ability to produce coloured images, and a wider selection of materials upon which to position the symbols. The disadvantage is that they are more expensive and therefore less widely available. In one application, coloured dot maps of the distribution of population were prepared on such an instrument.

More interesting are the continuous-curve plotters, which plot symbols and draw lines. The basic principle is simple. One stores, in sequential order, a large number of closely spaced points, identified by co-ordinates. The machine draws straight lines connecting the points. The lines will look smooth if the points are sufficiently close together. Of course these instruments are employed for many purposes other than

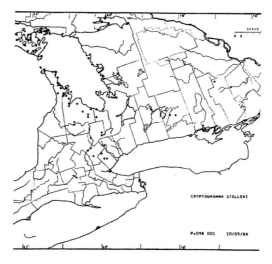

FIG. 2.   Part of a tabulator plotting of botanical observations from punched cards onto a preprinted continuous-form base map. *Courtesy of J. Soper, University of Toronto.*

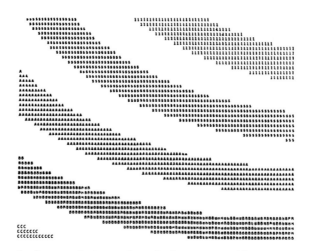

FIG. 3.   Detail of one technique of producing contour-like images on a printer.

making maps. Cartographic applications have included drawing various types of isarithms, map projections, census tract boundaries, street maps, and coastal outlines. In the United States, these graphic plotters have been employed by the Weather Bureau, the major oil companies, large regional planning offices, highway departments, some of the major cartographic agencies, and a few universities. An interesting use by the Weather Bureau has been the preparation of a motion picture showing

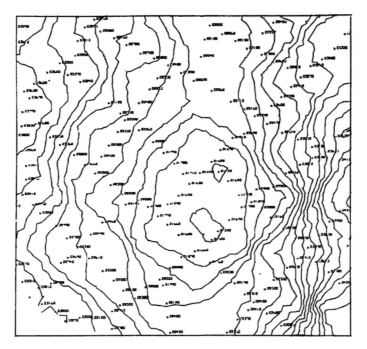

FIG. 4.   Contour map produced by a digital co-ordinate plotter.
Courtesy of California Computer Products, Inc.

Fig. 5.   Cathode ray tube map of the United States (reduced). *After Hershey.*

(interpolated) five minute changes in the position of barometric pressure lines for a
period of forty-eight hours. This requires 576 map drawings, all machine produced,
and gives a truly dynamic cartography, with the isolines moving about on the base
map.

14

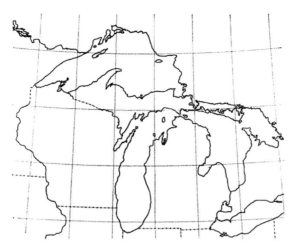

FIG. 6.   Great Lakes region drawn by a cathode ray tube from the same magnetic tape as
fig. 5. (reduced).

The cathode ray tube can also be used as a map-drawing device. This has the advantage of very high speed, which results in lower costs. The image is not quite as perfect as might be desired but certainly is legible. With such a system the information content, projection, and scale can be changed almost instantaneously. The Weather Bureau expects it will be able to produce weather-forecast maps at the rate of one per minute using such a system. Another current application is the superimposition of coastlines on to Tiros satellite pictures. The city of Chicago has employed a specially built cathode ray tube, the cartographatron, to prepare maps of over three million automobile trips. Certainly no human cartographer would want such a task!

PROBLEMS IN AUTOMATION

Several problems have been encountered in the attempts to automate thematic carto-graphy. The first of these is a requirement for geographical information identified by locational co-ordinates rather than by place name. All of the plotting equipment operates on co-ordinate principles so that if one does not have co-ordinates along with the information there is no possibility of automating map production. This point must be stressed strongly. There are many co-ordinate systems available, and almost any system will do. The most advantageous in the long run is latitude and longitude. A typical procedure is to convert these values to map projection co-ordinates and then to plotter co-ordinates. Having geographical information by co-ordinates also facili-tates analysis of the information, so that maps of the results of complicated analyses can be produced easily. From the point of view of cartographers, the statistical agen-cies of the world should report all of their geographical information by co-ordinates. The United States Bureau of the Census has begun to do this, but the international adoption of this practice is probably many years away.

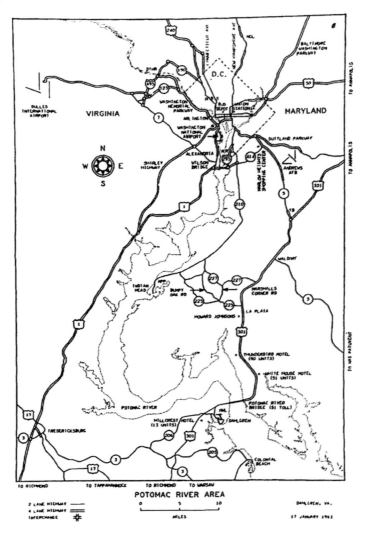

FIG. 7. Cathode ray tube map of Washington DC and vicinity. There is nothing on the map not drawn by machine. *After Hershey.*

Another difficult problem has been computer isolining. Typically one has scattered observational information and wishes to interpolate to obtain values for every point on the map (i.e., a continuous field). In cartographic circles this is known as logical contouring; to a mathematician it represents a problem in multi-dimensional interpolation. In one sense the problem has been solved; that is, there are now several computer programmes available which will draw isarithms when given observational information at only scattered locations. Of the several dozen such programmes now in existence, however, no two seem to use the same mathematical logic. One wonders which will give the more correct map. Almost everyone claims that their procedures are an improvement (in accuracy as well as speed) over the manual methods. Objective tests have not been published, but the majority of users seem satisfied. A short outline is given here for the production of an isarithmic map on a computer printer. The method is crude but has the advantage of extreme simplicity (there are more elegant methods).

To begin, one must reject the methods described in standard cartographic texts, since they are generally too ambiguous, and adopt a more rigorous procedure. This can take advantage of the computational facility now available. The computer printer has printing positions which are approximately 2.5 mm in width and slightly more in height. There are roughly 5,000 such positions on a page. The entire operation given below is repeated for each of these 5,000 positions, but this is still only a matter of seconds on a fast computer.

1. Compute the distance from the printing position to all observation points (which are assumed to be randomly distributed).
2. Out of all of these points find the nearest three which surround the position in question.
3. Fit a plane, $Z=AX+BY+C$, through the three points by solving the system of simultaneous linear equations necessary to fit a plane through points whose x, y, and z co-ordinates are known.
4. Calculate the estimated value at the desired printing position by inserting its co-ordinates into the equation.
5. If the value thus obtained differs by one or more contour intervals from the value of the printing position immediately previous, store a symbol, otherwise store a blank.
6. When the first row of (120) printing positions has been evaluated, print all of the stored symbols for that line, then begin on the next line. Stop when all lines have been examined.

The few steps presented above will yield a set of isarithms printed as a more or less continuous string of symbols. The method is not completely unambiguous as there may be ties in step two, but this can be resolved. It also gives a linear interpolation which is quite inaccurate at the edge of the map (outside the convex polygon which minimally surrounds the observation points). The extension to curvilinear interpolation can proceed by fitting a quadratic to the nearest six points, and so on. Variations and elaborations of this method are numerous in the extreme, some going as far as to represent the entire set of observational information with one equation and taking the sphericity of the earth into account. The complexity of these methods increases rapidly and the advantages of one system over another are rather obscure, even to mathematicians. The more attractive maps consist of smooth curves rather than strings of symbols. The procedures required are similar, with the additional step of rearranging the information to store the co-ordinate locations of each isarithm as an ordered sequence of points.

AUTOMATION AND GENERALIZATION

Another type of problem involves map generalisation. The question is not whether a computer can be used to generalize a map, but rather how it can be done. It may be that the level of generalization possible by computers is not economical. Certainly the general problem is difficult, particularly since (a) the process of generalization has not been formalized to a great extent, and (b) the methods of evaluating how well a computer has generalized a map are not at all clear. Generalization, and its evaluation, are quite subjective. Even as simple a step as not printing two symbols in the same position on a complicated and crowded map is difficult to programme without an extremely large computer memory.

As one example of computer generalization, consider the drawing of coastlines from one of the magnetic tapes of world outlines now available. Generalization of coastlines is, of course, only a special case of the more general problem, and a simple case at that. In this case we distinguish between a cognitive generalization and a statistical generalization. Perhaps this distinction will be useful in the more general case. In terms of coastlines, a cognitive generalization would, on the basis of topical information, designate and rank certain categories of points as being essential not to be eliminated-with deletion of other points being optional, whereas the simplest statistical generalization might randomly eliminate points while maintaining certain statistical properties of the parent population. The cognitive generalization obviously requires more carefully prepared information.

In terms of computer rules for generalization one could perhaps:

(a) Eliminate every nth point, where n is a function of the reduction in scale from the original compilation scale.
(b) Randomly eliminate $1/n$th of the points, where n is chosen as before.
(c) As above except that points of priority n (or greater) are retained.
(d) Eliminate points if, after conversion to scaled map projection co-ordinates, their distance apart is less than some function of the line width produced by the drawing instrument. This is the approach employed with considerable success by Hershey.[3] It takes advantage of rapid computational facilities and requires that the sometimes complicated map projection transformations be applied to each point before it is decided whether or not to include the point. Hershey's maps also raise the question of whether an extrapolation procedure might not be desirable if the points are too far apart.
(e) Plot all points.

A study of these questions is in progress, as are numerous other questions relating to generalization. Specific results can be expected in the next few years.

CONCLUSION

Finally we can ask what the result of the use of all of this complicated equipment will be. After considerable fumbling and misapplication it can be expected that some economic level of application will be achieved. We can expect more maps, cheaper maps, perhaps better maps, more useful maps, and more imaginative maps; possibly less artistic maps. Cartographic styles, like all styles, are subject to change. One of the

3. Hershey, A. V. (1963) 'The Plotting of Maps on a CRT Printer', NWL Report 1844, U.S. Navy, Dahlgren.

important implications of automated methods in cartography is that they disrupt the traditional attitudes and force· a re-examination of many of the conventional cartographic procedures. This necessary analysis of current cartographic practices should lead to many changes. The computer era is just beginning and one wonders what the maps of the twenty-second century will look like.

BIBLIOGRAPHY

Bedient, H. and Neilon, J. (1962). 'Automatic production of Meteorological Contour Charts', **Conference on Data Handling, Reduction, and Interpretation in Geophysics,** Yorktown Heights, processed, 8.

Bengtsson, B. E. and Nordbeck, S. (1964). 'Construction of Isarithms and Isarithmic Maps by Computers', **BIT,** 4.

Brending, D. O. (1964). 'Computers and Technical Documentation', **Graphic Science,** VI, 5 (May), pp. 23–31.

de Bronuner, M. S. (1963). 'Les Applications Cartographiques de L'Automation', **Bulletin,** Comite Fran~ais de Cartographie, 16 (Feb), pp. 134–139.

Bryson, R. A. and Kuhn, P. (1956). **Half-Hemispheric 500 MB Topography Description by Means of Orthogonal Polynomials,** Report No.4 (Madison: University of Wisconsin, Department of Meteorology), 15 pp.

Cain, J. and Neilon, J. (1963). 'Automatic Mapping of the Geomagnetic Field', **Journal of Geophysical Research,** 68, 16, pp. 4689–4696.

Chicago Area Transportation Study, **The Cartographatron,** Chicago (no date), 20 pp.

Creighton, R., Carroll, J., and Finney G. (1959) 'Data Processing for City Planning', **Journal, Am. Institute of Planners,** XXV, 2 (May), pp. 96–103.

Cude, W. C. (1962) 'Automation in Mapping', **Surveying and Mapping,** XXII, 3 (Sep.), pp. 413–436.

Dayhoff, M. O. (1963) 'A contour-map program for X-Ray crystallography', **Communications, Assn. for Computing Machinery,** VI, 10 (Oct.), pp. 620–622.

Dodge, H. F. (1964) 'Automatic Mapping System Design', **Surveying and Mapping,** XXX, 2 (Mar.), pp. 238–242.

Forgotson, J. M. (1963) 'How Computers Help Find Oil', **The Oil and Gas Journal,** March 18, pp. 100–109.

Garfinkel, D. (1962) 'Programmed Methods for Printer Graphical Output', **Communications,** Assn. for Computing Machinery, 5, 9 (Sep.), pp. 477–479.

Greenberg, G. L. (1962) 'Manifold Map Base Forms for Electronic Data Processing', paper presented at the 1962 St Louis conference of the Am. Congress on Surveying and Mapping.

Hägerstrand, T. (1955) 'Statistiska primaruppgifter, flygkartering och "Data Processing" maskiner: Ett kombination projekt', **Svensk Geografisk Arsbok.**

Harbaugh, J. W. (1962) 'Direct Printing of Computer Maps of Facies Data by Computer', **Bulletin, Am. Assn. of Petroleum Geologists,** 46, 2, p. 268.

Hauswitz, B., and Craig, R. (1952) 'Atmospheric Flow Patterns and their Representation by Spherical Surface Harmonics', **Geophysical Research Paper** No. 14 (Cambridge Research Center, MA), 80 pp.

Herndon, R. Jr. 1962) 'Modem USAF Cartography', **Surveying and Mapping,** XXII, 1 (Mar.), pp. 31–34.

Hershey, A. V. (1963) 'The Plotting of Maps on a CRT Printer', NWL Report 1844, U.S. Navy, Dahlgren.

Horwood, E., and Rogers, C. (1961) 'Research and Development of Electronic Mapping', **The Trend in Engineering,** 13, 4 (Oct.), pp. 9–12.

International Business Machines Corp. (1963) **Automatic Grid Contouring,** 1620 program library, 1620-MP-09X, May.

—— Film 'By the Numbers', 1962.

Jones, W. B., and Gallet, R. M. (1962) 'Methods for Applying Numerical Maps of Ionospheric Characteristics', **Journal of Research,** National Bureau of Standards, Vol. 66D, NO.6 (Dec.), pp. 649–662.

Kao, R. C. (1963) 'The Use of Computers in the Processing and Analysis of Geographical Information', **The Geographical Review,** LIII, 4 (Oct.), pp. 530–547.

Kogbetliantz, E. (1957) 'Interpretation of Magnetic and Gravimetric Surveys with the Aid of 704', paper presented at the Houston meeting of the Assn. for Computing Machinery.

Krumbein, W. C. (1959) 'Trend Surface Analysis of Contour-type maps with irregular control-point spacing', **Journal of Geophysical Research,** 64, 7 (Jul.), pp. 823–834.

Mach, R. E., and Gardner, T. L. (1962) 'Rectification of Satellite Photography by Digital Techniques', **IBM Journal of Research and Development**, 6, 3 (Jul.), pp. 290–305.

MacKay, J. R. (1962) 'Some Cartographic Problems in the Field of Special (Thematic) Maps', **Canadian Cartography**, 1, pp. 42–47.

Merriam, D., and Harbaugh, J. (1963) 'Computer Helps Map Oil Structures', **The Oil and Gas Journal**, 25 Nov.

deMeter, E. R. (1963) 'Latest Advances in Automatic Mapping', **Photogrammetric Engineering**, XXIX, 6 (Nov.), pp. 1027–1036.

Moser, F. (1963) **A Computer Oriented System in Stratigraphic Analysis**, Ann Arbor, Institute of Science and Technology.

Noma, A. A., and Misulia, M. G. (1959) 'Programming Topographic Maps for Automatic Terrain Model Construction', **Surveying and Mapping**, XIX, 3 (Sept.), pp. 355–366.

Nordbeck, S. (1964) 'Framstalling av kartor med hjalp av siffermaskiner', **Avhandlinger**, 40, Lund, Geografiska Institutionen, 99 pp.

Ostrow, S. M. (1962) **Handbook for CRPL Ionospheric Predictions based on Numerical Methods of Mapping**, National Bureau of Standards Handbook 90 (Washington, G.P.O., Dec.), 40 pp.

Pennsylvania Research Associates (1963) 'Investigation of the Compilation of Digital Maps', manuscript, (Philadelphia).

—— (1963) 'Statistics on Approximating Contours by Linear Segments', manuscript, (Philadelphia).

Perring, H., and Walters S. M., eds. (1962) **Atlas of British Flora**, Botanical Society of the British Isles, London.

Prey, A. (1922) 'Darstellung der Hohen-und Tiefenverhaltnisse der Erde durch eine Entwicklung nach Kugelfunktionen bis zur 16. Ordnung', **Abhandlungen der Koniglichen Gesellschaft der Wissenschaften zu Gottingen**, Mathematische Physikalische Klasse, Neue Folge, Ed. XI, 1 (Berlin: Weidmannsche Buchhandlung), pp. 1–42.

Sherman, J. C. (1961) 'New Horizons in Cartography', **International Yearbook of Cartography**, I, (London: G. Philip), pp. 13–19.

Simpson, S. M., Jr. (1954) 'Least Squares Polynomial Fitting to Gravitation Data and Density Plotting by Digital Computers', **Geophysics** XIX, pp. 250–257.

Slack, H. A., et al. (1963) 'Now-Map Making Made Accurate, Objective', **The Oil and Gas Journal**, 5 August, 7 pp.

Soper, J. H. (1964) 'Mapping the Distribution of Plants by Machine", **Canadian Journal of Botany**, 42, 8, pp. 1087–1100.

Thomas, E. N. (1960) **Maps of Residuals from Regression: Their Characteristics and Uses in Geographic Research**, Department of Geography Publication, 2 (Iowa City: State University of Iowa), 60 pp.

Thompson, P. D. (1961) **Numerical Weather Analysis and Prediction**, New York; MacMillan, pp. 140–146.

Tobler, W. R. (1959) 'Automation and Cartography', **The Geographical Review**, XLIX, 4 (Oct.), pp. 526–534.

—— (1963) 'Map Projection Research by Digital Computer', Paper presented, Assn. of Am. Geographers, Denver, Sep.

—— (1963) 'A Polynomial Representation of Michigan Population', **Papers and Proceedings**, Michigan Academy of Science, Arts, and Letters, 1964.

—— (1963) 'Geographical Ordering of Information', **The Canadian Geographer**, VII, 4 (1963), pp. 203–205.

Verzuh, F. (1958) **Contour Plot Program**, MIT Computing Center (SHARE program 506).

Wentworth, C. K. (1930) 'Average Slopes of Land Surfaces', **Am. J. of Science**, series 5, 20, pp. 184–194.

Wittke, H. (1958) 'Elektronische Schreib-und Kartiermaschinen', **Vermessungstechnische Rundschau**, June to September.

Wipperman, F. (1959) 'Kartenmassige Darstellung atmospharischer. Felder auf dem Schirm einer Kathodenstrahlrohre', **Tellus**, 11, 2 (May), pp. 251–256.

White, B. J. (1962) 'Studies of Perception', in **Computer Applications in the Behavioral Sciences**, H. Borko, ed., Englewood Cliffs, Prentice Hall, pp. 280–307.

Reflections on 'Automation in the Preparation of Thematic Mapping'

KENNETH FIELD

Editor, The Cartographic Journal / Esri Inc

In 1961 at the age of 31, Waldo Tobler received his PhD in Geography from the University of Washington at Seattle. Some four years later he published his paper entitled 'Automation in the preparation of thematic maps' in *The Cartographic Journal* after presenting the work at a Technical Symposium of the International Cartographic Association in Edinburgh in 1964. The paper was published in the first Issue of Volume 2 of the Journal and we can surmize that he might have been encouraged by the then fledgling Editorial Board and newly formed British Cartographic Society to submit his work for publication in what was then a brand new publication.

By the time of this paper, Tobler was already carving a very successful academic career. His publications to date had been broadly concerned with questions of automation, transformations and projections and he had himself derived a number of novel map projections and cartograms. Indeed, his PhD thesis explored a range of issues relating to transformations of geographic space (Tobler, 1961) which became one of the central themes of his remarkable career. Tobler's enthusiasm for such work was borne out of his participation in what is referred to as geography's quantitative revolution which characterized much of the discipline's theoretical and practical discourse in the late 1950s. He was, as befits most of us, a product of his time and went on to shape a career which saw over 150 published papers, with an emphasis on mathematical modelling and graphical interpretations. His use of computers as a central component of his research underpinned many of these works and the early paper we saw in *The Cartographic Journal* was an indicator of what was to come both from Tobler himself but also for geographical analysis and cartography more widely.

The article itself deals with an emerging aspect of cartography in the early 1960s — computerized design and production and specifically the creation of small-scale one-off maps. These are often used to illustrate a socio-economic dimension of the accompanying written work but less likely to be found in atlases, other compendia or topographic map series. Tobler explains that many of the steps involved in making a small-scale thematic map have either already been automated or can be, recognizing that what it is to be mapped is the first step, being the most important and from which all other steps emerge. Measuring the phenomena, giving the coordinates of the results and applying a map projection can in the most part be automated. The steps of generalizing and assigning symbols to the mapped phenomena remain 'difficult'. These observations remain interesting, as nearly 50 years on very little has changed. While speed of processing and more impressive algorithms have made certain steps in map-making much easier or quicker, there still remains an elusive 'make map' button that can properly and unequivocally handle the processes of generalization and symbolization. These are steps that require understanding on the part of the map maker and which even today algorithms struggle to execute convincingly in many situations since subjectivity remains a vital component of the cartographic toolbox. Tobler points to a number of important issues revealing his deep thinking about how best the emerging computer revolution might be practically applied to cartography. This was all taking place in parallel with developments at a company in Ottawa in the early 1960s where Roger Tomlinson was beginning to explore ways of using automated computer technologies to handle large geographical datasets. Tomlinson, of course, is credited as the 'father of GIS' for his early work but Tobler was part of this explosion too. While Tomlinson was forging his work in a practical sense, Tobler's career put in place much of the thinking that shaped practical endeavours. Indeed, even in the paper reproduced here, Tobler postulates how lines may be generalized by automated techniques — almost certainly the catalyst for more prominently applied work by the likes of Douglas and Peuker (1973) and others almost 10 years later.

In many respects we can trace several current and emerging fashions in cartography back to the early development of the computer as a tool to produce one-off products. Most of what we see on the Internet these days are single-use maps. They are generally produced to illustrate a single theme or a collection of related facts. The communication paradigm in cartography in the 1970s and 1980s grew from the idea that maps were vehicles for communication. In many ways this was a formalized result of the upsurge in single-use products that shared an idea. Despite the emergence of alternate theories in the way we think about cartography (MacEachren, 1995) there is still a very strong argument for the simple premise that a map is made to communicate an idea. This remains evident today as the Internet and the rise of so-called information graphics give people a mechanism to express their own ideas in graphical form.

The concluding paragraph in Tobler's paper is perhaps particularly apposite. In it, he ponders the utility of 'all of this complicated equipment'. He suggests that misapplication is inevitable and while we will see 'more maps, cheaper maps, perhaps better maps, more useful maps, and more imaginative maps; possibly less artistic maps' will also become evident. He was right. He still is right and the same could be said for much of modern cartography as we shift from one style in mapping to another. He considers automation as a disruptor. The same could be said of Google Maps and other transformational mapping technologies in recent years. They force a re-examination of convention and they lead to change. These are changes that technology, with its rapid pace, are still enforcing. Have we met the challenges and tamed technology? Possibly not. That, perhaps, is the ever-present curse of cartography; that technology moves so fast we need to stop every once in a while and look around. Tobler did, with clarity in this paper as he took stock of the remarkable changes that automation was bringing to cartography and some of the challenges that are to be faced. What he perhaps cannot have foreseen in 1965 was that many of his comments would still be relevant today and that we have yet to meet many of the challenges he notes.

Tobler's fascination with automation has continued throughout his career. Many of his later publications and presentations remain focused on dealing with how we might overcome some of the problems in developing satisfactory solutions to spatial and cartographic problems through the application of computer technologies. Tobler's most famous contribution to geography is his oft-quoted 'first law of geography' which states succinctly that 'Everything is related to everything else, but near things are more related to each other'. While formally the quote is supposed to have come out of work in 1970 during the production of a computer movie, in recent conversation Tobler himself, rather amusingly, claims that he could not actually ever recall where he mentioned this or, indeed, if he ever had at all (Tobler, 2013 pers. comm.). Whether this utterance is a geographically-related urban myth or whether he did in fact state it, you would be hard pressed to find anyone with a geographical or cartographic background that does not know of this phrase and what it means. Tobler has received numerous honours for his work and held a number of senior positions on various editorial and society boards. He published just once more in *The Cartographic Journal*, on the subject of map projections (Tobler, 1966), often preferring North American publications as a forum for disseminating his work. His contribution to cartographic thinking has been crucial in guiding us through the computer age. Much of what we do today stands on the shoulder's of his vast contribution to geography and cartography of which the paper reviewed here is but one small, albeit important, component.

REFERENCES

Douglas, D. and Peucker, T. (1973). 'Algorithms for the reduction of the number of points required to represent a digitized line or its caricature', **The Canadian Cartographer**, 10, pp. 112–122.
MacEachren, A. (1995). **How Maps Work, Representation, Visualization and Design**, The Guilford Press, New York.
Tobler. W. R. (1961). 'Map Transformations of Geographic Space', PhD Thesis, University of Washington, Seattle University Microfilms No. 61–4011.
Tobler, W. R. (1966). 'Notes on Two Projections: Loximuthal and Two Point Equidistant', **The Cartographic Journal**, 3, pp. 87–89.
Tobler, W. R. (2013). Personal communication.

Visual Perception and Map Design

M. WOOD

Originally published in *The Cartographic Journal* (1968) 5, pp. 54–64.

Visual perception is reviewed in relation to the search for principles in map design. Various cues of 'depth' are examined as aids in the specification of the qualities and dimensions of map symbols. Maps are analysed on the basis of 'receding planes', each plane containing a specific selection of graphic information. A design procedure is outlined in this content, and finally an example of a thematic map, the Residential Land Use of Glasgow-1965, is presented in detail.

Over recent years much has been written on the aesthetic principles and the technical aspects of designing maps. Since the Second World War, and certainly since the Esselte Conference on Applied Cartography in 1956,[1] there has been an increasing flow of information between map-producing countries. It is without doubt that cartographic technology has benefited from this exchange. The adoption of scribing on glass or plastic and the development of photoset lettering have contributed towards the clarity and crispness of the map image, and the development and refinement of tri-colour processing and printing has given the map designer a palette of hues, tints and shades, which is more than adequate for most of his needs. In the realms of more modest, but no less problematic monochrome cartography, pre-printed symbols and letters have taken much of the labour and uncertainty out of map preparation. However, as in some other fields, the concepts and processes of design have not kept pace with techniques and materials. In response to this imperfectly realised problem, some attempts have been made to investigate certain aspects of the use and function of maps in relation to design. These can be considered in two groups, consumer research and map evaluation.

Consumer Research

As in other fields this entails the collection of the reactions of a cross-section of potential users to new or established products, e.g. topographic maps or atlases. Results have varied, but perhaps the most interesting were received from school children (aged 8–10 years), who, in one study, rejected the brightly coloured maps of one atlas in favour of the subtler tints of another,[2] thus contradicting the oft-quoted assertion that younger children prefer strong colours. The uncontrolled nature of these investigations and the variety of suggestions presented by adults and children alike, has caused

1. **The Esselte Conference in Applied Cartography.** Stockholm, 1956.
2. J. S. Keates. 'The Perception of Colour in Cartography', a contribution by J. J. Klawe to the ensuing discussion. Page 27, **The Cartographic Symposium**, Edinburgh, 1962.

most map editors to resort once again to their own better judgement. Naturally mapping agencies and atlas publishing houses have their own approach to the subject of design. The appearance of contemporary editions of their maps may derive from the experience and judgment of directors or senior editors, may reflect the inertia of tradition and historical origins, or may have even more suspect and arbitrary roots. Of course, in many map-using situations there is no consumer choice since only one product is available. In few fields is there real competition, and where it exists, as in educational atlases, it is mainly a price war. Fashion, house style and the quest for a 'new image' play a large part in the considerations of design. David Pye's declaration that 'economy is the mother of most inventions, not necessity' certainly applies to commercial cartography.[3]

Map Evaluation

In various countries cartographers began to question and examine the maps they were making. Some university departments, with 'captive' classroom subjects, proceeded to make tentative visual tests of the effectiveness of such things as pre-printed area symbols, and especially the perception of proportional symbols for thematic maps.[4,5,6,7] Although valuable, these experiments have been few in number and often cover a statistically insignificant sample of subjects. In the U.S.A. and notably at the University of Wisconsin, where A. H. Robinson has laid special emphasis on cartography, a liaison with psychologists has been initiated, and a movement towards more refined psychological investigations has ensued. In the sphere of special purpose mapping, acknowledgement must be made to the work of the military departments in certain countries. Many and exhaustive tests have been carried out on personnel of all ranks to test their ability to read topographic maps and air charts. The U. S. Air Force, in particular, has devoted time to human factors' analysis, and the results will no doubt influence the future design and employment of maps in the cockpit environment.[8,9,10] Apart from this, research is still in the embryonic stage. A body of experimental evidence must first be accumulated. This may be original in nature or adapted from the myriad non-cartographic studies which exist in the literature of psychology.

The importance laid upon this aspect of design was underlined recently by the devotion of one session to 'Map and Colour' at the I.C.A. Technical Conference in Cartography at Amsterdam, 1967 (International Yearbook of Cartography VII 1967).[11] Many of the excellent contributions to this meeting will help to lay a foundation for future research.

3. David Pye. **The Nature of Design**, Studio Vista, London, 1964. Page 35.
4. R. L. Williams. 'Equal Appearing Intervals for Printed Screens', **Ann. Assoc. Amer. Geog.** Vol. 51, 1961.
5. R. L. Williams. **Statistical Symbols for Maps: Their Design and Relative Values.** Yale University, Connecticut, 1956.
6. Jenks and Knos. 'The Use of Shaded Patterns In Graded Series', **Ann. Assoc. Amer. Geog.** Vol. 51, 1961.
7. H. H. McCarty, N.E. Salisbury. **Visual comparison of isopleth maps as a means of determining correlations between spatially distributed phenomena.** Iowa State Univ. 1961.
8. S J. Skop. 'Modern Warfare and the Map', **Proceedings of the Second International Cartographic Conference**, Chicago 1958. Institut far Angewandt Geodiisie, Frankfurt, a.M. 1959.
9. R. E. Harrison. 'Art and Commonsense in Cartography', **Surveying and Mapping**, Vol. 19 (1) 1959.
10. Human Factors Research Inc. Santa Barbara Research Park, Goleta, California.
11. **International Yearbook of Cartography VII**, George Philip & Son Ltd, London, 1967.

It would be wrong to assume that certain principles of map design have not been sought. Early studies by Robinson[12] and Saunders[13] for instance, attempted to lay down some theoretical and empirical rules. Robinson's book, *The Look of Maps* is a worthy landmark in this difficult quest for good design. These principles, however, were sometimes negative in approach: e.g. 'do not juxtapose large areas of bright colour', 'broad line patterns disturb and attract, creating imbalance'; or very general, e.g. 'a map should have complexity and interest, but also possess unity.' If based on the experience of good cartographers, such rules simply stated can be valid and valuable to the map designer. Experience in teaching cartography and designing thematic maps brings the problem of design and comprehension into sharp focus. The lack of clear principles in map design at all levels seems to be responsible for many misapprehensions.

The Central Problems of Design

As the present avenue of research into this subject lies in visual perception, it is intended to examine some of the fundamental concepts of this science and then formulate and argue a theory which might, in certain map situations, provide a flexible guide for the inexperienced map maker. Before doing so it should be realised that the failure of many maps cannot always be totally attributed to the designer. The fault may lie in the method of communicating his specifications, or perhaps in his incomplete knowledge of the printing process. Maps for reproduction must be designed accordingly. At the outset one must acknowledge the obvious complexity of the whole concept of design, be it technical or psychological. Referring to non-cartographic problems, David Pye asserts that one determines the use or function of a thing, to be designed, in an arbitrary way, in spite of available information on its intended uses. 'Design of every kind is always a matter of trial and error'.[14] How much more does this apply to maps where, often, the specific purpose or intended uses cannot be clearly anticipated. A. H. Robinson has rightly said, 'there will always be the more sensitive cartographic designer who can erect a design structure that is more appropriate than that of the other designers',[15] but, as he states elsewhere, 'It is not necessary to be an artist to learn design effectively. The basic elements of good design lend themselves to systematic analysis, and their principles can be learned'.[16] However, it is often the lack of clear practical logic behind many such principles and the apparent range of alternatives to the 'ideal' answer, which frustrates the student. Can suitable logical bases be derived? As with any language, cartography can be learned empirically, but only after years of imitating good and bad practice, with no classical textbook to guide. This is a serious handicap to the cartographer. A knowledge of grammar, although not necessary in the learning of a language, is essential to good expression and effective communication. Something comparable to the flexible rules

12. A. H. Robinson. **The Look of Maps. An Examination of Cartographic Design**, University of Wisconsin Press, Madison, Wisconsin, 1952.

13. B. G. R. Saunders. 'Design and Emphasis on Special Purpose Maps', **Cartography**, Vol. 2. 1957–58.

14. David Pye. op. cit. p. 26.

15. A. H. Robinson. 'Psychological aspects of colour in cartography', **International Yearbook of Cartography VII**, George Philip & Son Ltd, London, 1967, p. 57.

16. A. H. Robinson. **Elements of Cartography**, John Wiley & Sons, Inc., London, 1960, p. 222.

of grammar and style should be possible, if not to provide the perfect formula, at least to point the way to perfection.

The Map as a Functional System

H. G. Alexander in his book *Language and Thinking*[17] describes a system of communication which is applicable to maps. It is best represented by way of his diagram, as follows:

$$\text{The Communicator} \qquad\qquad \text{The Communicatee}$$

$$E_1 - CC_1 \;\big\langle\!\!\begin{array}{c} S_1 \longrightarrow S_2 \\ R_1 \text{-----} R_2 \end{array}\!\!\big\rangle\; CC_2 - E_2$$

$$\text{Encoding} \qquad\qquad\qquad \text{Decoding}$$

Explanation: (note that capital letters with inferior numerals$_2$ refer to the communicatee).

E. Background, experience and attitude of communicator (map author).

CC. Concept of the Communicator.

S. Symbol used by the Communicator (this is where the designer may join the author to help him express his ideas graphically).

R. Referent as perceived or imagined by the Communicator.

Failure may occur at any or every stage in this system, giving rise to incorrect communication to the map user.

Many maps are unsuccessful in transmitting all the data they contain, either because they are too crowded, or because the visual presentation of points, lines, areas and names does not allow for separation or disentanglement of the superimposed patterns. This facility of separation, and the means whereby it can be induced will be examined at a later stage. As was mentioned above, a 'grammar' of design is lacking in cartography. In the ensuing review certain aspects of visual perception will be mentioned which may provide certain guides to the cartographer. The satisfactory manipulation of line and tint will always partly serve to please the eye of the beholder, but a more important function by far is to render, in the maps, an effective and accessible source of information.

AN OUTLINE OF VISUAL PERCEPTION

Vision is just one of several interacting senses which gives the observer an awareness of the external world. Visual perception includes every stage of the process of seeing, from original stimulation to final comprehension.

An object in the external world is observed. The light, received as electro-magnetic waves, passes through the eye lens and impinges on the retinal cells, the 'cones' and 'rods' which line the back of the eye. These cells are effectively sensitive only to some of this radiation, referred to as the visible spectrum. The resulting image is translated into electrical impulses and passed along the optic nerve to the brain, where it may or may not be received or understood. 'The mediating processes that determine

17. H. G. Alexander. **Language and Thinking**. D. Van Nostrand Co. Inc., London, 1967, p. 14 if.

the final response must first be set up ... and this preliminary adjustment is perception.'[18] However, there can be looking without seeing. It has been proved, using special viewpoint detection apparatus, that an observer may look directly at the object which he is seeking without recognising it.[19]

The action of seeing has three stages:

1. Physical: the stimulus. The wave lengths of light as received from the object.
2. Physiological: the reaction of the eye nerves to the stimulus. The retinal image may not be like the object owing to the particular qualities of the individual eye or the inability of the mosaic of retinal cells to resolve detail.
3. Psychological: The response. The measure of the ability of the brain to receive and interpret this retinal pattern.

The first stage can be measured, the second is a field of major research activity, but the third, although undergoing investigation, is still largely a mystery. Meaning must be given to the objects, and this requires knowledge, which can be gained through education or direct experience.

One reacts to light from all directions, and over a period of time. Light has the qualities of hue (red, green, etc.); saturation (or chroma), the degree of colourfulness of a given hue or the extent to which it departs from an achromatic scale; and brightness (or value), which is the apparent amount of light coming from an object and may be compared with lightness which is the continuum running from white surface to black surface. Colour in the general sense can encompass all tonal variations, including grey. Reduction of saturation produces a 'tint' of the hue and reduction of brightness gives a 'shade.' These dimensions of colour have been studied and organized into model form by certain colour scientists. In 1905 Munsell produced a system based on the characteristics mentioned above, namely hue, chroma (saturation) and value (brightness). This system has ten hues and ten steps of each hue, and each hue, tint or shade is given a unique number, which fixes it relative to the so-called 'colour solid'. Ostwald's method is based on twenty-four hues, black and white, but has been overtaken by Munsell in the western world. The layer colours in the International Map of the World, for instance, are standardized in Munsell notation.

Although it is known that 125 just noticeable differences (JNDs) of hue and in certain hues 200 JNDs of saturation can be detected,[20] most people can differentiate about fifteen hues across the visual spectrum. This can be extended with training.[21] The number of JNDs of brightness and saturation depends on the hue (e.g. yellow has fewer than blue) its shape and environment. The common map environment is white paper, but this need not always be the case. These steps are also limited to about ten and often only to only six or seven in a complex environment of juxtaposed areas. Colour exists in objects and areas, which in tum, are identified by shape, size, location, texture and number. All these factors affect the perception of colour and similarly the relative hue, brightness and saturation of the colour of objects may affect the perception of the objects.

18. D. O. Hebb. **Psychology**. W. B. Saunders Co., London, 1958. p. 179.
19. W. F. Floyd. 'On the Line of Sight', **Design Mag**. No 124 April 1959.
20. R. M. Evans. **An Introduction to Colour**, John Wiley & Sons, Inc., New York, 1951. Chap. VIII.
21. R. M. Hanes and M. V. Rhoades. 'Colour Identification as a function of extended practice', **J. Opt. Soc. Am.** Vol. 41, No. 11, 1959.

Apart from the limitations of colour vision, there also exist lower limits of visual acuity. If light from two small dots falls on two adjacent receptors in the retinal mosaic at the back of the eye, no gap will be seen to separate the dots. For perception of a space, one intermediate receptor must remain unstimulated. In general a separation between two points must subtend an angle of at least one second in the eye. Objects in so called 'hard colours' such as red can easily be seen, but the contours of pale tints and yellowish hues are difficult to define at small sizes, tending to merge into an off-white background.

There are two distinct regions of study within this science-colour and shape,-both of which are affected by size.

Some Aspects of Colour Perception

A group of predictable results has been gained from experiments involving the discrimination of meaningless isolated patches of colour, but when colours have shape and these shapes are in complex association, there occur some variations in normal colour response which are most significant to map designers. These variations reduce the range of observable steps of discrimination in every way. The main factor is colour contrast, which produces erroneous perceptions of brightness, saturation and hue. This gives rise to induction of the complementary hue and perhaps enhancement of that hue along its edge.

This can best be expressed by some examples:

(a) The effect of contrast on brightness alone: a grey ring appears light if seen on a dark background, or dark if seen against a light background.

(b) The effect of the size of an area on the hue: as an area becomes larger, brightness and saturation increase. A small area can render similar hues, like blue and green, impossible to distinguish.

(c) The effect of contrast on saturation: two identical yellow rings look different if surrounded by a pale buff on the one hand and a grey ring on the other.

(d) The effect of contrast on hues: the apparent hue of an object is affected by the hue of its surroundings.

(e) In some complex patterns the changes that occur among the colours evolved by small spacially juxtaposed areas, make them more alike, rather than more contrasted. For example, small areas interwoven with fine black lines usually appear darker than the same areas interwoven with fine white lines. This could be carried further; for instance, an identical grey area may appear yellow if interlaced with yellow, or bluish if interlaced with blue lines.

Other factors influence perceived colour: spectral relationships in the visual field, the length of time a colour is observed, the adaptation of the eye to certain qualities of light, psychological factors like the figure/ground relationship, and even the age of the observer may sometimes have to be considered. Manifestations of such contrast phenomena observed by makers and users of maps have been noted and included in the aforesaid empirical rules. What is lacking here is not an exhaustive list of visually incorrect combinations, but a theme or analogy which can be employed in a logical and methodical fashion by the designer. It is hoped to show that some aspects of depth and distance perception may provide such a guide in certain circumstances.

The Perception of Form

The eye sees colour, but colour cannot exist without form and form can only be observed if it contrasts with its background. This contrast may be a combination of brightness and hue, or brightness alone as in black and white photography. There are many map situations where a limited palette may be very effective, using contrast alone for quite simple effects. The edge, where the pale tint meets the strong hue of the object is called the 'contour'. The two shapes which exist on either side of this line cannot be viewed simultaneously. One is the 'figure' i.e., it stands out from the background or 'ground'. Hence the term 'figure/ground effect'. The extent to which one area or shape appears to stand out in front of another is mainly a function of the contrast. The greater the contrast, the more pronounced the effect, and the greater the recession of the ground. The figure may have texture which attracts the eye, while the background appears filmy. If brightness contrast is insufficient, and this could be intentional, lines may have to be introduced to enhance hues, and make objects stand out. Good manipulation of the figure/ground phenomenon aids the clarity of perception and simplifies complex patterns of lines and shapes, such as maps. It minimizes the effort required to attend to the important shapes and overall distribution and facilitates examination of the details.

Specific Shape and Meaning

In our field of view there exists an array of forms where clarity and degree of prominence is a result of contrast, but how do we see a particular form as figure? There are two main theories for this, which have been examined by psychologists. The earlier 'structuralists' proposed that all objects were merely the sum of their smallest parts which were seen individually. *Gestalt* psychologists, however, believed that people perceived whole objects and organizations, and that forms were regions of meaning bounded by contours. When an object is being scanned its form can be built up in the mind although the retinal image is constantly changing.* This becomes more important with increasing complexity, and many maps are among the most complex images we use.

1. The Attributive Factors of Form

The first stage of the investigation concerns the size, brightness and association of meaningless shapes. From cumulative experimental results Gestaltists have suggested some laws of organization to help predict which of two possible shapes will be seen.

(a) *Area.* If a closed region is small it is more likely to be seen as figure.
(b) *Proximity.* Dots or objects which are close together tend to be seen as groups. The theory of the half-tone printing image illustrates this principle.
(c) *Closedness.* Areas with closed contours tend to be seen as figures more than those with open contours.
  These three principles can be tested, but those which follow are less easily proved.

---

* *'Gestalt' means 'whole', 'configuration' or 'form', and the psychologists associated with this revolutionary theory (launched in the 1920's) were Wertheimer, Koffka and Kohler.*

(d) *Symmetry*. The more symmetrical a closed region the more it tends to be seen as figure.

(e) *Good Continuation*. That arrangement of figure and ground tends to be seen which will make the fewest changes or interruptions in straight or smoothly curving lines or contours.

More recent research has tried to summarize all these parts under the law of 'Simplicity and Minimum Principle' which states that we see an object by the simplest possible arrangements of line. Also we classify unknown objects by relating them to known objects. We do this by the shortest and most obvious route. This is termed 'Series and Form Generalization' in psychology.

2. *Cognitive Aspects of Shape*

Cognition lends further clues to anchor the figure/ground relationship. This is the de-coding stage in the communication process, and is dependent on the attitude and experience of the receiver. Normal learning enables him to identify everyday objects, but specific training is generally required to give an understanding of many symbols.

A map differs from a photograph in that the latter is meant to look like the object, while a symbol expresses an idea by standing for it and is thus the medium for the expression of an idea. Symbols are common groupings or categories, representations are not. The concept behind the symbols has referents (associated ideas) which help to complete the intended information. Referring back to Alexander's diagram, failure of the communicatee to provide the correct referents will reduce the efficiency of the system.

Map symbols are combinations of artificial devices,[22] naturally associated with the referents by way of

1. Imitation, e.g. cliff drawing.
2. Contiguity, e.g. oil derricks for oilfields.
3. Synesthesia, association of two senses e.g. red/orange for hot deserts.

Many of these connections may be abstractions in that one aspect or quality is taken for the whole. They may also be 'primary', e.g. a detailed plan outline of a town, or 'substitutes', e.g. a small circle replacing the city plan, but perhaps still giving it value or significance. As scale is reduced so the value factor increases and such symbols predominate on the map. The symbols may be obvious or easy to decode. Here no attention would be wasted on the symbol; one would think at once of the idea expressed. On the other hand they may be obscure or difficult to decode.

Although some conventions exist in maps the symbols are not as standardized as they are in written language, and a legend is always required for each map to allow for complete interpretation. If a symbol is a picture of the object, immediate recognition should follow. Artificial devices, however, rely on the user supplying extra information himself. The complexity of the idea may create further doubt. For example, whereas an area depicting 'housing' can be easily assimilated, a zone of 'morphological change' may require further or specialized knowledge for its complete comprehension. Each class of information will be assimilated at different rates.

22. H. G. Alexander. op. cit. p. 63 ff.

The eye and brain, when faced with a new map, are only party prepared to receive and understand what it contains. Well-known general terms such as 'sea' and 'river' will be obvious, although, being unspecific they may be singularly uninformative. New symbols are either learned before hand or continuously referred to in the legend. This oscillation of attention is bound to affect the rate of comprehension. Above all, however, the map user must be able to extract the required symbol or group of symbols from the map, with maximum facility.

3. Attention to the Symbols
S. A. Hempius states that

> perception is not a passive registration of incoming signals, but an active adaption of the visual process by activating certain channels between cortex and retina, and blocking up other routes. The whole process functions like a selective filter which is adjusted to the expectations on hand and provides data for a first checking. This filtering procedure is of both a physiological and mental nature and may be optimised by training.[23]

Maps are designed to communicate information. General pattern or specific parts of the pattern may have to be studied. An overall awareness will persist as the eyes scan the perceptual field, one part after another becoming figural. This could give rise to a continuous alternation from figure to ground. However, attention as well as awareness must be directed to the map, and this vital factor depends on personal interests, or specific direction from outside, through education. Location of points on a map is, perhaps, the easiest thing to achieve, but relationships and distributions are much more difficult to appreciate. This is especially true when the design does not afford rapid extraction of the relevant points or areas. Original experiments of the *gestalt* phenomena were based upon the brief viewing of single shapes. Examining this, Rudolf Arnheim refers to Japanese 'long period' viewing experiments where 'the figure abandoned its good gestalt and produced new configurations ... sometimes secondary details left their context and became the centre of attention'.[24] 'Time and reflection change the sight little by little', says Cezanne, 'till we come to understand'.

Since a general awareness of the whole field renders little information, specific instructions must be given to find certain symbols. In this way the observer is mentally 'set', having a task or *aufgabe*. He can imagine the object of his search and proceeds to seek something which matches this mental image. If this image is wrong, or differs from the symbol, he may search in vain. Rewards such as passing an examination, or reaching a journey's end by the quickest route, can improve the attention of the map user.

Of course, should the search for symbols be unsuccessful, this does not mean that the activity has been wasted. Familiar symbols, like those for rivers, roads and cities have been observed, their presence noted, and if the symbols originally sought e.g. 'oil pipelines' are now located, they will take their place within this framework of knowledge of the rest of the map. As De Haas put it 'the purpose of education is to give the human being an adequate repertoire of expectations'.[25]

23. S. A. Hempius, W. G. L. de Haas, A. P. A. Vink. **Logical Thoughts on the Psychology of Photo-interpretation**. International Training Centre Publication, series B-No 41, Delft, 1967.

24. R. Arnheim. **Art and Visual Perception**, University of California Press, Berkeley, 1954.

25. S. A. Hempius *et al.*, op. cit.

Gestalt psychologists, while investigating the idea of good configuration, discovered that if an observer sees an incomplete image, he tends to complete it: e.g. three spaced dots make a triangle.[26] Further it was shown that if selected patterns were thoroughly memorized by the observers, they were much more easily recognized when embedded in complex figures, than unknown patterns would have been.[27] The more assymetrical the embedded figure, the longer it remained undetected within a complex form.[28] Map patterns are very complex, but some small tests have shown that if students 'know' the general form of a distribution (e.g. main railways in Scotland), or if they can examine, briefly, a colour 'pull' of one or two colours of a multi-coloured map (e.g. water/contour pull of an Ordnance Survey one inch sheet) they can more readily see this 'gestalt' through the other complex patterns on the complete map. They can then proceed to associate the latter distributions with this first one. Naturally, prior colour separation of map elements cannot be regarded as practicable-so the next best solution should be to create a visual image whereby patterns can be seen as a whole and be extracted without undue difficulty.

Another important quality of a map is its permanence; it can be examined over a period of time. Unlike writing and speech, which are sequential, a whole map is visible at once and there is no obvious order in its reading. This is why many people fail to 'get into' a map at all.

The intention in the above account has not been to comprehend the whole of visual perception, but merely to outline the main stages in the process of seeing: stimulus, reaction and response. It is against this background that the specific problems of the map will be examined.

THE PRINCIPLES OF PERCEPTION IN MAP DESIGN

The cartographic problem of separating the layers of graphic information which comprise the map was referred to initially. So far, attention has been devoted to colour and form perception in any objects without specific reference to maps. Three-dimensional space has an almost infinite potential of visual planes, receding from the eye, which can contain the objects of the external world. As these planes are separate and have certain qualities which enhance this illusion of depth, attention can be focussed on selected parts of the field. Although the map is two dimensional having only one plane to store information, there are certain monocular cues of depth which can be employed. They are cited by Vernon[29] as follows:

'Relative sizes of the retinal images of different parts of the field'.
'Relative brightness, clearness, and colour saturation of different parts of the field'.
'Perspective effects, overlapping planes, shadows and the general structural arrangement of the field'.
'Previous knowledge of and familiarity with the spatial arrangement of the field'.

26. L. L. Thurstone. **A Factorial Study of Perception**, University of Chicago Press, 1944.
27. S. S. Djang. **J. Exper. Psychol.** Vol. 20 (1937) p. 29.
28. I. G. Campbell. **J. Exper. Psychol.** Vol. 28 (1941) p. 145.
29. M. D. Vernon. **A Further Study of Visual Perception**, Cambridge University Press, 1962. p. 94.

A number of experiments have been carried out which suggest that if a familiar object is decreased in size in the visual field, it appears to move away from the observer, and vice versa. In the presence of other depth cues, this can be a compelling illusion.

The effect of depth or distance on the clearness, brightness and colour of objects, referred to as aerial perspective, makes distant objects appear to tend towards violet as well as become less distinct. Although accurate estimates of distance cannot be made the relative impression is effective. In one experiment[30] observers saw bright colours, white and yellow, as relatively nearer than neutral grey; and darker colours, green, blue and black as relatively farther away than neutral grey. The observation of texture in proximate objects can reinforce these depth effects.[31]

Perspective effects are not so relevant to the present problem, but overlapping planes and objects are significant. Overlapping is inevitable in maps and the depth effect can often be induced this way.

Kopferman[32] showed that if patterns or configurations on two planes were very similar and seemed to fit together, then they were seen on the same plane. This is a warning to designers of one-colour maps especially, to separate different line patterns, for instance, by obvious change of character and thickness.

Previous knowledge of the landscape gives the map user an extra facility to increase the effect of depth, e.g. by seeing the names of mountain peaks as being closer to him, or by neutralising the effect, for instance, of roads 'floating' above ground.

Gibson[33] proposed that many of the above cues can again be understood under a more general concept of 'stimulus gradient'. Surfaces in the visual world typically have textures, and the greater the distance away the greater the density of texture in the retina. This can be imagined as a textured plane extending away from the eye with a regular pattern of dots which become more dense with distance.

By applying the cues described above the map designer can create the illusion of depth, carefully adjusting the qualities and relative sizes of the symbols he employs. If analogy is made with a landscape on a hazy day, and at the same time the physical variables of colour and line available to the map maker are considered, the following observations can be made. The foreground has the richest colours (maximum saturation/tint) while colours in shadows have the darkest appearance (minimum brightness/shade). The foreground also has strong contrasts and good definition of detail. With increasing distance, contrast is reduced and colours take on paler tints (less saturated) with duller shadows: applying this, decreasing scales of magnitude or importance could be suggested in the map. It is also necessary to consider the saturation and brightness of boundary lines for such areas so that they lie in the correct plane. If it is necessary to place a number of hues in the same visual plane it may be necessary to judge them achromatically, e.g. to equate a red to a blue, a tint or a shade of the red may be required so that it does not stand out unduly.

This 'depth-cue' approach can be applied to colour, tint, symbol size, line thickness, textures and even names, in fact to all the elements which comprise a map image. Each part may be relatively adjusted in size, etc., until it takes up a desired visual

30. E. H. Johns and F. C. Sumner. **J. Psychol.** Vol. 26 (1948) p.25.
31. M. D. Vernon. **Brit. J. Psychol**. Vol. 28, p. 1 and 125.
32. H. Kopferman. **Psychol. Forsch**. Vol. 13 (1930) p. 293.
33. J. J. Gibson. **The Perception of the Visual World**, Houghton Mifflin Co., Boston, 1950.

level of importance. In certain map design problems, therefore, it may be helpful for the cartographer to attempt to create that effect artificially (still retaining maximum definition of detail) and thus achieve a separation of distributions which can accelerate the map reading process.

The idea of visual planes in cartography is not a new one, having been suggested by Robinson[34] and Heath[35] on previous occasions. Saunders, after some investigations of preference by users of certain map designs, concludes that 'the three most favoured maps all created the impression of different visual planes, which clearly indicated the appeal of the effect and its significance for cartographic procedure'.[36]

Maps consist of elements which are conveniently classified under four headings: areas, lines, points and names. If this is the alphabet of graphic language, how can it be manipulated to give cartographic symbols?

1. *Analysis*
Areas: Areas suggest the continuous distribution of a feature within their outlines. This assumption is one of the fundamental weaknesses of the map, as the real world can seldom have this homogeneity. Nevertheless, the areas should be perceived in their entirety and be easily matched with isolated patches of identical definition. Areas can have hue and pattern.

(a) *Hue*: Hue gives a unique response which can be emotional and affect legibility and acuity. This response can be modified by adjusting the associated tints or shades to the required level. When hue is introduced into a map the differing appeals of certain hues may make them difficult to assign to any specific visual plane. However, provided that the aim of the map is appreciated, adequate adjustments should be possible.

(b) *Pattern*: Pattern is of prime importance in one colour maps. Textures of lines, dots etc., also produce reactions unique to each, but not in quite the same way as hue. The contrasts of the internal structure of the pattern can not only attract attention to itself in a stable uniform way, (as would red) but also (through illusion) cause disturbing side effects which may reduce the definition of the area contour. In a region of diagonal lines, if thick lines are moderately spaced, a thin contour line may be ineffective as a boundary, especially in places where it runs sub-parallel to the lines. Even when a thicker, more appropriate line is used (equal to the thickness of the lines in the pattern) maximum effectiveness is only reached when it is perpendicular to the line pattern, thus acting as a 'stop' to the eye, and defining the edge more clearly. When the lines are parallel to the contour, minor frustration may result, and the only way of reducing this effect is to employ a pattern of closely spaced lines. It is just as important to consider the contrast of adjacent patterns, as it is with hue. Two sets of opposite, widely spaced diagonal lines may disturb the eye, as would adjacent areas of red and green of the same brightness. Such contrasts impair the smooth passage of attention from one area to the next.

34. A. H. Robinson. **Elements of Cartography**. op. cit. p. 224.
35. W. R. Heath. 'Cartographic Perimeters'. **International Yearbook of Cartography VII**, 1967.
36. B. G. K. Saunders 'Map Design and Colour in Special Purpose (Geographic) Cartography'. **Cartography**, Vol. 4. 1961–62.

The first improvement is crossed lines, a new pattern, where the line effect is replaced by a stability which reaches its maximum in the regular dot pattern. Coarse, open patterns (like parallel lines) have two major disadvantages: (i) unless shapes are large and simple, small variations of contour will not be enhanced, and the broad bands will tend to be seen as additional structures within the outlines: (ii) It may be impossible to depict small areas of the same category which fall within one of the band widths.

As the pattern texture becomes finer, and limitations of acuity are reached, increasing distance may be suggested, and the impression of tone results. The pattern is now only perceived on close inspection. Experiments conducted by Jenks and Knos[37] showed that users preferred dot styles and especially those of fine textures.

Regular patterns of any style or textural dimensions can be modified in tint by increasing the thickness of line or size of dot, i.e. by reducing the white paper content. Contrast may thus be produced without introducing new textures.

Confining attention to 'black and white' maps we can again examine the nature of depth cues and the arrangement of textural patterns in graded series from dark to light. This can be achieved as stated above, by increasing the size of the textural unit, dot or line. Referring back to depth perception, it is noted that decreasing retinal angle (i.e. the angle subtended in the eye by the object) is another cue of depth: the reduced sizes of dots or lines in a tint pattern produce a lower light percentage. The lightness scale can also be created by selecting a shading pattern unit size or thickness and adjusting the spacing of the units, e.g. the spacing of lines. The dominating theme is still one of reduced saturation, i.e. ink/paper ratio, but this time a secondary precept is introduced. Whereas one texture, varying in spacing and size suggests the visual gradients of texture referred to by Gibson,[38] (where a texture of known size appears to recede into the distance, thus becoming finer) in this case, while the main effect is one of increasing lightness of tint, (as in graded steps for decreasing rainfall, and hence decreasing importance), there is a subtler and opposite effect of increased texture or graininess, and hence increased attraction. (Incidentally this coarseness gives rise to the problems of misrepresenting small areas). This may help to explain why the former approach to graded series is more acceptable. It also explains why a normal series may be interrupted or reversed if a pattern is inserted which has the correct ink/paper ratio for that step but where texture is inappropriately attractive or coarse. The effect would be of gradually receding steps with one which vibrated from foreground by texture to background by value.

Area patterns in maps exist at all degrees of textural coarseness but as the 'grain size' increases and becomes more widely spaced, adequate contrast along the whole edge is lost or must be reinforced by suitable outlines. The finer the texture, the more homogeneous does the area appear. Finer textures are also more suited to receive other superimposed distributions like lines.

The afore-going remarks are levelled largely at conventional one-colour maps and so far no mention has been made of the reproduction process and the technique used to produce hues, tints and shades. In early maps each hue and tint required a separate

37. Jenks and Knos. op. cit.
38. J. J. Gibson. op. cit.

printing plate, with the colour adjustments made at the ink stage. However this can be prohibitively expensive for multi-coloured maps, especially thematic maps, and contact screens and later, film screens, employing a fine pattern of lines or dots, came into use. These patterns were incidental, serving only to satisfy the printing process. Following the theory that any hue can be created by combining the correct proportions of cyan, magenta and yellow it has become common practice for a cartographic office to have a tri-colour printed chart displaying ranges of hues and tints available through the combination of the dot or line screens employed by their processing department. Special departments, like soil survey, may require an even wider selection than is normally available. The screens to be used in this way had fairly coarse line patterns at around 70–80 lines per inch. For example the Directorate of Overseas Surveys used screens with 75 lines per inch for many large scale thematic maps, and the Netherlands Soil Survey Institute used line screens of from 40–80 lines per inch. The latter have now adopted dot screens for improvement of appearance. Even in combination, such patterns are fairly coarse and did not satisfactorily replace the flat tint-although they did aid, sometimes unintentionally, distinction between hues. When dot screens of suitable density (133 lines per inch and finer) became available, the trichromatic chart began to have the effect of numerous single printings, as the screen texture was too fine to be seen. Such screens, ranging from 90 lines per inch to 300 lines per inch are now most popular in the map printing world. A noteable exception is that of the Swiss Land Survey Department which prepared a new set of line screens (120 lines per inch) to employ in the new Swiss National Atlas.[39] The specifications for a Regional Economic Atlas of the Philippines contains an excellent account of a comprehensive system of tint production.[40] Here dot screens are employed in a so-called 'Color- Trol' system, but symbol patterns may be substituted for the magenta or cyan printers to extend the range to cover the 800 classes required for soil maps. The new fine screen techniques certainly improve the overall appearance of a map, but by removing textures completely, they present the reader of complex thematic maps with certain colour-matching problems, which result from colour contrast phenomena.

Williams[41] reported in 1958 on experiments designed to discover the correct screen percentages (black dots) to give 10 equal appearing intervals. These were, 2.5, 8, 17, 28.5, 43, 58, 73, 85, 94 and 100 per cent. In 1961 Jenks and Knos[42] compared this and seven other attempts to achieve the same goal. They discovered that the Williams scale, that of the U.S. Aeronautical Chart and Information Center (5, 10, 17, 28, 35, 42, 58, 76, 100 per cent) and another empirical scale of their own were most popular with the subjects. However, it is significant that these refer only to black, where the total range is maximum. It is obvious that if yellow only was being used, fewer and probably different values would be chosen. In spite of this, and of former experiments, the screens in general use are standardized either to a selection of the following: 10,

39. E. Spiess and H. Stump. 'Experience in Three-Colour Printing for Coloured Areas in Thematic Maps', **International Yearbook of Cartography IV**, 1964.

40. E. Spiess and H. Stump. 'Specifications for a Regional Economic Atlas (Philippines)', Proceedings, Vol. 2, **4th U.N. Regional Cartographic Conference**, Manila. New York 1966.

41. R. L. Williams. 'Equal Appearing Intervals for Printed Screens', op. cit.

42. Jenks and Knos. op. cit.

20, 30, 40, 50, 60, 70, or as with the Swiss Atlas maps 5, 10, 25, 50, 75. If these are applied with good judgment the results should certainly be adequate, although it would be interesting to observe the effects on colours of specifically prepared screen percentages. To return briefly to area boundaries, all screens, especially when printed in dark colours, can be detected below 150 lines per inch. It is thus desirable for practical purposes to mask the rough edges of areas with a line, which may also provide some leeway in the registering of two tints.

Before examining the next category of symbols, note that lines, as used above, act either together or as substitutes for hue stimuli or individually to reinforce poor figural contrasts, or rough edges between areas.

2. Lines

Lines are employed to join points or to separate areas. They should have sufficient distinction to allow for easy tracing on the map, and, since they often comprise a distribution, they should be appreciable as a whole. The nature or importance of a line feature is brought out by:

(a)   Hue. Certain conventions use hue for distribution (e.g. blue rivers).

(b)   Character. The appearance (pecked or dotted) of a line can be changed, instead of the hue, if the latter is not available. Hue and character are often combined to reinforce ideas.

(c)   Thickness. The thickness of a line can suggest importance and is equivalent to either an increase of tint or shade. However, simultaneous contrast will affect lines as if they were very narrow areas, and thus subject to the same rules. Rivers are often misrepresented for a technical reason: if they are printed over yellow or greenish areas, they will appear dark green. This physical factor is difficult to overcome. When narrow coloured lines are used on a white background, there is a tendency for all hues, except yellow and bright red, to be seen as black. Changing the background colour can reduce this effect. Equally the criss-crossing of lines over an area can create the effect of a pale tint over the whole region. Clear figural distinction must be made between lines employed to separate areas and lines depicting specific features, and special problems of contrast are introduced when line symbols have to cross patterned areas. This, once again, is where the designer can apply his depth cue formula, if, in one colour, a well selected set of line thicknesses for items of differing importance, can be employed. The task is more difficult in colour, when one is limited to the printing colours available.

3. Points

These symbols are used to depict physical objects or associated ideas at specific locations. The points may be studied individually or in groups along with other symbols. The rules which govern line variations in hue, character and weight or thickness, apply equally to point symbols, and small sizes severely limit discrimination of hue. The 'depth cue' approach may also be applied in this case.

4. Names

These are symbols of another language which have many other associations Although they can be analysed in a different way they may also be treated graphically, as were points, as having physical attributes such as hue, size and boldness. These may be adjusted to the correct visual level in the map.

II. *Synthesis*

It is insufficient to examine symbols in isolation, they must be studied in the context of the map, where meaning is given to the shapes. The function of a map must be clear and the -more limited this is the more easily will it be executed.

Maps may be divided into:

1. Those which describe and record spatial information with or without relative emphasis.
2. Those which emphasise the relative importance or value of the spatial ideas.

When the topics are noted in order of importance this is converted to a visual scale of comparable range. The next: stage is to superimpose these so that they may be seen in correct order when the map is viewed for the first time. This can be achieved by consciously creating the visual levels alluded to above, which can be attended to without confusion. If topics are placed on an imaginary scale of distance planes, closely spaced if the data is similar in quality and importance, and spaced more widely if relative emphasis is required, each plane could provide for focussed attention and a good 'gestalt' be procured. This technique is not expected to provide a complete answer, but is a method of approach in constructing a clear map.

The last consideration is that of meaning. Meaning will affect the relative acceptance of each category. This de-coding stage is the most variable and unpredictable, when subjective reactions, colour and pattern, knowledge of the conventions, the region, the topic and the ideas being expressed, all play a part. These may have a flattening, influence on the prepared visual levels in the map. For instance, black names, although standing out from a pale background, may be seen to lie in the valleys to which they refer. This precept could be enhanced by making the name closer in quality to the valley tones.

To illustrate some of the thought processes envisaged here, consider the following example of a 'political' atlas map containing three separate states. In addition, there are rivers, railways, roads, state and county boundaries and names for states, 'counties, inlets, rivers, seas, islands and towns.

Theoretical approach:

If no obvious order of reading is implied, it is necessary to erect an order, if only to aid the design process. The minimum requirements of a map of this nature are clarity and legibility, arid certainly a facility to concentrate on anyone of the contained symbols or distributions without difficulty.

Major Levels or Planes:

1. The sea is the least important.
2. The land surface with its associated physical features and names requires more attention. The colours chosen at this stage will help to enhance the symbols below.
3. Cultural features, towns and communications, have an obvious importance in such a map.
4. Political and administrative units with their names are naturally the most significant aspects of this map.

It might appear that in carrying out this separation, we are introducing certain conceptional errors; for example if culture and communications are designed to lie on a visual plane, closer to the observer than the land surface, we are misrepresenting the facts. However, this impression of separation is not intended to be a vivid one and should be neutralized by the simple effects of seeing the paper as a flat surface, and knowing that the physical and cultural features are in the same plane in real life.

Construction of the separate planes

1. Sea. This is blue, but, at the greatest distance from the observer, it will be reduced in saturation (tint) until the sensation of hue is just present. Blue is one of the Gestalt 'soft' or 'film' colours, suited to backgrounds, and which can also display the recession illusion through chromatic abberation in the eye. Sea names should also be blue; their style, size and weight being governed by the importance of the features. Above all, the names should be visually associated with the sea.

2. Land and associated features. This visual plane is considerably closer to the observer, and contrast at the land/sea edge should be such that even a quick glance at the map should render the land as a good sharp 'figure'. Inadequate contrast-not uncommon-reduces definition and thus increases required attention by an undesirable degree. If, on the other hand contrast is too great through the choice of a dark saturated hue for the land, the 'distance' left between this plane and the observer may be insufficient to provide for the 'closer' more significant distributions. In other words, if the background is 'shouting' at the reader, the foreground may not be able to compete successfully. The resulting map will be overpowering, with saturated hues, thick lines and large, bold names. A tint of green could be selected as the general hue for the land. It is 'harder' than blue, through the addition of yellow, a 'hard' colour. If two other colours must be chosen, merely to add emphasis to the areas of states, these should not be of too great contrast and less than that between land and sea. Small hue changes could be introduced towards red, or even yellow, but the impression of a narrow visual plane must be maintained. Water features and names could be blue/green. If hill names must be distinguished in grey this can be suggested by employing light weight black lettering, the thinness maintaining the impression of relative distance. Choice of type style is a critical consideration in all design problems.

3. Cultural Features. Again there is a step upwards in the scale of significance and a movement closer of the visual plane. If red was available, it could be used for one or all of these symbols, the character and thickness being chosen for adequate contrast with plane 'two'. Black would have to be manipulated with care if chosen to be linked with red in this plane.

4. Political Features. This is the last and most important aspect of the map. Black lines and names can be made relatively thicker. Black is a good 'form' colour and this final plane should be easy to create.

Technical Aspects

Assuming that the map is being designed for economical production, using four colours, blue, red, yellow and black, it may be necessary or desirable to modify the primaries. This done, attention may be turned to the resultant tri-colour chart of tint combinations. The selection and matching of hues will be aided by punching circular or square holes in a second chart which can then be superimposed. Choice of colours is now carried out by comparing tints through the holes, and the effects of simultaneous contrast can be anticipated. There is no reason why most of the theoretical design features described above cannot be initiated, although certain colour changes arising from the superimposition of, for example, blue over green areas should be anticipated.

Other Examples

Examples, displaying the simple effects of visual planes and depth, can be found in many atlases and maps, although this may not have been the specific intention of the map maker. The following, selected from the Atlas of Switzerland, illustrate the impression which is sought:

> *Types de Maisons Paysannes*, sheet 36
> *Climat et Temps* I   „   II
> *Bale: Travailleurs se rendant chaque*
>   *jour a Bale en* 1960. sheet 44
>   also: Fig. 2 *Ports de Bale Ville.*
> *Geophysique*   sheet 10

Even with this condensed examination of map design it appears that many principles of perception could be more consciously applied to maps. Moreover there are analogies in our surrounding environment which can be adopted to link many of the fragmented and unassociated rules which exist at present.

A SPECIFIC EXAMPLE:

THE RESIDENTIAL LAND USE MAP OF GLASGOW-1965

This is an example of a map showing all classes of thematic data on approximately one 'plane'. It is not presented as the epitome of all the ideas given before, although some have been employed. It does however, highlight a particular problem of colour and tint identity, and suggests a means of overcoming this.

The source material for this map was a survey carried out by students of the Department of Geography, University of Glasgow, and directed by Mr D. R. Diamond, who prepared the original classification. The object of a small scale derived map was to show two selected aspects of residential housing:

1: types of houses, 2: age of houses. All that remained was to design a map which displayed both in an equally legible manner. The geographical distribution of these classes -which appear in semi-continuous sub-concentric zones extending outward from the centre with decreasing age and change of type-had to be considered. Any design should take advantage of this pattern and enhance each zone for maximum comprehension.

The first compilation and generalization was executed on a scale of 1 : 25,000. The main housing 'blocks' were established at this stage and secondary features like roads, railways and canals were selected. Industrial and commercial land was included, but only to complete the picture of the built-up area of the city. Ambitions to print the map at a scale of 1 : 25,000 were abandoned on the grounds of cost, and instead, a fully coloured manuscript was carefully prepared, at the compilation scale, for display purposes. Standardization of tints was achieved with Letraset Instant Dry Colour and photoset names were affixed in the normal way.

Design and Content
The first solution of the design of two topic scales was to separate house types with distinct hues; red for tenements, the most dominant house type in Glasgow; blue for terrace houses which must have three in a row; yellow/ green for cottages-which are villas, or semi-detached; grey for flatted houses, or 4-in-a-block dwellings; yellow for multi-storey blocks.

Then three tints of each were introduced to cover the necessary age periods: pre-1880, 1880–1914, 1914–1945, post 1945. This catered for two psychological reactions, one to hue and one to decreased saturation or tint. In the legend box, these scales appeared to work satisfactorily and suitable colour steps were selected from the Letraset range. However, one of the features of a city is its complexity, and carto-graphically, this manifests itself in house blocks and zones of many sizes, intermingled and juxtaposed in such a way that simultaneous contrast causes several colour-matching ambiguities, especially in the paler tints. This was overcome by adapting a method used in the advertising literature of paints and inks. A replica of the legend was prepared on card, holes were punched in each hue or tint rectangle, and the identification of a selected map area was made by visual comparison through the appropriate holes. One could have adopted the standard procedure of adding a letter or figure to each tint, as on geological maps, but at manuscript stage this was not regarded as necessary.

Publication, having been abandoned at the large scale was later investigated for a scale of 1 : 50,000. It was realised that the manuscript could be photographically reduced and the map produced at the smaller scale, without generalization. This scale also proved to be economical, especially if designed for four printing colours.

Reduction intensified the problem of identification of separate tints in small areas and the printed map had to contain an unambiguous means of tint reference. By employing a normal 133 line dot screen, and selecting combinations from a tricolour chart, once again a pleasing legend was prepared, but it did not embody the positive tint check required. The method of index letters may be suitable for a working docu-ment like a geological map, but in this case, numerous letters or figures would have marred the map image. Also many of the areas were too small to contain other than the barest patch of the hue concerned. The only alternative was to introduce a second-ary stimulus, that of pattern. As stated, the fine dot screens do not provide sufficient pattern changes and the line screens commonly employed in soil and geological maps are altogether too coarse for this situation. It was not until screens of 120 lines per inch became available that the solution was found. It was then possible, not only to introduce a series of tint steps within each colour, but to apply a particular screen angle for each age period, cutting across the hues.

In order to maintain a light appearance in the map, fairly desaturated hues were chosen. This, in effect, pushes back the visual plane so that the colours do not 'shout'. One outcome of this particular decision was that the 25 per cent tints (originally selected for the most recent age period) lost much of their hue individuality, and certain confusions could result between green and blue in small areas. Hence when attention was directed at the distribution of house types, the lower end of the tint scale failed to register as a definite part of the pattern. The ideal solution would have been to raise the percentage tint at the lower end to 30 per cent or 35 per cent but, as described earlier, these did not exist for the lines screens being employed. Fifty per cent was the nearest approximation available. This decision inevitably restricted the number of useable steps to three (50 per cent 75 per cent and solid) and so, to extend the range, a 50 per cent black screen was printed over a repeat of the solid hue at that end of the range, hence reducing its brightness. This, incidentally reflects a feature of many of the older houses built of sandstone, which have been blackened by smoke over the years. The specifications for the legend are as follows:

Pre-1880	1880-1914	1914-1945	Post-1945
solid hue +	solid hue	75 *per cent* tint	50 *per cent* tint
50 *per cent* black line		screen at	screen
screen at 135°		45°	horizontal

These specifications apply to tenements, terraced houses and 'cottages', but since both 'flatted houses' and blocks of flats are psot-1914 in age only two steps, namely 1915–1945 and past-1945 are required. Restriction to four colours required 'flatted houses' to be composed of a combination of red and blue.

1914-1945	Post-1945
75 *per cent* red	50 *per cent* red
45 *per cent* line screen +	horizontal screen +
25 *per cent* blue	25 *per cent* blue
135° line screen	135° line screen

This predominance of red is designed to associate these areas with the tenement, rather than the terrace house. The yellow printing, representing blocks of flats on the original manuscript, was omitted. Emphasis was given to high blocks by increasing the line thickness round each area and applying a horizontal line screen in black, again associating it with the post-1945 period.

Grey, being a neutral colour, enhances the main hues in this map. Thin black area outlines were used for the main colours. The industrial and institutional land and the open land (black 25 per cent and 10 per cent screens) being secondary in emphasis are differentiated by pattern as well as tint, but no outlines were used to avoid drawing attention to the boundary. This also induces a sense of 'film' or background colour to increase the figural effect of the main topic-housing. The latter decision, although theoretically valid, is more difficult to carry out successfully as the two patterns do not fit perfectly, one being of lines and the other of dots. Also a line pattern does not define edges with consistent quality as the best effect is gained when the screen lines are at right angles to the contour edge.

Univers 690 (medium condensed) was selected for district names to raise the level of this information above the chorochromatic plane. The lines in this style 'of letter, at 5 point size, are thicker than the lines bounding the area. The area surrounding the city is in 40 per cent black to help concentrate the attention on the housing. The latter was also the reason for differentiating the centre with a coarser but lighter dot screen.

The nature and range of categories in this map provided an interesting problem for design.

The state of psychological research into the map image is such that one cannot avoid concluding with the remark that 'more must be done in this field'. There is little chance of this research progressing quickly and so it would benefit cartographic designers to do selective reading into the relevant branches of psychology. In this way they may develop design aids, perhaps like the one of induced visual planes, which although not experimentally confirmed could prove successful with experience.

ACKNOWLEDGEMENTS

The author wishes to acknowledge the advice and cooperation given by Mr J.S. Keates and Mr G. Petrie of the Department of Geography, University of Glasgow, and to Mr D. R. Diamond of the Department of Town and Regional Planning, who conceived the original idea for the map, and to Dr A. Shirley of the Department of Psychology. He also wishes to express thanks to Mr A. Kelly of the Department of Geography, who did most of the original drawing for the land use map.

Grateful acknowledgement must also be made to The Carnegie Trust for the Universities of Scotland and to the Court of the University of Glasgow, without whose generous aid the coloured map could not have been produced.

Reflections on 'Visual Perception and Map Design'

AMY L. GRIFFIN

University of New South Wales

Michael Wood's (1968) article entitled 'Visual Perception and Map Design' aimed to develop a flexible grammar of map symbols that cartographers could use to guide their design of maps and that was based on the limits and capabilities of visual perception. Wood wanted to provide a positive counterpoint to the 'how not to design maps' approach Robinson (1952) often took in *The Look of Maps*. Working within the paradigm of the cartographic communication model of map design, Wood argued that when maps fail to communicate accurately, this failure is often the result of a set of design decisions that 'prevents the separation or disentanglement of superimposed patterns' (p. 55). He drew upon psychological literature to review aspects of the visual perception of colour and form. Then, applying his findings to a cartographic context, he advocates for the use of visual planes and depth cues to remedy the problem of superimposed patterns, relegating the least important map elements to the lowest planes in the map. In so doing, he echoes several earlier arguments presented by Robinson regarding the use of visual planes within cartography (1960) and the need for coincidence between the map's intellectual and visual logics (1952). Finally, he makes a set of practical suggestions about symbolization decisions, such as the choice of visual variables, which will help the cartographer to create visual planes within the map and provides what he terms an example of applying his guidelines to a particular dataset, though without an accompanying graphic illustration. The title of the article notwithstanding, Wood also ventures briefly into the realm of the cognitive and explores the making of meaning from symbols for map readers.

Dent (1972), obviously influenced by Wood's (1968) article — even using some of the same illustrative examples — applies the concept of visual planes specifically to thematic maps and borrows several of Wood's suggested depth cues: overlaps among map elements and a variation in the level of contrast between map symbols to induce the appearance of different distances from the map reader; a form of aerial perspective. Through his textbooks, Dent later popularized the use of the term 'visual hierarchy' to explain the concept of visual planes within a map. Wood himself further developed the relevance of the concept of Gestalt and the character of visual marks that produce figure/ground relationships, especially in the context of how different map readers see and interpret these marks in different ways, highlighting the increasing importance of human factors research to our field (Wood, 1972). Later, other researchers explored the role of variations in contrast in promoting or hindering the perception of clear figure/ground relationships through experiments with map readers (MacEachren and Mistrick, 1992).

The group of cartographers whose practice has been most directly influenced by Wood's idea of promoting depth cues has been those designing visual displays in operational environments such as air traffic control towers and power plant control rooms (e.g. Taylor, 1985; Reynolds, 1994; Van Laar, 2001). In such environments, visual displays are often designed to support specific tasks, which must be conducted efficiently as well as correctly. The basic principle that their designers have borrowed from Wood is that the most task-relevant information should appear higher in the visual hierarchy. Experimental testing has demonstrated that using visual layers based on an intellectual hierarchy within displays produces more efficient task completion, a lower perceived workload, and task error rates that are not significantly different from other ways of displaying the same information (Van Laar and Deshe, 2002; Van Laar and Deshe, 2007).

Another way in which Wood's 1968 article had impacts upon cartography can be seen when researchers took up his call for flexible grammars that could provide guidance to less experienced cartographers to help them to make design decisions. While a range of cartographers sought to codify cartographic grammars through expert systems during the 1990s, these systems often proved to be insufficiently flexible to handle the wide variety of situational constraints that occur during map symbol design (e.g. Buttenfield and Mark 1991; Forrest 1993; Forrest 1999). However, at least some of the flexible guides that Wood envisioned have finally emerged, though in perhaps somewhat different form. ColorBrewer was the first of a series of focused support systems designed to help novice cartographers make better design choices (Harrower and Brewer, 2003). In addition to providing example colour schemes, the ColorBrewer tool aims to improve the map designer's understanding of the perceptual effects of different design choices. Other, similar efforts followed in ColorBrewer's wake, and there is now a series of simple yet flexible tools to make more aesthetically pleasing and more effective maps (e.g. Schnabel, 2005; Sheesley, 2007). When Wood wrote his article, cartographers had no choice but to deliberately make symbolization choices; today's mapping systems have many defaults that may allow a cartographer to abstain from making a choice and thereby produce an ill-designed map. Wood's desired flexible guidance systems are therefore perhaps needed more than ever today.

REFERENCES

Buttenfield, B. P. and Mark, D. M. (1991). 'Expert systems in cartographic design', in **Geographic Information Systems: The Computer in Contemporary Cartography**, ed. by Taylor, D. R. F., pp. 129–150, Pergamon Press, Oxford.

Dent, B. D. (1972). 'Visual organization and thematic map communication', **Annals of the Association of American Geographers**, 62, pp. 79–93.

Forrest, D. (1993), 'Expert systems and cartographic design', The Cartographic Journal, 30, pp. 143–148.

Forrest, D. (1999). 'Developing rules for map design: A functional specification for a cartographic-design expert system', **Cartographica**, 36, pp. 31–52.

Harrower, M. and Brewer, C. A. (2003). 'ColorBrewer.org: An online tool for selecting colour schemes for maps', **The Cartographic Journal**, 40, pp. 27–37.

MacEachren, A. M. and Mistrick, T. A. (1992). 'The role of brightness differences in figure-ground: Is darker figure?', **The Cartographic Journal**, 29, pp. 91–100.

Reynolds, L. (1994). 'Colour for air traffic control displays', Displays, 15, pp. 215–225.

Robinson, A. H. (1952). **The Look of Maps**, University of Wisconsin Press, Madison, WI.

Robinson, A. H. (1960). **Elements of Cartography**, 2nd ed., Wiley and Sons, New York.

Schnabel, O. (2005). 'Map Symbol Brewer — A new approach for a map symbol generator', in **Proceedings of the 22nd International Cartographic Conference**, A Coruña, Jul 9–16. http://icaci.org/files/documents/ICC_proceedings/ICC2005/htm/pdf/oral/TEMA24/Session%205/OLAF%20SCHNABEL.pdf (accessed 15 August 2013).

Sheesley, B. C. (2007). 'TypeBrewer: Design and evaluation of a help tool for selecting map typography', PhD Thesis, University of Wisconsin-Madison.

Taylor, R. M. (1985). 'Colour design in aviation cartography', **Displays**, 6, pp. 187–201.

Van Laar, D. L. (2001). 'Psychological and cartographic principles for the production of visual layering effects in computer displays', **Displays**, 22, pp. 125–135.

Van Laar, D. L. (2002). 'Evaluation of a visual layering methodology for colour coding control room displays', **Applied Ergonomics**, 33, pp. 371–377.

Van Laar, D. L. and Deshe, O. (2007). 'Color coding of control room displays: The psychocartography of visual layering effects', **Human Factors** 49, pp. 477–490.

Wood, M. (1968). 'Visual perception and map design', **The Cartographic Journal**, 5, pp. 54–64.

Wood, M. (1972). 'Human factors in cartographic communication', **The Cartographic Journal**, 9, pp. 123–132.

The Map as a Communication System[1]

ARTHUR H. ROBINSON AND BARBARA BARTZ PETCHENIK

Originally published in *The Cartographic Journal* (1975) 12, pp. 7–15.

Only recently have cartographers devoted much attention to the study of the map as a communication system. It is now generally accepted that the study of how maps convey knowledge to a recipient, and therefore the study of the percipient as well as the process, are important areas of research. Various models of communication systems are analysed from the point of view of the inputs and transformations involved. The element of 'noise', characteristic of all systems, is examined as it relates to the map. The problem of assessing the amount of information included in a map and the efficiency with which geographical knowledge is evoked in a percipient is reviewed. Because a map is a two-dimensional presentation it is a very different form of communication from the one-dimensional linear forms that depend upon a temporal sequence, and consequently, the techniques of information theory are judged not to be directly applicable.

Just as most writing assumes a reader, most map making assumes a viewer to whom the map will convey information. The author-reader transfer and the cartographer-map viewer transfer are each systems of communication, although they differ in essential ways as noted by Moles[2] and Bertin.[3] Whereas the operation of language as a communication system has received a great deal of study, the way a map functions has not. It is important that we examine the basic characteristics of the communication process as it operates with maps for several reasons: cartographers can hardly employ their communication system efficiently unless its essential attributes are known; maps differ sufficiently from other methods of communication, so that seemingly simple, parallel concepts, for example 'noise', may actually be rather different; and the recognition of the various cognitive components involved will help to clarify problems of error in and measurement of the efficiency of the system.

At the outset it is necessary to select and define a term for the 'receiver', so to speak, in the communication process, who insofar as maps are concerned is one who, by viewing a map, augments his understanding of the geographical milieu, that is his previous conception of the real world. We use the name 'percipient'. Presumably,

---

1. Editor's note: This article is a slight modification of one in a collection of essays by the authors planned for publication under the title *The Nature of Maps, Essays Toward an Understanding of Maps and Mapping*.

2. A. A. Moles, (1964), 'Theorie de l'information et message cartographique', **Sciences et Enseignement des Sciences** (paris), Vol. 5, No.2, 11–16.

3. J. Bertin, (1968), 'Cartography in the Computer Age', paper presented at **Technical Symposium S 40**, ICA-IGU, New Delhi. J. Bertin, (1970), 'La graphique', **le Seuil** (Paris), Vol. 15, 169–185.

there are more than a few people, in addition to very young children, who for one reason or another, can look at a map without it having any noticeable effect on their geographical understanding. These would simply be map viewers, not percipients. Also, a percipient is to be distinguished from those we designate by the more restricted terms 'map reader' and 'map user'. The term 'map reading' implies a rather specific and limited action, such as looking up the name of a city or country, or finding out how high a particular hill is. Similarly, the term 'map user' connotes the employment of a map for a specific purpose, such as that of the farmer who obtains the data needed for contour ploughing, or the engineer who lays out a road with the help of soil and topographic maps. Neither the map reader nor map user are necessarily adding to their spatial knowledge. Both terms suggest operations similar to that of using a dictionary to find out simply how to spell or pronounce a word which adds little if anything to one's understanding of the meaning of the word and therefore of the language to which it belongs.[4]

When one surveys the history of cartographic thought one cannot help but be surprised at the fact that, until recently, very little concern, either practical or theoretical, has been focused on the map as a communication system. Koláčný[5] summarized the state of affairs by observing:

> In fact, however, cartographic theory and practice have almost exclusively been concerned with the creation and production of cartographic works, so far. One would hardly believe how little thought literature has, until recently, given to the theory and practice of map using. Not until in the last few years have the authors of some handbooks complemented their definitions of cartography by extending the term to the employment of maps.

> On the present level of cartographic theory and practice, the work of the map user is therefore largely determined by the cartographer's product. There prevails the tacit assumption that the user will simply learn to work with any map which the cartographer makes. In other words, the map user is expected to submit, more or less, to the cartographer's conditions.

Only since the burgeoning of cartography in the 1940s has the percipient, that is, the receiver of the information prepared by the cartographer, become a subject for investigation. The earlier general analyses, such as Robinson,[6] Keates,[7] and the more

4. The term 'percipient' is a generally accepted appelation employed by those who study perception. For several years the authors used the term 'cartoleger' (kar-tol-ə-jər) but decided against employing it, mainly to escape the accusation of creating jargon. 'Cartoleger' has etymological justification: 'cart' is from Fr. *carte*, (map), fr. L. *charta*, Gr. *khartes*, (sheet, leaf) and 'leg' is from L. *legere*, Gr. *legein* (gather, speak, choose, read), cf. legend, legible. In French, cartoleger would be *cartolege* and in German *Kartoleg*, coordinate with *cartographe* and *Kartograph*. A natural cognate of cartoleger would be cartology, the study of the map as a medium of communication and has been proposed in the literature (L. Ratajski, (1970), 'Kartologia' (Cartology), **Polish Cartographical Review**, Vol. 2, 97–110; L. Ratajski, (1973), 'The Research Structure of Theoretical Cartography', **International Yearbook of Cartography**, Vol. 13, 217–228, pp. 220–226). Cartology has the obvious advantages of simplicity and a readily apparent meaning over another term sometimes used, 'metacartography'. Cartology has also been suggested several times as a term to cover the study of maps as documents (their origins, history, quality, *etc.*), but it has not been generally accepted.

5. A. Koláčný, (1968), **Cartographic Information-a Fundamental Notion and Term in Modern Cartography**, Czechoslovak Committee on Cartography, Praha, (in English), p.1.

6. A. H. Robinson, (1952), **The Look of Maps**, University of Wisconsin Press, Madison.

7. J. S. Keates, (1962) 'The Perception of Colour in Cartography', **Proceedings of the Cartographic Symposium**, Edinburgh, 19–28.

specific psychophysical investigations into stimulus-response relationships, such as Flannery,[8] Williams,[9] Clarke,[10] Ekman, Lindman and William-Olsson,[11] and Jenks and Knos,[12] dealt primarily with particular characteristics of the map percipient. This explicit concern with the reactions of the viewer to the graphic components of the map was not, of course, the beginning of such interest. Occasionally, earlier cartographers had wondered about the effectiveness of what they did, but that interest was not considered a part of cartography. It was not until the development of formal communication theory and investigation into information processing systems that cartography incorporated this aspect. In his review of the subject and method of cartography, Salichtchev[13] remarks on the 'Strong ties [that] have emerged between cartography and the theory of scientific information' and he incorporates 'the theory and methods of map use' as one of the elements comprising the content of contemporary cartography.

A review of the literature of cartography during the past two decades reveals a persistent increase in the concern for the problems of cartographic communication. Numerous Master's theses and Doctoral dissertations have delved into various facets of the broad subject, ranging from the general in which the entire system is of concern (e.g. Dornbach[14]) to the specific in which selected elements of the map are examined (e.g. Wright,[15] Pearson,[16] Dent,[17] Cuff,[18] Yoeli and Loon[19]). Published research which relates various characteristics of the percipient to specific questions of cartographic design has appeared regularly, such as, Castner and Robinson,[20] Crawford,[21]

8. J. J. Flannery, (1956), **The Graduated Circle: A Description, Analysis and Evaluation of a Quantitative Map Symbol**, unpublished Ph.D. dissertation, University of Wisconsin.

9. R. L. Williams, (1956), **Statistical Symbols for Maps: their Design and Relative Values**, Yale University Map Laboratory, New Haven, Conn.

10. J. I. Clarke, (1959), 'Statistical .Map Reading'. **Geography**, Vol. 44, 96–104.

11. G. Ekman, R. Lindman and W. William-Olsson, (1961), 'A Psychophysical Study of Cartographic Symbols', **Perceptual and Motor Skills**, Vol. 13, 355–368.

12. G. F. Jenks and D. S. Knos, (1961), 'The Use of Shading Patterns in Graded Series', **Annals**, Association of American Geographers, Vol. 51, 316–334.

13. K. A. Salichtchev, (1970), 'The Subject and Method of Cartography: Contemporary Views', **The Canadian Cartographer**, Vol. 7, 77–87. (Translation by J. R. Gibson), pp. 83–85.

14. J. E. Dornbach, (1967), **An Analysis of the Map as an Information Display System**, unpublished Ph.D. Dissertation, Clark University.

15. R. D. Wright, (1967), **Selection of Line Weights for Solid Qualitative Line Symbols in Sel'ies on Maps**, unpublished Ph.D. dissertation, University of Kansas.

16. K. Pearson, (1970), **The Relative Visual Importance of Selected Line Symbols**, unpublished Master's thesis, University of Wisconsin.

17. B. D. Dent, (1971), **Perceptual Organisation and Thematic Map Communications: Some Principles for Effective Map Design with Special Emphasis on the Figure Ground Relationship**, unpublished Ph.D. dissertation, Clark University.

18. D. J. Cuff, (1972), **The Magnitude Message: A Study of the Effectiveness of Color Sequences on Quantitative Maps**, unpublished Ph.D. dissertation, The Pennsylvania State University.

19. P. Yoeli and J. Loon (1972), **Map Symbols and Lettering: A Two Part Investigation**, European Research Office, United States Army, London, England (NTIS No. AD 741834).

20. H. W. Castner and A. H. Robinson, (1969), **Dot Area Symbols in Cartography: the influence of Pattern on their Perception**. Technical Monograph No. CA-4, Cartography Division, American Congress on Surveying and Mapping, Washington, D.C.

21. P. V. Crawford, (1971), 'Perception of Grey-tone Symbols', **Annals**, Association of American Geographers, Vol. 61, 721–735.

Flannery,[22] Dent,[23] Head,[24] Cuff,[25] and Meihoefer.[26] Special sessions at meetings of cartographic organizations as well as separate symposia dealing with problems of cartographic design in relation to the intended use of maps have proliferated, and in 1972 the General Assembly of the International Cartographic Association created Commission V, Communication in Cartography, with the following terms of reference.[27]

(1) The elaboration of basic principles of map language.

(2) The evaluation of both the effectiveness and efficiency of communication by means of maps with reference to the different groups of map users.

(3) The theory of cartographic communication, i.e. the transmission of information by means of maps.

Numerous attempts to enlist the aid of Information Theory have been reported, a subject to be dealt with explicitly later in this paper. All in all, there seems no doubt that the field of cartography has opened wide its arms to welcome the concept that it is a communication system. To be sure, there are always those who recoil from such things as being theoretical and impractical ('let's get on with the job of making maps'), but it is doubtful that the trend can be resisted. The dean of European cartographers, Eduard Imhof stated it clearly and vigorously in a closing section of *Kartographische Geliindedarstellung* dealing with the Reform of Map Graphics where he observed:

A good part of the labours of topographers and map editors remains fruitless because of graphic defects. A very important user of graphics once condemned maps simply as 'graphic outrages'. It is in order, therefore, to clean out the Augean stables. Many excellent cartographers have already begun this Sisyphean [-like] task. Map graphics must be reformed. Only the simple is strongly expressive. A map should contain nothing that a normally gifted user cannot easily see. The 'laws of vision' and the experience of map readers are to be followed and observed.[28]

As an accomplished artist and teacher, Imhof lays great stress on good training and gifted personnel, but it seems fully as important that the cartographer clearly understands his role in the larger system of communication of which he is but one part.

One may study communication systems in a variety of ways. At one end of the range is the attempt to model the organization simply on the basis of the energy

22. J. J. Flannery, (1971), 'The Relative Effectiveness of Some Common Graduated Point Symbols in the Presentation of Quantitative Data', **The Canadian Cartographer**, Vol. 8, No.2, 96–109.

23. B. D. Dent, (1972), 'Visual Organisation and Thematic Map Communication', **Annals**, Association of American Geographers, Vol. 62, 79–93.

24. G. Head, (1972), 'Land-Water Differentiation in Black and White Cartography', **The Canadian Cartographer**, Vol. 9, No.1, 25–38.

25. D. J. Cuff, (1973), 'Colour on Temperature Maps', **The Cartographic Journal**, Vol. 10, 17–21.

26. H. J. Meihoefer, (1973), 'The Visual Perception of the Circle in Thematic Maps: Experimental Results', **The Canadian Cartographer**, Vol. 10, 63–84.

27. L. Ratajski, (1974), 'Commission V of I.C.A.: The Tasks It Faces', **International Yearbook of Cartography**, Vol. 14, 140–144, p. 140.

28. E. Imhof, (1965), **Kartographische Geliindedarstellung**, Walter de Gruyter and Co., Berlin, p. 400.

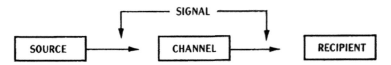

FIG. 1.   Fundamentals of a communication system.
(University of Wisconsin Cartographic Laboratory)

inputs, flows and losses in the various parts of the system. At the other extreme one can attempt to analyse the information content itself and the relation between the substantive input and the output of the system. Most analyses of the mapping communication system will incorporate some aspects of both and will thus lie somewhere between the extremes. We will begin our analysis of the cartographic communication system by first looking at the simplest possible form of a general communication network.

Singh[29] shows that a typical communications network consists fundamentally of a. source (or transmitter), a channel which conveys the message, and a receiver (Figure 1). In everyday terms a speaking person may be the source, the air which carries the sound waves of his voice would be the channel, and a listener the receiver. A more detailed analysis of the basic system will usually include the insertion of an encoder between the source and the channel and a decoder between the channel and the receiver, so that the system will be as diagrammed in Figure 2. The function of the encoder is to improve the efficiency of the system. In our original illustration the voice mechanism of the speaker constitutes the encoder, taking the thoughts of the source and transforming them into sound waves, while the hearing mechanism of the listener is the decoder, transforming the sound waves back into thoughts. The elements of other systems, such as telegraphy, the hand sign language of the deaf, television, and so on, are easy to fit into the same sort of model.

An unwanted but apparently inevitable component of every communication system is the element called 'noise'. Noise consists of interference with the signal, such as incorrect voice sounds in speech, static in radio, and distortions of appearance in television. In the modeling of the general communication system the noise is shown as entering the system at the channel (signal) phase (Figure 2). As we shall see later,

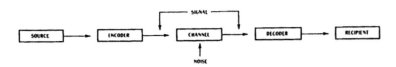

FIG. 2.   The place of encoders and decoders in a simple communication system
(University of Wisconsin Cartographic Laboratory.)

29. J. Singh, (1966), **Great Ideas in Information Theory, Language and Cybernetics**, Dover Publications, Inc., New York.

FIG. 3.   A generalized communication system.
(After Board: 1967. University of Wisconsin Cartographic Laboratory.)

the concept of noise is considerably more elusive in the cartographic communication system.

With a little squeezing and broad analogy the cartographic communication process can also be made to appear analogous to the generalized communication system. For example, Board,[30] in his discussion of the map-model analogy employs the following generalized communication system drawn from Johnson and Klare.[31] Board makes use of a map which shows on a nominal scale the occurrence of the older 'ridge and furrow' cultivation patterns in the Midland counties of England.

Board likens the source (Figure 3) to the real world, the encoding to the black symbolism on the map employed to show the presence of ridge and furrow, the signal to the graphic pattern provided by the black patches and blocks, the receiver to the eyes of the reader of the *Geographical Journal* in which the map appeared, and the decoder to the mind of the reader who is, of course, the destination. No transmitter is indicated and noise is assumed to be 'distracting information', a component not identified, but by implication consisting of anything other than the primary signal, and which thus presumably includes 'essential names and insets', although how anything can be both essential and noise is not made clear.

It was not Board's purpose to investigate the degree of correspondence between the cartographic system and the general model of communication systems, but the example shows that clearly they are similar. The world and the cartographer constitute the source, the map is the coded 'message', the signal is made up of the 'light waves' which make the message visible, the channel is space, and the decoder and recipient are the receiver-destination. Nevertheless, as Board and others realize, when the cartographic communication process is analysed in detail it is evident that it departs significantly from the general model.

The general model simply assumes a given message, its transmission and its receipt. Any specialized system, such as the cartographic, will incorporate complex processes of selection and interpretation in both the source (the cartographer) and the destination (the percipient). These, along with whatever corresponds to noise in the general system, produce a discrepancy between the real world on the one hand and the image developed by the percipient on the other. Part of this difference is due to the personalized, and therefore inevitably 'distorted', views of reality held by the cartographer

30. C. Board, (1967), 'Maps as Models', pp. 671–725, in R. J. Chorley, and P. Haggett, **Models in Geography**, Methuen and Co., London, pp. 672–675.

31. F. C. Johnson, and G. R. Klare, (1961), 'General Models of Communication Research, a Survey of the Development of a Decade', **Journal of Communication**, Vol. 2, No.1, 13–26 and 45. p. 15.

and the percipient and part is due to the methods of coding the message, i.e., the map made by the cartographer and to its decoding by the percipient. To reduce this discrepancy is, of course, a fundamental aim in cartography.

The first step in this direction is the realization that the map maker and the percipient are not independent of one another, and that an increase in the efficiency of the system depends upon an active feedback, so to speak, from the decoding, percipient phase, to the encoding, mapmaking phase. This was clearly realized in the earliest penetrating analysis of the problem made by Koláčný.[32] This stemmed from an elaborate research programme on a variety of aspects of the efficiency of cartographic symbolism carried out at the Research Institute for Geodesy and Cartography in Praha (Prague) during the period 1959–1968. Koláčný reported:

> ... cartographic work cannot obtain its maximum effect if the cartographer looks upon the production and the consumption of the map as on two independent processes. That maximum effect can only be obtained if he considers the creation and utilisation of works of cartography to be two components of a coherent and in a sense indivisible process in which cartographical information originates, is communicated and produces an effect. ...
>
> The creation and communication of cartographic information is actually a very complex process of activities and operations with feedback circuits on various levels.[33]

To illustrate the basic system as it operates in cartography Koláčný constructed a diagram which has been reproduced several times.[34] Koláčný's analysis of the cartographic communication system appears to have served both as the prototype and as a catalyst for other analyses.

He summarized the implications of the system as follows:

> This is a combination of very complex relations which have not been conceived in their mutual relations, so far. Various new questions are thus added to the traditional problems of cartography, questions which hitherto have been either neglected or solved separately. This refers in particular to research into the map user's needs, interests and inclinations, the study of his work with maps, the check on the function and effectiveness of maps, the promotion of 'map-consciousness' and cf. desirable habits of work with maps, the encouragement of 'map-fancying', etc. Moreover, the construction of a complete theoretical system of cartography and the preparation of methods for the solution of the prospective tasks of cartography also call for light to be shed on the problems mentioned.[35]

32. A. Koláčný, (1968), **Cartographic Information-a Fundamental Notion and Term in Modern Cartography**.

33. A. Koláčný, (1968), **Cartographic Information-a Fundamental Notion and Term in Modern Cartography**, pp. 3–4.

34. A. Koláčný, (1968), **Cartographic Information-a Fundamental Notion and Term in Modern Cartography**. A. Koláčný, (1969), 'Cartographic Information-a Fundamental Term in Modem Cartography', **The Cartographic Journal**, Vol. 6, 47–49. A. Koláčný, (1970), 'Kartographische Information-ein Grundbegriff und Grundterminus der modemen Kartographie', **International Yearbook of Cartography**, Vol. 10, 186–193. D. Woodward, (1974), 'The Study of the History of Cartography: A Suggested Framework', **The American Cartographer**, Vol. 1, 101–115.

35. A. Koláčný, (1968), **Cartographic Information-a Fundamental Notion and Term in Modern Cartography**, p. 7.

Much the most detailed model of 'cartographical transmission' is that of Ratajski.[36] It was constructed for the purpose of illustrating his conception of the research structure of theoretical cartography. Very similar in essentials to the model of Koláčný, and rather like a diagrammatic plan of an urban mass transportation system, it charts the routes of transmission and characterizes them in terms of input and output and their efficiency. Like Koláčný, Ratajski makes no attempt to analyse the fundamental nature of the percept itself other than to observe that the map is a product of selective observation and symbolization of reality. The map when observed by a 'receiver (map user)', results in an 'imagination of reality' more or less at variance with reality because of losses in information and in efficiency occasioned by a variety of factors including the input of both the cartographer and the percipient. It is to be expected that both Koláčný and Ratajski's formulations of the transmission of cartographic information will receive general approval as basic outlines of the essentials of the communication process in cartography; Salichtchev has already so employed Koláčný's in simplified fashion.[37]

Others have found it useful to simplify the model of cartographic communication to focus attention on either the mechanics involved or the substantive characteristics.

Muehrcke[38] characterizes the cartographic process as a series of transformations, and he diagrams their relationship as shown in Figure 4. His description of the process is as follows:

Data are selected from the real world (T_1), the cartographer transforms these data into a map (T_2), and information is retrieved from the map through an interpretive reading process (T_3). A measure of the communication efficiency of the cartographic process is related to the amount of transmitted information, which is simply a measure of the correlation between input and output information. The cartographer's task is to devise better and better approximations to a transformation, T_2, such that output from T_3 is equal to input to T_2; i.e., $T_3 = (T_2)^{-1}$.[39]

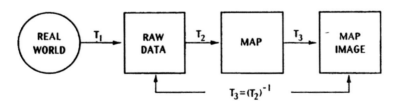

FIG. 4. Muehrcke's diagram of the cartographic processing system. *(University of Wisconsin Cartographic Laboratory.)*

36. L. Ratajski, (1972), 'Cartology', **Geographica Polonica**, Vol. 21, 63–78. L. Ratajski, (1973), 'The Research Structure of Theoretical Cartography'.

37. K. A. Salichtchev, (1973), 'Some Reflections on the Subject and Method of Cartography after the Sixth International Cartographic Conference', **The Canadian Cartographer**, Vol. 10, 106–111. K. A. Salichtchev, (1973), 'Some Features of the Modern Cartography Development and their Theoretical Meaning', **Moskorskiy Universitet, Vestnik, Seriya V Geografia**, No.2, 3–11. D. Woodward, (1974), 'The Study of the History of Cartography: A Suggested Framework'.

38. P. C. Muehrcke, (1969), **Visual Pattern Analysis, a Look at Maps**, unpublished Ph.D. dissertation, University of Michigan, p. 3. P. C. Muehrcke, (1970), pp. 199–200.

39. P. C. Muehrcke, (1970), p. 199.

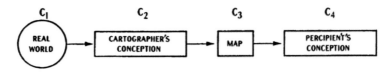

FIG. 5.   A diagram of the cartographic communication system with emphasis upon the conceptual aspects. *(University' of Wisconsin Cartographic Laboratory.)*

Muehrcke's model points up clearly the important activities in cartography. Each of the transformations is an appropriate area for research and development and provides a useful framework into which to fit these activities. Muehrcke[40] himself has so employed it with respect to thematic cartography, but no such review has been undertaken in the area of general mapping, that is non-thematic, although there have been a few significant studies such as the penetrating analysis by Keates[41] of the shortcomings of the transformations of real-world facts to the generalized symbolism of topographic maps.

Instead of concentrating attention on the operational aspects of cartography, so to speak, as Muehrcke did, one can focus on the substantive aspects employing some of the stages of Koláčný[42] and Ratajski,[43] but in a simplified manner. If we diagram this, we obtain the array shown in Figure 5. This comes much closer to the idealized information processing system discussed at the beginning of this chapter in the sense that the cartographer's (selective) conception of the real world (C_2) is the message to be transmitted, the map (C_a) is the coded signal, and the percipient's conception (C_4) is the message received. The arrows between each of the stages constitute Muehrckes' transformation stages.

Instead of modelling cartographic communication in a linear fashion by tracing the flow of a geographical conception and observing the operations performed upon it, it is also instructive-as well as being perhaps more appropriate-to 'map' the relations among the significant cognitive fractions of the communication. To do this we can employ a rectangular version of the Venn diagram (Figure 6).[44] In the simplified diagram shown in Figure 6 the entire area, S_c+S_e represents the total conception of the geographical milieu held by mankind, A represents one cartographer's conception or image of a subset of that milieu, and B, another subset, comprises the image of the, milieu held by a percipient-to-be of a map prepared by the cartographer. The dashed line separating S_c and S_e divides the conception of the milieu into error-free (S_c) and erroneous (S_e) segments. The definition of what constitutes error in this context need

40. P. C. Muehrcke, (1972), **Research in Thematic Cartography**, Resource Paper No. 19, Commission on College Geography, Association of American Geographers, Washington, D.C.

41. J. S. Keates, (1972), 'Symbols and Meaning in Topographic Maps', **International Yearbook of Cartography**, Vol. 12, 168–181.

42. A. Koláčný, (1968), **Cartographic Information-a Fundamental Notion and Term in Modern Cartography.**

43. L. Ratajski, (1972), 'Cartology'.

44. The authors are indebted, for suggestions and refinements, to colleagues and students, notable among whom was Mr Peter Van Demark at the University of Wisconsin who pointed out the necessity of including error in the system.

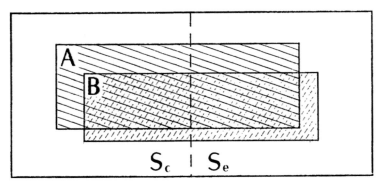

FIG. 6. A Venn diagram symbolising man's conception of the geographical milieu. *(University of Wisconsin Cartographic Laboratory.)* Diagram elements: S_c-correct conception of the milieu; Se-erroneous conception of the milieu; A-a cartographer's conception of the milieu; B-a percipient's conception of the milieu

not now concern us. The relative sizes of A and B in the diagram symbolize an assumption that the cartographer has a more extended image of the milieu than the percipient, a generally desirable state of affairs, but also immaterial to our purpose here. Both A and B are coincident for a portion of S showing that the images held by the cartographer and the percipient are the same in some respects.

In Figure 7[45] the shaded rectangle, M, superimposed on Figure 6 symbolizes the conception of the milieu put in the graphic form of a map by a cartographer. It is

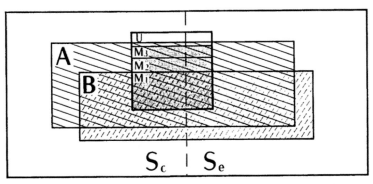

FIG. 7. A Venn diagram summarising the cognitive elements in cartographic communication. *(University of Wisconsin Cartographic Laboratory.)*
Diagram elements: C–cartographer; P–percipient; S_c–Correct conception of the milieu; S_e–Erroneous conception of the milieu; A–Conception of the milieu held by C; B–Conception of the milieu held by P; M–Map prepared by C and viewed by P; M_1–Fraction of M previously conceived by P; M_2–Fraction of M concerning S newly comprehended by P, a direct increment; M_3–Fraction of M not comprehended by P; U–Increase in the conception of S by P not directly portrayed by M but which occurs as a consequence of M, an unplanned increment.

45. Ibid.

the 'coded message' in the general communication system previously discussed. The rectangle, M, lies on both sides of the dashed line separating the error-free from the erroneous conceptions of the milieu since it is reasonable to assume that any map will contain some component of error arising not only from incorrect conceptions of the milieu but from error introduced by the cartographer or percipient.

The area of M is composed of three subdivisions: M_1, M_2, and M_3. The first, M_1, represents a kind of redundant fraction of the map consisting of that information which was already a part of the percipient's understanding. In other words, M_1 added nothing in the sense, for example, that a first-rate geographer would ordinarily find nothing new to him in an outline map of the United States. The fraction M_2 symbolizes those elements included in the map not previously comprehended by the percipient and thus constitutes a direct increment to his spatial understanding. The portion, M_3, represents the fraction of M not comprehended by the percipient, and accordingly it symbolizes the discrepancy between input and output in the communication system. This breakdown in the information transfer can occur for a variety of reasons, but whatever the causes it is a matter of critical concern to cartography. This has been recognized by cartographers who have studied the process,[46] but generally speaking no systematic study of the deficiency represented by M_3 has as yet been undertaken, mainly because the investigation of why something does not occur is both a never-ending and unrewarding task. One element contributing to the discrepancy between input and output in the cartographic system is what is called noise. Although the term is widely used it is not easy to comprehend the concept.

The unwanted signals that enter an electronic communications system, such as static in radio or humming in a telephone, came to be called noise for obvious reasons. When broadcasting systems added a graphic dimension it was only natural for the electronically produced snow on a television screen also to be called noise. The term is now widely used to refer to anything in a graphic display that interferes with the percipient obtaining the desired message. There is, however, a marked difference between the concepts of electronic noise and graphic noise, assuming each must be a part of that which is perceived. A radio blaring close at hand while one is trying to talk on the telephone may be noisy, but it is not noise in the communication system. Consequently, those deterrents to graphic communication, such as a severe headache or external distractions, which interfere with the percipient's concentration are not properly called noise because they are not a part of the system. But even here one gets on shaky ground; it is difficult to assert that one's head, even an aching one, is not part of the system.

Generally, when one refers to the graphic noise inhibiting cartographic communication, one points the finger at a variety of distracting elements, such as prominent patterns, eye-catching configurations, dense or overpowering lettering, or simultaneous contrasts of hue and value.[47] There is an almost unlimited number of such occurrences which can qualify as graphic obfuscations, but they are not easy to define

46. C. Board, (1967), 'Maps as Models', pp. 682–683, 698–704; A. Koláčný, (1968), **Cartographic Information-a Fundamental Notion and Term in Modern Cartography**; P. C. Muehrcke, (1969), **Visual Pattern Analysis, a Look at Maps**, pp. 2–5; P. C. Muehrcke, (1972), **Research in Thematic Cartography**, p. 3.

47. Cf. M. Merriam, (1971), 'Eye Noise and Map Design', Cartographica, Monograph No.2, 22–28.

and classify. When an odd geographical shape, or any other element difficult to see past, draws the eye but is part of that which is being communicated, it is illogical to call the same thing both message and noise in the same situation. To define noise as simply that which the percipient finds graphically distracting would be unsatisfactory because it ignores the message component. It is not our purpose here to define noise, but it does seem reasonable, in analogy with the original meaning of the term, to suggest that graphic noise may have to be limited to those delineations that are not necessary to the communication of the message. In any case, it is a significant part of the deficiency represented by M_3 in Figure 7.

Two other sections of M in Figure 7 symbolize additional consequences of the preparation and perception of a map. The area designated as U represents an increase in spatial understanding by the percipient that may occur because of his viewing the map but which was neither intended nor symbolized in any way by the cartographer. For example, if the percipient were quite familiar with the bedrock structure of a mapped area and the map showed the surface hydrography of the region, including the occurrence of springs, the existence of a significant correlation between the frequency of springs and the intersection of the ground surface with the contact between pervious and impervious beds might well come to mind, although this relationship is not actually mapped. Area U signifies a kind of unplanned increment, and it is probably a fairly common occurrence for geographically sophisticated percipients who integrate the limited symbolization of the milieu on a map with their previously acquired understanding.

The areas of M and U in Figure 7 which fall in the region of Se stand for a significant but negative component. Collectively they represent the 'error term' in the cartographer-percipient equation. The error can occur simply as a consequence of 'normal' incorrect conceptions of the milieu such as arise because of instrumental and observational error in data gathering or processing or it can come about as a consequence of symbolism.[48] As a specific example, numerous studies, such as those of Flannery,[49] Clarke,[50] Ekman, Lindman, and William-Olsson[51] and many others, have clearly shown that, when graduated circles are used as symbols and scaled so that their areas are in a linear proportion with the real world magnitudes they represent, then the 'message' conveyed is incorrect. Numerous other discrepancies between the cartographer's intent and the actual result due to map symbolism have been documented, primarily in the area of magnitude scaling.

There has been no definitive investigation of the extent of this 'error term' but numerous studies have all shown that it is quite large. These have ranged from

---

48. J. K. Wright, (1942), 'Map Makers are Human. Comments on the Subjective in Mapping', **The Geographical Review**, Vol. 32, 527–544; D. I. Blumenstock, (1953), 'The Reliability Factor in the Drawing of Isarithms', **Annals**, Association of American Geographers, Vol. 43, 289–304; A. H. Robinson and R. D. Sale, (1969), **Elements of Cartography**, John Wiley and Sons, Inc., New York.

49. J. J. Flannery, (1956), **The Graduated Circle: A Description, Analysis and Evaluation of a Quantitative Map Symbol**; J. J. Flannery, (1971), 'The Relative Effectiveness of Some Common Graduated Point Symbols in the Presentation of Quantitative Data'.

50. J. I. Clarke, (1959), 'Statistical Map Reading'.

51. G. Ekman, R. Lindman and W. William-Olsson, (1961), 'A Psychophysical Study of Cartographic Symbols'.

general studies of the nature of the problem, such as Jenks'[52] review of the extent of the error in thematic mapping to the more numerous investigations of specific components of the cartographic assembly, such as, for example, in typographic legibility,[53] the reading of statistical symbols,[54] the judgment of line weight,[55] or colour symbolism.[56] No attempt will here be made to apportion responsibility among the cartographer, the map percipient, or any outside factors for the portions of M and U falling in Se. The problem of the strategy to be employed to minimize the error quantities in the cartographic communication system[57] is a matter of prime concern in cartography. Among other things it may well involve the construct of an 'average map percipient' similar in concept to a kind of humanized economic man.[58]

The final matter yet to be treated in this survey of the cartographic communication system is the problem of measuring the quantity of geographical data encoded and transmitted by a map. Such a concept can encompass quite a range, from the simple notion of 'there is more on this map than on that one' to measures of the 'complexity' of a map, which presumably would be useful in studying questions of comparative map generalization and map design.[59] The problem of the quantitative assessment of map content has come under review only in recent years and promises to be a difficult one. Because it involves the measurement of the information provided by a map, a seemingly reasonable possibility would be that a solution to the problem might be found in the methods of Information Theory. On the contrary, it is the authors' conviction that, although something conceptually similar to the techniques of Information Theory may well have some limited use in special cartographic analyses, the fundamental concepts upon which Information Theory rests make impossible its direct application in cartographic communication.

Information Theory had its beginnings in the late 1920s at the Bell Laboratories when Hartley,[60] an electrical engineer, developed a mathematical formulation for the capacity of various systems of telecommunications to transmit electronic signals representing verbal language. These basic ideas were extended by Wiener[61] and by

52. G. F. Jenks, (1970), 'Conceptual and Perceptual Error in Thematic Mapping', **Technical Papers**, 30th Annual Meeting of the American Congress on Surveying and Mapping, 174–188.

53. B. Bartz, (1969), 'Type Variation and the Problem of Cartographic Type Legibility', **Journal of Typographic Research**, Vol. 3, No.2, 127–144; B. Bartz, (1970), 'Experimental Use of the Search Task in an Analysis of Type Legibility in Cartography', **Journal of Typographic Research**, Vol. 4, No.2, 147–167.

54. J. J. Flannery, (1971), 'The Relative Effectiveness of Some Common Graduated Point Symbols in the Presentation of Quantitative Data'.

55. R. D. Wright, (1967), **Selection of Line Weights for Solid Qualitative Line Symbols in Sel'ies on Maps.**

56. D. J. Cuff, (1972), **The Magnitude Message: A Study of the Effectiveness of Color Sequences on Quantitative Maps**; D. J. Cuff, (1973), 'Colour on Temperature Maps'.

57. A. H. Robinson (Moderator), (1971), 'Discussion: An Interchange of Ideas and Reactions', **Cartographica**, Monograph No.2, 46–53; Muehrcke, (1971).

58. A. H. Robinson, (1970), 'Scaling Non-numerical Map Symbols', **Technical Papers**, 30th Annual Meeting, American Congress on Surveying and Mapping, 210–216.

59. P. C. Muehrcke, (1973), 'Some Notes on Pattern Complexity in Map Design', unpublished paper prepared for panel discussion **Map Design for the User**, A.C.S.M. Annual Meeting, 15th March.

60. R. V. L. Hartley, (1928), 'Transmission of Information', **Bell System TechnicalJournal**, Vol. 7, 535–563.

61. N. Wiener, (1948), **Cybernetics**, John Wiley and Sons, Inc., New York.

Shannon and Weaver[62] culminating in laws having to do with matching the coding of messages to the noise characteristics of the channel. As Weaver clearly observed, the word 'information' in Information Theory is used in a special sense not to be confused with meaning, and two messages, one meaningful and one nonsense, can be equivalent in terms of their information content.[63] The measure of information in Information Theory is based upon the principle of uncertainty in the sense that a greater number of possible choices indicates more uncertainty, and the greater the uncertainty the greater the information capacity simply because more choices are available. The unit of uncertainty, i.e., information, is based upon the two equally likely choices in a binary system and is called the bit, a contraction of 'binary unit'.[64] Logarithms to the base 2 are employed to express the amount of information, so that the measure of the information content of a simple binary system would be unity, i.e., $\log_2 2 = 1$. All elements of a coding system are not necessarily equally probable, however, so that the measure is determined by summing the products of the logarithms of the probability, p, of every element each multiplied by that same probability, as in

$$H = -\sum_{i=1}^{n} p_i \log_2 p_i$$

This summation, is designated by H and called entropy, because of its analogy to the quality of random disorganization called entropy in statistical mechanics. Since the maximum value for H for a system would be attained when all its elements are equally probable,[65] His therefore a simple measure of uncertainty, and the greater the freedom of choice (uncertainty), the more the information. It is apparent, then, that meaning, in the ordinary sense, and information in Information Theory are essentially the inverse of one another. This has been recognized by Bertin who writes:

> ... if 'information' is the improbable [uncertain], maximum information is maximum improbability, that is, total spontaneity. But meaningful perception is a relationship (Fechner); and the relationship between two complete spontaneities, that is, between two completely unknown things, has no meaning.[66]

It is surprising that this 'essential paradox', as Bertin calls it, has not received more attention from those interested in applying the techniques of Information Theory in contexts other than electronic communication.

The lure of Information Theory as a possible method for quantifying the content of a variety of communication systems has been great, partly because of its tremendous importance and impact on the understanding of the processes connected with the transmission of signals.[67] Psychology was immediately attracted to the apparent potential of Information Theory, especially in the area of perception. Numerous

62. C. E. Shannon and W. Weaver, (1949) **The Mathematical Theory of Communication**, University of Illinois Press, Urbana.

63. C. E. Shannon and W. Weaver, (1949) **The Mathematical Theory of Communication**.

64. J. Singh, (1966), **Great Ideas in Information Theory, Language and Cybernetics**, p. 14.

65. J. Singh, (1966), **Great Ideas in Information Theory, Language and Cybernetics**, pp. 15–21.

66. Personal communication, in French, 4th January 1972.

67. Y. Bar-Hillel, (1964), **Language and Information**, Addison Wesley, Reading, MA and the Jerusalem Academic Press, Ltd., p. 283.

papers and books appeared, but by the 1960s doubts about its direct applicability were being voiced. For example, Pierce remarked:

> *It seems to me, however, that while information theory provides a central, universal structure and organisation for electrical communication it constitutes only an attractive area in psychology. It also adds a few new and sparkling expressions to the vocabulary of workers in other areas.*[68]

One field of research in psychology, analogically close to map perception, is figure perception. The inability of Information Theory to contribute much in this area is spelled out in detail in a paper by Green and Courtis[69] with the expressive title *Information Theory and Figure Perception: The Metaphor that Failed*. They point out[70] that for Information Theory to be employed in any perceptual investigation among the requirements that have to be met are two: (1) an agreed-upon alphabet of signs with known and constant probabilities of occurrence, and (2) the probabilities must be objective. They demonstrate clearly that in the perception of two-dimensional graphic arrays of marks, neither requirement can be met when one attempts to measure the 'information content', even though they are 'readily justified in the original context in which information theory was developed ...'.[71] They go on to emphasize that Information Theory is concerned with components (and the probabilities of their occurrence) which are to be perceived in linear sequences, and this is totally different from the perception of a two-dimensional array of marks in a figure (map).

The assumption of linearity in Information Theory, that is, the sequential pattern in telecommunication systems, is clearly stated even in Hartley's pioneering paper:

> *In any given communication the sender ... selects a particular symbol and ... cause, the attention of the receiver to be directed to that particular symbol. By successive selections a sequence of symbols is brought to the listener's attention.*[72]

It is generally agreed that we do not perceive the elements of two- or three-dimensional displays in any strict sequence, but in the unified and interrelated form of a Gestalt. To be sure, there will be some sequential pattern of eye movement involved in the total view, but research on eye movements so far indicates that the patterns are relatively varied and unpredictable.[73] Green and Courtis point out that one can impose a linear sequence on the perception of a figure, but that this completely distorts the normal situation, since:

> *Perception of a sequence of relationships between points [or signals] is a very different process from that of the perception of simultaneous relationships between parts [of a figure].*[74]

68. J. R. Pierce, (1961), **Symbols, Signals and Noise: the Nature of Communication**, Harper Torchbooks, The Science Library, Harper and Row, New York, p. 249.

69. R. T. Green and M. C. Courtis, (1966), 'Information Theory and Figure Perception: The Metaphor that Failed'. **Acta Psychologica** (Amsterdam), Vol. 25, 12–36.

70. R. T. Green and M. C. Courtis, (1966), 'Information Theory and Figure Perception: The Metaphor that Failed', p. 12.

71. R. T. Green and M. C. Courtis, (1966), 'Information Theory and Figure Perception: The Metaphor that Failed', p. 13.

72. R. V. L. Hartley, (1928), 'Transmission of Information', p. 536.

73. G. F. Jenks, (1973), 'Visual Integration in Thematic Mapping: Fact or Fiction', **International Yearbook of Cartography**, Vol. 13, 27–35.

74. R. T. Green and M. C. Courtis, (1966), 'Information Theory and Figure Perception: The Metaphor that Failed', p. 31.

Also, they note that one can break up a figure (abstract picture, design, map, etc.) into cells for a summary analysis, unit by unit, but they note that ' . . . this is not how it is *seen* by the percipient'.[75]

Another very significant element in the problem is that the amount of knowledge conveyed by a map (figure) is not only a function of the cartographer but of the percipient as well. Furthermore, even with respect to language Green and Courtis make the very important point that:

> . . . *the alphabet of signs is not one of letters or words, but of meaning units, which are . . . peculiar to the individual. What is transmitted may be regarded by an engineer as simple, a string of words, or letters and spaces for that matter, but what is* communicated *is a set of meanings . . . the essential property of language is not to be found in Markov chains but in syntactic constituents . . . which leaves information theory right back where it started — a mathematical tool in communication engineering, particularly useful for dealing with the technical problems of channel capacity.*[76]

Another parallel between figural perception and map perception is that of blank spaces. In a caricature drawing often what is left out, so to speak, is very meaningful because of the input of the percipient. Green and Courtis refer to this in a very perceptive section analysing cartoon drawings.[77] In cartography, a blank space can contain a great deal of information by inductive generalization, such, for example, as the 'empty' area shown between a meandering stream and its (floodplain) edge shown by the sharp rise of the valley wall. Yet by being blank and homogeneous that area would have to be assessed as containing zero bits of information.

The possibility of measuring the information content of a map, and by subtraction, assessing the efficiency of its transmission, lured cartographers to attempt to apply the techniques of information theory in mapping. For example Roberts[78] reported that a standard USGS quadrangle contains between 100 and 200 million bits of information, and a variety of other assessments 2nd applications have been made.[79] Sukhov's study illustrates the basic problems referred to above. In an analysis of a map made by superimposing a grid of cells he determined the 'information content' to be 1025.1 bits per cm^2. Disregarding the problem of equating the 'information content' of one symbol as opposed to another, the practice of summing the 'information content' in the cells of an arbitrary grid totally disregards the positional factor. In other words, if the cells were rearranged so that the resulting array was absolute geographical

75. R. T. Green and M. C. Courtis, (1966), 'Information Theory and Figure Perception: The Metaphor that Failed', p. 31.

76. R. T. Green and M. C. Courtis, (1966), 'Information Theory and Figure Perception: The Metaphor that Failed', p. 32–33.

77. R. T. Green and M. C. Courtis, (1966), 'Information Theory and Figure Perception: The Metaphor that Failed', p. 20–27.

78. J. A. Roberts, (1962), 'The Topographic Map in a World of Computers', **The Professional Geographer**, Vol. 14, No.6, 12–13.

79. E.g. V. M. Gokhman, M. M. Meckler and A. P. Polezhayer, (1970), 'Special and General Informative Capacity of Topical Maps', unpublished paper prepared for the 5th International Cartographic Conference of the I.C.A.; V. I. Sukhov, (1970), 'Application of Information Theory in Generalisation of Map Contents', **International Yearbook of Cartography**, Vol. 10, 41–47; V. M. Gokhman, M. M. Meckler and A. P. Polezhayer, (1970), 'Special and General Informative Capacity of Topical Maps', unpublished paper prepared for the 5th International Cartographic Conference of the I.C.A.; V. Balasubramanyan, (1971), 'Application of Information Theory to Maps'. **International Yearbook of Cartography**, Vol. 11, 177–181.

nonsense, the 'information content' would still be the same. Furthermore, the probabilities must be assigned in some quite arbitrary way for there is no simple geographical equivalent to such things as word and letter frequency or redundancy in a language.

Because the communication of the 'information' on a map is in no way like that of a coded sequential message consisting of signals, the measurement of the 'information content' of a map and of the amount transmitted by a cartographer to a map percipient cannot be effectively obtained with the techniques of Information Theory. We will have to search for other ways to assess M in Figure 7.

Although the direct application of Information Theory seems not to provide a method for assessing the actual 'information content' of a map, some analytical method similar in concept may be useful in quantifying some of the characteristics of a map of concern to the cartographer. For example Muehrcke[80] in a summary of approaches to analysing 'pattern complexity' in maps, suggests the utilization of the general concept of redundancy as a measure of this attribute of a map. Because he retains a fixed 'alphabet' by limiting the application to a specific class of maps and because the 'alphabet' varies along only one dimension (numbers of class intervals) the application is legitimate. Even though the redundancy sums themselves may not have any absolute validity in the sense that they do not actually assess the total 'information content', the general approach of Information Theory can provide a useful basis for comparison.[81] It seems reasonable that a quantity similar to that of entropy (H), arrived at by some as yet undevised method, may be developed to quantify such attributes as 'systematic regularity' or its opposite 'visual disorganisation', 'figural dominance', and other such qualities not now well understood and subsumed by the term 'pattern complexity'. Presumably, such attributes should be related to the quantity of information communicated by a map. What is needed, however, is not simply the direct application in cartography of the mathematical-statistical techniques of another field, but the development of techniques, perhaps through adaptation, to the unique conditions of cartography.[82]

In this paper we have seen that mapping is basically an attempt at communication between the cartographer and the map percipient. Although there may be fundamental differences between the kinds of 'messages' being conveyed by various classes of maps, all maps have as their aim the transfer of images of the geographical milieu. Analyses by cartographers of the communication process have tended to concentrate on the flows and summary character of the operations involved in the communication, primarily as a means of identifying the components and interrelationships among the various operations which make up the field of cartography. It seems entirely possible thus to identify the basic components of the field of cartography in a 'macro-sense' by surveying its fundamental activities as described briefly in this article. On the other hand, any thorough understanding of the field must involve a much greater penetration in which man's perceptual and cognitive processes are probed.

80. P. C. Muehrcke, (1973), 'Some Notes on Pattern Complexity in Map Design'.

81. K. A. Salichtchev and A. M. Berliant (1973), 'Methodes d'utilisation des cartes dans les recherches scientifique', **International Yearbook of Cartography**, Vol. 13, 156–183, pp. 162–163.

82. I. Kádár, M. Agfalvi, L. Lakos and F. Karsay, (1973), 'A Practical Method for Estimation of Map Information Content', unpublished paper presented at **Meeting of Commission III**, I.C.A., Budapest, August; A. Molineux, (1974), 'Communication Theory and its Role in Cartography', unpublished paper presented at **A.C.S.M. Annual Meeting**, St. Louis, March.

Reflections on 'The Map as a Communication System'

GEORG GARTNER

Vienna University of Technology

Cartography can be explained, analyzed and understood in several ways. Attempts have been made to emphasize the artistic element of cartography while others have focused more on its technological characteristics. The role of maps was very often explained from a perspective of how they function as artefacts, and papers such as 'The Map as a Communication System' by Arthur H. Robinson and Barbara Bartz Petchenik, first published in *The Cartographic Journal* in 1975, were among the first to offer a more holistic explanation of the function of maps and thus a new view towards defining cartography as a communication science. The authors sought to integrate former important contributions like those from Bertin (1968), Salichtchev (1970), Ratajski (1973) and especially Koláčný (1969), which all dealt with the application of communication science to cartography.

Robinson and Petchenik aim to define the 'big picture' on all the models being proposed in this regard. They also try to use a comparative methodology by applying formal descriptions and aim to evaluate how the communication models differ, especially in respect to the 'input' and the communication transformation through maps. When assessing the amount of information being included in a map they confirmed that the more linear information theory models are not applicable to the processes of cartographic communication.

The merits of this paper lie in the authors' differentiated discussion of the elements of the cartographic communication models, thus confirming their validity and importance. Although concepts and terms like 'noise' (in the sense of possible threats to transmitting a code via graphical symbols), 'percipients' or 'communication process' were not introduced for the first time in their paper, by re-evaluating them and by putting them in an analytical framework, Robinson and Petchenik contributed significantly to the acceptance of their importance within the domain of cartography.

The understanding of the map as an instrument of communication between a cartographer and a map percipient still carries some value today. The communication paradigm is useful, but it also has its limitations, which become clearer when digital cartography, and especially interactivity with maps, are considered. The domains of communication processes can not only be reduced to syntactical and eventually semantic aspects of transmitting information, but rather the pragmatic aspects become more important. Freitag (2008) suggested that the cartographic communication model needs to be enhanced with a new dimension that allows social aspects of the communication process to be included, while Gartner (2009) showed, within the context of Web Mapping 2.0, that most academic cartographic research has focused on controlling the syntactical dimension of communication and trying to provide better maps by exploring the useful semantic aspects of semiotics. The collaborative and participatory nature of Web Mapping 2.0 will lead to a change in research priorities, with pragmatics likely to receive much more attention. It is the user's behaviour and interests that determine the communication process in Web 2.0; semantics and especially symbol and sign syntax are usually beyond the control of collaborative users.

Today, maps can be created and used by any individual possessing only modest computing skills from virtually any location on Earth and for almost any purpose. In this new map-making paradigm, users are often present at the location of interest and produce maps that address needs that may arise instantaneously. Cartographic data can be delivered digitally and wirelessly in finalized form to the device in the hands of the user or they may derive the requested visualization from downloaded data *in situ*.

While the above advances have enabled significant progress on the design and implementation of new ways of map production over the past decade, many carto-graphic principles remain unchanged; the most important one being that because maps are an abstraction of reality, they need to communicate spatial information to the recipient via an efficient depiction that perpetuates its use. The visualization of selected information means that some features present in reality are depicted more prominently than others while many features might not even be depicted at all. Abstracting reality makes a map powerful, as it helps us to understand and interpret very complex situations very efficiently. Papers like this one by Robinson and Petch-enik hold especial value in that the insights they gained are still valid in the era of today's modern cartography.

In their summary the authors point out that for further understanding, 'a greater penetration in which man's perceptual and cognitive processes are probed' is needed. It can be noted that such efforts have been made and are still ongoing, as the newly founded ICA Commission on Cognitive Visualization proves. However, there are still numerous unanswered questions and we are still far from having 'the' theory of cartography.

It is interesting to read this paper nearly 40 years after it was first published. It seems to be borne out of a particular era of cartography where the major proponents worked in a certain 'atmosphere' by helping to lay a strong theoretical foundation to what they believed necessary for a bright and important future of the discipline of cartography. You can nearly smell some kind of 'pioneer' spirit in the contributions of that era. The discipline was noble, clearly defined and respected. This has changed over time. A paper like this could be published under a variety of headings today; it could still be seen as a contribution to cartography but also as a contribution to geoinformation science, geovisualization, geospatial analysis, geodata infrastructure, geomatics, or several other terms. It is the framework we are operating in today that has changed and makes reading a paper like this so rewarding, as it reminds us that we might need eventually another era of pioneers in cartography. The noble overall aim of Robinson and Petchenik's paper of 1975 — to contribute to build upon a theoretical foundation of a discipline — requires a similar endeavour nowadays.

Credit should go not only to the authors of this specific paper but also to *The Cartographic Journal*. It is the activities of the leading journals which more or less shape a discipline. In this respect, *The Cartographic Journal* was and is a spearhead of the scientific discipline of Cartography and all of us owe our thanks. It is my sincere pleasure to wish the Journal ongoing success in attracting trend-setting papers for the next 50 years to come.

REFERENCES

Bertin, J. (1968). 'Cartography in the computer age', in **Technical Symposium S 40**, ICA-IGU, New Delhi, December 1968.

Koláčný, A. (1969). 'Cartographic information — a fundamental concept and term in modern cartography', **The Cartographic Journal**, 6, pp. 47–49.

Freitag, U. (2008). 'Von der Physiographik zur kartographischen Kommunikation — 100 Jahre wissenschaftliche Kartographie', **Kartographische Nachrichten**, 58, pp. 59–66.

Gartner, G. (2009). 'Web mapping 2.0', in **Rethinking Maps**, ed. by Dodge, M., Kitichin, R. and Perkins, C., Routledge, New York.

Ratajski, L. (1973). 'The research structure of Theoretical Cartography', **The International Yearbook of Cartography**, 13, pp. 217–228.

Salichtchev, K. (1970). 'The subject and methods of cartography', **The Canadian Cartographer**, 7, pp. 77–87.

Topographical Relief Depiction by Hachures with Computer and Plotter

PINHAS YOELI

Originally published in *The Cartographic Journal* (1985) 22, pp. 111–124.

The introduction of lithographic reproduction methods for printing of maps instead of copper engraving techniques almost 150 years ago, has supplanted the use of hachures for cartographic relief depictions. The article explains. How computers and plotters can be used to produce hachures based on an initial input of digital terrain models. Instead of using hachures, as in the past, as a possible means of topographic relief presentation only, it is suggested to utilize them as a source for the computer assisted treatment of morphometric topics.

HISTORICAL BACKGROUND

One of the earlier methods of topographical relief depiction on maps was the use of 'hachures'. They were applied in two forms. The 'slope hachures', as first introduced by]. G. Lehman,[1] were short lines drawn in the direction of the slope of the terrain. Their widths were made proportional to the gradient of the slope according to the principle 'the steeper the darker'. The density of the strokes, i.e. the number of hachures per centimeter in a row of hachures was kept constant over the map except for horizontal areas which were left blank. A system of horizontal contours at constant vertical interval served as the base of the hachuring. The length of the individual hachure was, consequently, supposed to be the horizontal projection onto the map plane of the length of the steepest slope between two adjacent contours.

As 'slope hachures' in the form described above created almost no direct visual impression of the third dimension (see Figure 1), 'shadow hachures' came to be used. An oblique direction of light-usually from the north-west direction-was assumed and the width of each individual hachure was drawn according to whether it came to be situated on an 'illuminated' or a 'shadowed' slope. This created a three dimensional effect. Some of the most beautiful and meticulously executed examples of this method of relief depiction are the topographical maps of Switzerland at the scale of 1:100 000 known as the 'Dufour Atlas' which were first published during the period of 1842–1864. A small section of such a map is shown-in enlargement, in Figure 2.

A comprehensive description of these methods and their various forms of applications can be found in E. Imhors book: 'Cartographic relief presentation'.[2]

1. Lehmann, J. G. (1799) **Darstellung einer neuen Theorie tier Bergzeichnung der schiefen Fliichen im Grundriss oder der Situationszeichnung der Berge,** Leipzig.
2. Imhof, E. (1982) **Cartographic relief presentation,** Walter de Gruyter, Berlin-New York.

Fig. 1.   Part of a map with 'slope hachures'

Fig. 2.   Part of a 'Dufour' map-twice enlarged

The development of modern topographical surveying techniques, the introduction of photogrammetric surveying methods-both terrestrial and aerial-and last but not least the replacement of copper engraving cartographic reproduction methods by lithographic processes, stimulated an ever increasing usage of topographic contours, supplanting rapidly all previous methods of topographical relief depictions. It should be pointed out, however, that similar to 'slope hachures', contours alone also lack the direct three dimensional visual impact. To overcome this detriment, oblique hill shading is often combined with contour presentations.

The aim of this article is to describe the logics of a computer program named 'HACHURE' as developed by the author. It computes hachures which can then be drawn by a plotter either in the form of 'slope hachures' or 'shadow hachures' according to the decision of the program user.

Perhaps the question may be asked: Why bother about a method of cartographic relief depiction which has become as obsolete as hachuring and why waste time and effort on such an enterprise?

The answer is, that we did it first of all as an exercise in experimental computer assisted cartography, prompted also by the beauty of the maps of the 'Dufour Atlas' and the copper engraving hachuring technique in general. (Out of admiration for Albrecht Durer, that great master of these techniques, we were even tempted at times to name our program 'ALBRECHT'.)

We wished to find out if it can be done with a machine and, when successful, to experiment with the method in such a fashion as would have been practically impossible when done manually.

The second reason for trying to produce hachures with computer assistance was, that the method of hachures and especially of 'shadow hachures' does perhaps not deserve to be as forgotten as it is. We wish to quote E. Imhof from his book 'Cartographic relief presentation' p.229: 'Hachures alone, without contours, are more capable of depicting the terrain than is shading alone since the stroke direction indicates the line of maximum gradient everywhere and in every direction and this gradient can be worked out from the length of the hachures. In small scale maps, where contours can no longer be used easily, the hachure is still a very good element for portrayal. Hachures also possess their own special, attractive graphic style. They have a more abstract effect than shading and perhaps for this reason are more expressive. The finely grained, changing play of black to white of paper shade and hue, increases the drama and brilliance of expression'.

The present obsoleteness of hachuring on maps of even small scales, where they could still fulfil a useful role in relief presentations, is due mainly to the often laborious efforts required for their manual drawing and the high professional skill needed to achieve results of good quality.

A third reason for the development of this algorithm was the possible applicability of hachuring techniques for the graphic depiction of morphometric topics such as slopes. Examples of this employment of hachures are shown in the *Figures* 35, 36, 37 and 38.

Program 'HACHURE', whose logics will be explained forthwith, tries to instruct a computer together with a plotter to do the job. We should, however, not delude ourselves to believe, that considering the present technical state of digital plotters one could achieve with them alone the same graphic excellence and finess of manually engraved hachures as produced by superbly skilled craftsmen about a hundred years ago.

As mentioned previously, both 'slope hachures' and 'shadow hachures' were constructed according to certain rules. There are 'five of them, of which the first four were identical for both forms of hachures.[3]

1. The direction of every individual stroke follows the direction of the steepest gradient (Figure 3).
2. The hachures are arranged in horizontal rows (Figure 4).
3. The density of the hachures, *i.e.* the number of strokes per centimetre of horizontal extent is constant over the whole area of the map except for horizontal planes which remain blank (see Figure 3 and Figure 4).
4. The length of each stroke corresponds to the actual local horizontal distance between assumed contours of a certain vertical interval. So far for both types of hachures. The fifth rule for 'slope hachures' is as follows:
5. The width of the strokes is proportional to the slope according to the principle: 'the steeper the darker' (see Figure 5).

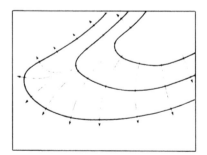

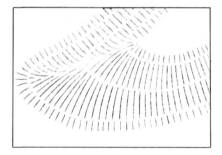

FIG. 3.   Direction of hachures (Courtesy of E. Imhof)   FIG. 4.   Arrangement of hachures in horizontal rows (Courtesy of E. Imhof)

3. Imhof, E. (1982) **Cartographic relief presentation**, Walter de Gruyter, Berlin-New York.

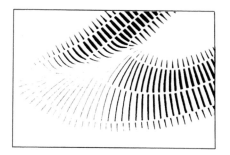 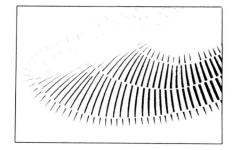

FIG. 5. 'The steeper the starker' — 'Slope hachures' (Courtesy of E. Imhof)

FIG. 6. 'Shadow hachures' (Courtesy of E. Imhof)

For 'shadow hachures' the analogous rule is:

The stroke width is a function of the amount of light falling from an assumed oblique light direction on the slope according to the principle: 'the darker the wider' (see Figure 6).

There is a contradiction between this rule and rule 3. The brightest spots on an obliquely illuminated relief are to be found on slopes facing the light direction. These receive even more light than the horizontal planes. According to rule 3, however, even the most illuminated slope must still be covered with the same number of strokes as the darkest declivity, although with minimum width. This leaves the horizontal planes of the terrain as the brightest parts of the map. The explanation of this obvious inconsistency is to be sought in the chronological development of the methods of cartographic relief depictions. First 'slope hachures' and afterwards 'shadow hachures' came into use, both before the invention of lithography (in 1796 by Alois Senefelder) came to be applied for cartographic reproduction purposes (since 1808) enabling the printing of shadow toning by nonlinear graphic elements. To achieve a toning with hachures consistent with the real shadow effects of an oblique illumination, one would have to cover all the horizontal planes of a map with thin hachures, leaving only the brightly illuminated slopes, facing the light, uncovered. Instead of this it was preferred, as illogical as it was, to remain with the 'Lehmann heritage' and draw hachures on every slope but not on horizontal planes.

The algorithms of program 'HACHURE' are based, in principle, on the computer compatible formulation of the five rules mentioned above. There is, however, no need to formulate different rules for 'slope hachures' and 'shadow hachures'. As we shall see later, the varying widths of the hachures as computed by the program are a function of the direction of light chosen at will by the program user. Vertical illumination will produce 'slope hachures' while oblique illumination results in 'shadow hachures'.

THE INPUT DATA FOR PROGRAM HACHURE AND ITS PRELIMINARY TREATMENT

As hachures, according to rule 2 and 4, are supposed to be based on a contour system, the input data for program 'HACHURE' consists primarily of a file of digitally stored contours of the area to be hachured. The source of these contours is irrelevant. They may be either the digitized contours of already existing topographical maps of the area, preferably at a larger scale than the scale of the planned hachure-map, or the

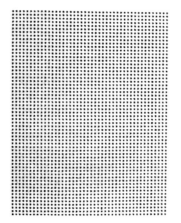

FIG. 7.   DTM of 2700 points
(45 X 60) 1:25 000

FIG. 8.   Contours threaded into the DTM.
Scale: 1:25 000, Contour interval: 5 metres

result of the contour threading process of a computer program using as primary input a digital terrain model.

The choice of the vertical interval of the contours used depends, according to rule 4, on the lengths desired for the hachures. On manually drawn hachured maps this length is at least 0.3 mm for the steepest slope. As an example, for a map in the scale of 1:20 000 of a terrain with possible steepest slopes of approximately 40° this would require a contour-system of a vertical interval of approx. 5 m.

For the specific examples illustrating this article, the contours used are the stored output of a program named 'CONTOUR' which threads contours into a regular digital terrain model at any required vertical interval.[4] The DTM used was surveyed photo-grammetrically and consists of 60 rows and 45 columns of height points. The distance on the ground between the rows and the columns is 30 metres and the ground surveyed covers therefore an area of 1770 × 1320 m² situated south of Haifa in Israel. Into this lattice of 2700 points (see Figure 7) program 'CONTOUR' threaded contours at 5 m interval as shown in Figure 8, in the scale of 1:25 000.

The algorithm of program 'CONTOUR' is based on a linear interpolation of the contour heights between neighbouring points of the DTM. In other words, it finds the intersection points of the required contours with the lines connecting the points of the DTM in the direction of the rows and the columns. These are the 'support points' of the contours as shown in Figure 9.

Their x and y coordinates are stored and catalogued on a special file used as input for program 'HACHURE'. Every branch of a contour is preceded by the registration of its height. To produce the drawing of Figure 8, spline curves were interpolated between the 'support points' resulting in continuous curves (see also 'Cartographic drawing with computers'[5]).

4. Yoeli, P. (1984) 'Cartographic Contouring with Computer and Plotter', The American Cartographer, Vol. 11, No.2, pp.139–155.

5. Yoeli, P. (1982) Cartographic drawing with computers, Computer applications, special issue, Vol. 8, Department of Geography, University of Nottingham.

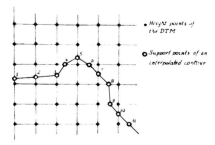

FIG. 9. Support points of a contour

Following rule 2 of the way hachures were produced manually, the computation and drawing of the hachures is executed by program 'HACHURE' in horizontal belts delineated by two contours whose height difference is one vertical interval. For this purpose the program finds in a first step the heights of the highest and the lowest contour levels stored on the external contour file.

After rewinding the file it then collects from the file the coordinates of the first highest contour branch encountered and stores them internally in the arrays XUP and YUP. This contour will now serve as a sort of 'washing line' from which the first row of hachures will be suspended. The density is dictated by a parameter named IDENS which indicates to the program the number of hachures to be produced per centimetre of 'washing line'.

As the support points of the contour stored in XUP and YUP have not necessarily the required proximity nor are they equidistant, the contour must first undergo what may be called 'cosmetic treatment', ensuring an array of points situated on the contour at equal distances determined by IDENS.

If the contours were digitized in distance mode from an existing contour map, there is, of course, no need to modify the point arrays, provided the proximity of the digitized points has been chosen in accordance with the required density of the hachures.

If the input are the support points of contours threaded by a contouring program into a DTM such as the type shown in Figure 7, one is often faced with the fact, that the resulting contours are afflicted with 'noise', i.e. small bulges and irregularities in the flow of the contours. If these are no dictated by the geomorphological character of the terrain, but are the result of the threading technique applied or caused by the use of a DTM too dense for the cartographic job at hand, they are better removed·
-by a preliminary smoothing process of the contours.

As mentioned before, the contours used for the demonstrations of this article are the result of a computer executed contour threading process as shown in Figure 9. The positioning of the equidistant points could, of course, be done on the lines connecting the given support points of the contours as shown in Figure 10a.

The angularity of such a polygon would, however, also influence the appearance of the rows of hachures strung along its linear segments. The original contour point array is therefore sent to a subroutine named 'MODSMO' (MODified SMOoth) (see

73

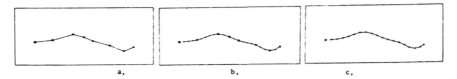

a,   b,   c,

FIG. 10a.   Original contour polygon
FIG. 10b.   Contour-polygon after 'MODSMO'
FIG. 10c.   Equidistant array of contour points

'Cartographic drawing with computers'[6]), which interpolates and stores temporarily an array of points lying on a smooth curve which passes through the given array. The distance between these interpolated points can be chosen at will, but even if we choose them in accordance with IDENS, this interpolated array will not yet be completely equidistant as it still contains also the original contour points (see Figure 10b). Therefore, in a next step the equidistant array is now established using the 'curved' polygon interpolated by 'MODSMO' (see Figure 10c). The algorithm for the creation of an equidistant array superimposed on a given polygon is simple and shown in Figure 11.

The first point of the polygon is chosen as the centre of a circular array of points defining a circle whose radius equals the required distance to the next point as dictated by IDENS. The intersection of this circle with a segment of the given polygon establishes the next point and also the centre for the next circle, etc. These intersection points are now considered to be the new array of contour points and their x and y coordinates are overwritten into the arrays XUP and YUP.

The algorithm applied above does not necessarily remove the bulges and irregularities mentioned previously. This task is executed by a subroutine name 'NOSPIKE' which functions as follows. Among the parameters of the program, to be chosen by the map editor, are two values—ARCMAX and ARCMIN which indicate to the program the maximal change of direction allowed between two adjacent sides of the contour polygon. Beginning from the second point of the given array, the polygon angles are computed, and if found to be bigger than ARCMAX or smaller than

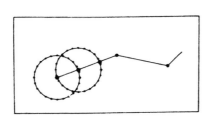

FIG. 11.   Explanation of subroutine EQDIST

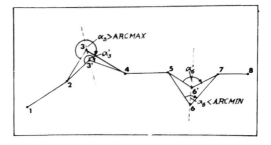

FIG. 12.   Smoothing of a contour (removal of bulges)

6. Yoeli, P. (1982) Cartographic drawing with computers, Computer applications, special issue, Vol. 8, Department of Geography, University of Nottingham.

ARCMIN the vertex of the angle is moved in the direction of the angle bisector to a position where the size of the new angle is inside the allowed margin. Figure 12 illustrates the procedure.

The polygon angle α_3 in point 3 was found to be bigger than ARCMAX. Point 3 was therefore moved to 3' where the new angle α_3 complies with the defined restrictions.

In point 6 the angle α_6 was found to be smaller than ARCMIN and the vertex was therefore moved to 6'.

This treatment is preferably applied to the original contour array before it is send to subroutine MODSMO but can, of course, be activated immediately afterwards. For the contours used for the illustrations in this article the following values were defined: ARCMAX = 225° and ARCMIN = 135°.

THE SEARCH FOR THE BELT TO BE HACHURED

After this preparation of the upper contour-the 'washing line'-the lower contour delineating the belt to be hachured must be collected from the external contour file. Its height is, as mentioned before, one vertical interval lower than the contour stored in XUP and YUP. But difficulties arise here. Not every contour branch of this height is suitable to serve as the lower border of the belt. To fulfil its role as the lower delineator of the hachures strung along the upper contour, the lower contour must encircle the higher one. Not all contour branches of the lower height answer this demand. Figure 13 illustrates this dilemma.

Let it be assumed, that the contour of height 50 m in Figure 13 is the upper contour stored in XUP, YUP. The vertical interval of the contour system is assumed to be 10 m. Of the four possible contour branches of height 40 m occurring inside the area of Figure 13, only branch III is suited to serve as the lower delineator for hachures to be strung along the contour 50 m. To enable the computer to reach this conclusion certain preparatory steps must be taken.

Once a certain contour branch of the lower height, consisting of ILOW points, has been stored internally in the arrays XLOW and YLOW, it is checked whether

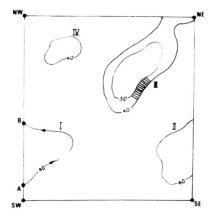

FIG. 13.   The search for the relevant contour belt

this contour closes on itself, i.e. if XLOW (1) = XLOW (ILOW) and YLOW(1) = YLOW(ILOW). In case of the affirmative, such as with contour branch IV in Figure 13, the control, if the contour encircles the upper contour, is straightforward. The program has to check if any point of the higher contour is situated inside the area enclosed by the lower contour. In program 'HACHURE' this control is executed by a subroutine named INOUT which determines if a certain point falls inside or outside a closed polygon (INOUT is part of a library of cartographic computer programs described and listed in P. Yoeli: 'Cartographic drawing with computers'). If this control proves to be negative the next contour branch of the relevant height is dealt with, overwriting the arrays XLOW, YLOW with the coordinates of the next contour branch of the lower height.

Not all contours close on themselves inside the map area or the area covered by the DTM. In fact, most contours start usually on the rim of the DTM area and must consequently end on the rim such as the contour branches I, II and III of Figure 13. In order to be able to use subroutine INOUT for these contours they must first be made to close on themselves, even if only artificially.

Let us assume, as an example, that the starting point of contour branch I in Figure 13 is the point A and B is its end point. This contour can be closed artificially, by connecting B to A by adding the coordinates of A once more to the end of the contour point array, making the first and last point of the contour array identical. Another possibility is the way around the border of the DTM, i.e. by adding after B the points NW, NE, SE, SW and A again. Which of the two possibilities is to be chosen depends on which side of the contour the terrain rises. The contour stored in XUP, YUP is on higher ground and only if the artificial closure of the lower contour will possibly encircle the upper contour can the activation of subroutine INOUT have any sense.

To be able to reach the right closure decisions the program needs information about the *rim profiles* of the DTM. It is therefore necessary that the DTM which served as the base for the creation of the contours is also available to the program as an additional input file. In a case where the contours were digitized from existing topographical maps, the border profiles of the relevant map area will also have to be digitized and stored on a separate file. As mentioned before, the contours used for the illustrations of this article were produced from a DTM. The program stores therefore, in an early stage of its algorithm, the four rim profiles of this DTM in the four arrays HLEFT HRIGHT, HDOWN and HUP, (see Figure 14).

Once the coordinates of a lower contour branch, composed of ILOW points, have been read off the external contour file and stored in the arrays XLOW and YLOW, the program finds out on which side of the four possible sides of the DTM the first point of the contour has come to be situated. For the contour to start on the left rim XLOW(I) must be zero, while a start from the lower rim would be indicated by YLOW(I) equalling zero. For a start on the upper rim YLOW(I) will have to be equal to YMAX and for a start on the right side XLOW(I) must equal XMAX (the values of XMAX and YMAX can be computed from: XMAX = DEL TAX *(N-l) and YMAX = DELTAY *(K-l), where DELTAX and DELTAY are the distances between the columns and the rows of the DTM and N and K the number of columns and rows respectively).

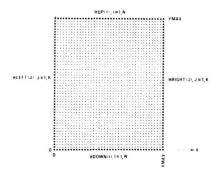

FIG. 14.   The rim profiles

Once the position of the starting point of a contour is determined, the program must discover between which pair of height points on the rim profile the starting point has come to be situated. After this it is easy to determine on which side of the contour the terrain ascends or descends.

In Figure 15 the first point of the contour line of height 30 m has been found as being situated on the first column of the DTM between the ninth and tenth points whose heights had previously been stored in HLEFT(9) and HLEFT(10) respectively. As the height of HLEFT(9) is found to be higher than the contour height, it indicates that pursuing the contour in the direction of increasing indices of the contour points, the terrain rises towards the right hand side. The program characterizes this contingency by allocating an indicator variable named IRIGHT the value 1. If HLEFT(10) would have found to be higher than the contour, the ascent of the terrain would have been towards the left hand side of the contour in which case IRIGHT would have received the value zero. Similar deductions can be made for contours starting on any of the other sides of the DTM.

Figures 16a–16e and 17a–17e illustrate the decisions made by the program for a fictitious contour closure once the value of IRIGHT for specific contour has been made.

Figures 16a–16e show the case for IRIGHT=O for different end points of contours, all starting on the left rim. Under each illustration the required point sequence of the correct closure is indicated. Both the cases of 16a and 16e (and 17a and 17e) show a

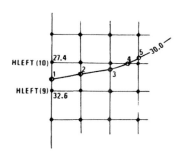

FIG. 15.   Determination of IRIGHT for open ended contours

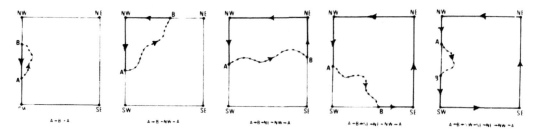

FIGS. 16a, b, c, d, e.   The artificial closure of a contour with IRIGHT = O

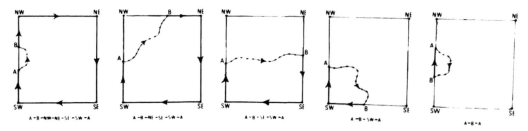

FIGS. 17a, b, c, d, e.   The artificial closure of a contour with IRIGHT = 1

contour starting and ending on the same, in this case left, rim, the only difference being that for case 16a (and 17a) the value of YLOW(I) is less than YLOW(ILOW) while for 16e (and 17e) the opposite is true.

Analogical relationship can be found for contours starting and ending on all the other rims of the DTM. The subroutine in program 'HACHURE' taking care of the correct artificial closure of open ended contours is named 'ENCIRCL' and it is, indeed, the most copious of all the subroutines of the program.

THE DIRECTION AND THE LENGTH OF THE HACHURES

Each hachure drawn in the belt between two contours is supposed to start at one of the points of the equidistant array stored in XUP, YUP, defining the upper contour. According to rule 1, the hachure is to follow the direction of the steepest gradient. The bisectors of the angles of a contour polygon are perpendicular to the tangents to

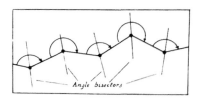

FIG. 18.   The direction of hachures

the interpolated curve at the contour points. These bisectors can therefore be regarded as indicating the directions of the steepest gradients at these points. Figure 18 shows a segment of a contour polygon with the bisectors of the polygon angles.

The hachure, starting on the upper contour, must obviously be drawn in the direction of the *descent* of the terrain. To be able to make this decision, the program must know the 'IRIGHT' value of the contour. Therefore, before the actual hachuring of the belt begins, the value of IRIGHT of the upper contour is determined in exactly the same way as explained previously for the lower contour. However, when we discussed the determination of the value of IRIGHT required for the closure decision for open ended lower contours, the case of a *a-priori* closed contour need not have been dealt with. The 'encirclement condition' for such a contour could be established without reference to an IRIGHT value. Not so for upper contours. For the decision on which side of a closed upper contour to apply the hachures the IRIGHT value must be found as for open-ended contours. The value depends on the answers to the following two questions:

1. Is the contour part of the contour depiction of a hill or a hollow?
2. In which sense of rotation were the stored support-points of the contour computed or digitized?

To answer the first question, a point of the DTM falling inside the area of the closed contour is found. If its height proves to be higher than the height of the closed contour the form of the terrain is an elevation, otherwise it is a hollow.

The procedure applied by the program to find the answer to question two, is explained with the help of Figures 19a and 19b.

Through a point P positioned inside the area of the closed contour a parallel to the ordinate axis y is laid and the intersection points of this line with the contour polygon are computed. The two points I and I+1 of the contour points array between which the lowest intersection point (point A in Figures 19a and 19b) has come to be situated are now compared. If the x coordinate of the point I+1 is larger than that of I, the sense of rotation of the indices of the contour point array is anticlockwise-as in Figure 19b. If x of I is larger than that of I+ 1, the sense of rotation is clockwise as in Figure 19a. The value of IRIGHT for Figure 19a would therefore be one for a hill or zero for a hollow. For Figure 19b IRIGHT would be zero for a hill and one for a hollow.

Once the IRIGHT value of the upper contour is known the following rule can be applied: *Pursuing the contour in the direction of increasing indices the hachures are to be drawn to the left of the contour if IRIGHT equals one and to the right if IRIGHT equals zero.*

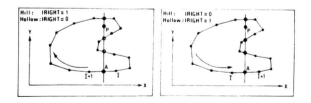

FIG. 19.   Determination of IRIGHT for closed contours

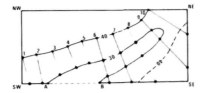

FIG. 20. Multiple intersections of hachure lines with the lower delineator of the hachured belt

The next problem to be solved in order to be able to draw the hachure, is to find its end point. According to rule 4 this should be the intersection point of the hachure line with the lower contour. Such a point, however, can not always be found-at least not with the real part of the lower contour in case it has been closed artificially. Besides, often more than one intersection will be found. Figure 20 illustrates the dilemma.

The figure shows the contours of the foot of a slope. Along the upper contour of height 40 m, ten hachures at 10 contour points have been drawn. As the IRIGHT value of this contour is obviously zero (following the contour in the direction of increasing indices, the terrain rises towards the left) the hachures have been drawn to the right of the direction of increasing indices. The lower contour of height 30 m is part of a valley contour which-in order to encircle the upper contour has been closed artificially by extending the original point array running from A along the contour to B, by the points SE, NE, NW, SW and back to A. Let us now examine what happens when the program intersects the hachure lines with the lower contour. For the hachure in point lone intersection point situated on the rim of the map is found. The same applies to the hachure in point 2 and both cases pose no problem. Each of the hachures in the points 3, 4 and 5 intersect the real part of contour 30 m once, and again, pose no problem. For each of the hachures in the points 6, 7, 8 and 9, however, three intersection points will be found by the program; two of these with the real part of the contour and one with the artificial contour extension. The hachure in point 10 poses still another problem. Although only one intersection point will be found-situated on the rim exactly like the intersection points of hachures 1 and 2-something is wrong. The hachure may have crossed another branch of the contour of height 40 situated on the opposite slope of the valley (in Figure 20 this possible branch is drawn with a dashed line). The program does not know, at this stage of the processing, of the existence of such a contour. The solution for the problem posed by the hachures 6, 7, 8 and 9 is easy. The program is instructed that in the case that more than one intersection point with the lower contour is found, the one nearest to the starting point of the hachure is to be selected as endpoint. The problems posed by cases demonstrated by hachure 10 can only be solved if we supply the computer with additional information about the valley and ridge lines of the terrain in the form of an additional data file. In fact, the cartographers who, manually produced the beautiful copper engraved hachures, had at their disposal a detailed layout of the structure lines of the terrain. These are the valley and ridge lines (watersheds) or lines of terrain discontinuities such as upper rims of cliffs etc. We can not expect that a machine will produce a similar product if we do not supply it with the same amount of information. For our purposes we found that only valley lines and lines of discontinuities

FIG. 21.   Valley lines of the terrain

were required. The file of these structure lines can be produced either by drawing them manually on the background of the contours followed by their subsequent digitization, or analytically with the help of an appropriate computer program, using as input the original DTM from which the contours were created. Figure 21 shows the valley line system as found by program STRUCT (see Yoeli[7]) in the DTM of Figure 7-superimposed on the contours of Figure 8.

Once a structure-line file is provided, the hachure lines are not only intersected with the polygon of the lower contour, but also with the structure line system, and if intersection points are found, they are added to the pool of the previously mentioned intersection points from which the one nearest to the starting point of the hachure is chosen to be the end point.

In addition to the problems illustrated in Figure 20, another three contingencies must be dealt with. At sharp turnings of the upper contour it may happen that a hachure will come to intersect a previously established one. The hachures of the belt are therefore stored, once their starting and end points have been established, in the temporary arrays XSTART, YSTART and XEND, YEND. Beginning from the second hachure in a belt, the program then checks if the hachure under investigation does not intersect any of the already stored ones (see Figure 22).

As the possibility of such an intersection is usually confined to the nearest neighbours, there is no need to store the entire assemblage of the hachures of a belt. The nearest 5–10 hachures are usually enough.

Another problem is posed by topographical saddles illustrated by Figures 23a and 23b.

Let us assume that the fully drawn contours of height 50 m and 40 m in the Figures 23a and 23b are the upper and lower contours stored in XUP, YUP and XLOW, YLOW respectively. Although the contour, drawn in the above illustrations with a dashed line, exists on the contour file, the program does not know about its existence at this stage. Hachures strung along the upper contour will therefore reach

7. Yoeli, P. (1984), 'Computer-Assisted Determination of the Valley and Ridge Lines of Digital Terrain Models', **International Yearbook of Cartography**, Vol. 24, Kirschbaum Verlag-Bonn-Bad Godesberg, pp. 197–206.

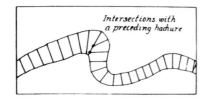
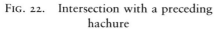
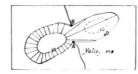
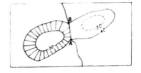

FIG. 22.   Intersection with a preceding
hachure

FIG. 23.   Interruption of hachures at topographical
saddles

right across the saddle and transvers the elevation on the other side of the saddle as shown in Figure 23a. The obvious way to prevent this, is to define-on the structure line file-the valley line right across the saddles i.e. to unite the two valley lines starting at both sides of a saddle into one continuous line (see points A and B in the above figures). This will cause a halt to the hachures as shown in Figure 23b.

The last problematic case to note, although it may occur rather seldom, is the closed contours of hollows. For the lowest of such contours no lower delineator can be found. Therefore, in cases like these, no hachures are drawn. This is similar to the reverse case of the area enclosed by the highest contour of a hill. There too, no hachures are applicable.

When the end point of a hachure is finally established, its length can be computed. The height difference between the start and end points of the hachure is one vertical interval of the contour system in a case where the end contour point lies on the lower contour. In a case where the end point is an intersection point with the structure line system, the height of the point will have to be established by interpolation from the height data of the structure line file.

In Figure 24, the hachure AB of the belt between the contours 40 m and 50 m has been found to intersect a valley line at point B. Its height is consequently found by a linear interpolation between the two valley line points whose height is given as 44.2 and 37.8 respectively.

From the length of the hachure and its height difference its slope can now be computed~ According to the second part of rule 3 no hachures shall be drawn on horizontal areas of the map. It is up to the program user to decide what the minimum slope still to be hachured should be. This slope is a parameter of the program named SLOPMIN. If the slope of the hachure under investigation proves to be smaller than SLOPMIN, it is not drawn and the next hachure in the belt is dealt with.

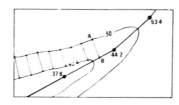

FIG. 24.   Intersection of a hachure with a valley line

FIG. 25. Various densities of hachures

The Figures 25a, 25b and 25c demonstrate the hachuring of the terrain shown in Figure 8 and in the same scale of 1:25 000 with a varying density of hachures. The contour interval used for the hachuring of the horizontal belts is 5 m.

In Figure 26, contours at 10 m interval were added to the map depicted in Figure 25b. This addition is optional, as well as the vertical interval of the drawn contour system, which must not necessarily be identical with the contour interval of the hachured belts.

The hachures of Figures 25 and 26 still lack, of course, the characteristics of slope or shadow hachures, i.e. their varying width.

THE WIDTH OF THE HACHURES

The wider one draws the hachures, the darker is the graphic effect. Their width is therefore a means to express the amount of illumination falling on the terrain from

FIG. 26. Hachures and contours

an assumed light direction. In order to be able to compute this illumination intensity, one can suppose that each hachure represents a small facet of the terrain, whose length equals the length of the hachure and whose width equals half an interval to the adjacent hachures on both sides. Figure 27 shows such a hachure-facet.

For a vertical illumination, as assumed by J. G. Lehmann for his 'slope hachures', the illumination intensity of a hachure depends on the cosine of the slope angle e (Figure 28) and the width of the hachure was determined as a function of this angle.

But this is also the angle between the normal to the slope and the light direction. In order to be able to determine the necessary width of the hachures for all cases we shall try, therefore, to compute this angle for any light direction.

Let us assume, that the hachure representing the hachure facet in Figure 27 is a vector $\vec{b}$ (leading from the point A situated on the upper contour to the end point B on the lower contour). If the coordinates of the points A and Bare XA, YA, HA and XB, YB, HB respectively, the components of vector $\vec{b}$ are

$$bx = XB\text{-}XA$$
$$by = YB\text{-}YA$$
$$bz = HB\text{-}HA$$

Let us define a second vector $\vec{a}$-also commencing in A perpendicular to vector $\vec{b}$. The end point of $\vec{a}$ is the point C. (see Figure 29), supposed to have the same height as point A. The length of $\vec{a}$ is half the hachure interval. This interval is, of course, a function of the parameter IDENS mentioned previously. HACHINT = IDENS/2.

According to Figure 29 the components ax and ay of vector $\vec{a}$ are:

$$ax = HACHINT.\cos \beta$$
$$ay = HACHINT.\sin \beta$$

$$where \; \beta = \arctan \left(\frac{bx}{by} \right)$$

and az = o.

The vector product $\vec{a} \times \vec{b}$ results in a vector $\vec{n}$ perpendicular to the plane of the hachure facet (Figure 30).

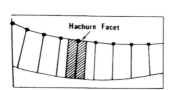

FIG. 27.   A hachure facet

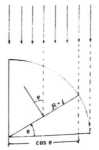

FIG. 28.   Vertical illumination

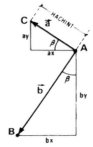

FIG. 29.   The vectors $\vec{a}$ and $\vec{b}$

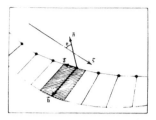

Fig. 30.   The vector $\vec{n}$

The components of $\vec{n}$, in general, are:

$$nx = ay \cdot bz - az \cdot by$$
$$ny = az \cdot bx - ax \cdot bz$$
$$nz = ax \cdot by - ay \cdot bx$$
but as $az = 0$
$$nx = ay \cdot bz$$
$$ny = -ax \cdot bz$$
$$nz = ax \cdot by - ay \cdot bx$$

Care should be taken that the program determines for the components of vector $\vec{a}$ such signs-that, according to the 'corkscrew rule' the vector $\vec{n}$ will point outwards, i.e. into the air and not into the ground.

If we put into the light direction a vector $\vec{s}$ whose components are sx, sy and sz, the scalar product of $\vec{s} \cdot \vec{n}$ can be computed.

$$\vec{s} \cdot \vec{n} = |\vec{s}| \cdot |\vec{n}| \cos e = sx \cdot nx + sy \cdot ny + sz \cdot nz \qquad (1)$$

where $|\vec{s}|$ is the absolute value of $\vec{s}$;

$$|\vec{s}| = \sqrt{sx^2 + sy^2 + sz^2}$$

and $|\vec{n}|$ is the absolute value of $\vec{n}$;

$$|\vec{n}| = \sqrt{nx^2 + ny^2 + nz^2}$$

The angle e is then the required angle between the light direction and the normal to the facet.

In order to simplify computations, the components of $\vec{s}$ can be defined in such away, that $|\vec{s}| = 1$. The parameters of the light direction are to be chosen by the program user and are therefore best expressed in angles of azimuth and inclination (named HORANG (HORizontal ANGle) and VERANG (VERtical ANGle) in the program). The program converts these two parameters into the vector components sx, sy and sz as shown in Figure 31.

The starting point of the light vector $\vec{s}$ is supposed to be situated in the origin of the coordinate system x, y, z, of the map. Angle ex in Figure 31 is then the azimuth of the chosen light direction and the angle β is its inclination from the positive direction of the vertical z axis. Let us regard the light vector $\vec{s}$ as the spatial diagonal of length unity of a cube (line MN in Figure 31) whose edges are parallel to the

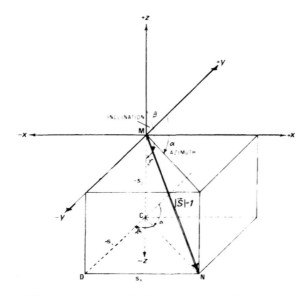

FIG. 31. The computation of the components of the light vector $\vec{s}$

coordinate system. The edges MC, CD and DN can then be regarded as the components sz, sy and sx respectively of the vector $\vec{s}$. If CN is the orthogonal projection of the diagonal MN onto the 'bottom' of the cube then:

$$-\,sz = \cos \gamma \; MN$$
$$\text{where } \gamma = 180° - \beta$$
$$\text{and } MN = 1$$
$$\text{therefore} \quad \boxed{sz = -\cos (180 - \beta)}$$

In the triangle NCD:

$$-\,sy = CN \cos \upsilon$$
$$\text{where } CN = \sin \gamma$$
$$\text{and } \upsilon = 180 - \alpha$$

$$\text{therefore} \quad \boxed{\begin{array}{l} sy = -\sin \gamma \cos \upsilon \\ sx = \sin \gamma \cos \upsilon \end{array}}$$

From equation (1) we have:

$$\cos e = \frac{sx \cdot nx + sy \cdot ny + sz \cdot nz}{|\vec{s}| \cdot |\vec{n}|}$$

Inserting the values for the components of $\vec{n}$ and the value of $|\vec{n}|$ from (2) we get

$$\cos e = \frac{sx \cdot ay \cdot bz + sy \cdot az \cdot bz + sz(ax \cdot by - ay \cdot bx)}{\sqrt{(ay \cdot bz)^2 + (ax \cdot bz)^2 + (ax \cdot by - ay \cdot bx)^2}}$$

This is the general equation for any light direction. For a vertical illumination, *i.e.* for 'slope hachures' the angle of inclination is 180° and the components of $\vec{s}$ are then

$$sx = 0$$
$$sy = 0$$
$$\text{and } sz = -1$$

and equation (3) is reduced to:

$$\cos e = \frac{-ax \cdot by + ay \cdot bx}{\sqrt{(ay \cdot bz)^2 + (ax \cdot bz)^2 + (ax \cdot by - ay \cdot bx)^2}}$$

For this special case there is, of course, a second way of computation. For a vertical illumination the angle e is equal to the angle of slope of the facet and

$$\tan e = \frac{HA - HB}{\sqrt{(XB - XA)^2 + (YB - YA)^2}}$$

where HA and HB are the heights of the points A and B respectively (for most hachures HA-HB is the vertical interval of the contour system except in cases where the hachure intersects a valley line or another hachure).

The functional relation between the width of the hachures and the cosine of the angle e can be formulated in sundry ways. As the options for the variation of the width with an incremental plotter are, a-priori, rather limited, a very simple relation was chosen for program 'HACHURE' (see Table 1). The Calcamp plotter 960 of the Computing Centre of Tel-Aviv University used for the demonstrations of this paper draws lines with a ball-pen at an approximate width of 0.02 cm. The varying width of hachures can be achieved with an incremental plotter by a repeated drawing of the hachure following a small displacement of the pen. The minimal displacement possible with the plotter at our disposal is 0.01 cm. Starting from the slimmest hachure of the width of 0.02 cm every displaced repetition increases therefore the width by not less than 0.01 cm.

In program 'HACHURE', the following relations between the size of the angle e and the width of the hachure and, consequently, the number of strokes per hachure (IREPEAT), were used.

For other plotters of different minimal line width and/or different minimal step other relations may have to be established.

Figure 32 illustrates how the required width of a hachure is achieved.

Let A and B in Figure 32 be the starting and end points respectively and the width of the hatched area the minimal width of the hachure attainable with one stroke of the plotter pen. Let this width be named d. If for example the value of IREPEAT was found to be 2, a second stroke would have to be added to the hachure AB. If XA, YA and XB, YB are the coordinates of the points A and B respectively, the displacements ΔX and ΔY for the starting and end points A' and B' of the second stroke can be computed from:

$$\Delta X = \frac{d}{2} \sin \alpha$$
$$\text{and } \Delta Y = \frac{d}{2} \cos \alpha$$

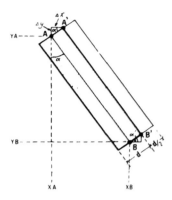

FIG. 32.   Displaced repetitions of the drawing of a hachure

$$\text{where } \alpha = \arctan \left(\frac{XB - XA}{YB - YA} \right)$$

For higher values of IREPEAT the strokes are applied alternately at both sides of the hachure. The maximum number of strokes per hachure allowed for in HACHURE is 5.

For the case that the slopes of the terrain are too gentle to produce a satisfactory scale of hachure widths, program 'HACHURE' is provided with a parameter named HEIINCR (HEIght INCRease), serving as a factor by which the vertical component bz of the vector b̄ is multiplied.

In the following illustrations the effects of various light directions and height exaggerations are demonstrated. In the Figures 33a, 33b and 33c the results of variations of HEIINCR are shown. The azimuth of the light direction and its angle of inclination are the same for all three figures and are both 135°.

In the Figures 34a, 34b, and 34c the parameter HEIINCR was given the value 2, the light inclination for all three figures is 135°, but the azimuth of the light direction was varied.

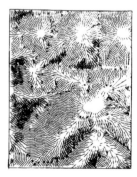  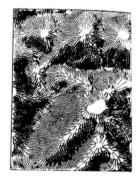

FIG. 33.   Variations of the height exaggeration

FIG. 34.   Variations of the azimuth of the light direction

The scale of all the six illustrations above is 1:25 000 and the density of the hachures is 14 hachures per centimetre for Figures 33a, 33b, 33c and 10 hachures per centimetre for Figures 34a, 34b, and 34c. As a further illustration Figure 35 shows the computer-assisted hachuring of the area in the scale 1:10 000.

e = arc(cose)	width of hachure (cm)	IREPEAT (number of strokes for hachure)
1° − 20°	0.02	1
20° − 45°	0.03	2
45° − 65°	0.04	3
65° − 75°	0.05	4
75° and above	0.06	5

THE DRAWBACKS OF COMPUTER PRODUCED HACHURES

If we compare the hachures as produced by our program with hand-drawn or hand-engraved hachures, we realize that the machine generated strokes lack the refinement and subtleness of a good manual product. This is mainly due to the fact that the direction of the gradient of slope changes continuously and not just on the upper delineator of the hachured belt. The strokes should be curved whenever required by the terrain, as seen in the Figures 3, 4, 5 and 6, where they are perpendicular not only to the upper contour of the belt but also to the lower one. This problem can, of course, be solved by a further sophistication of the program, leaving ample space for further experimentation with the method. By choosing a very small contour interval the effect of this drawback can be reduced.

A second reserve concerns the limitations put on the quality of the drawing by incremental plotters. In comparison with the graphic finess attainable with a hand-held drawing pen, plotters--even expensive ones-are soulless, uncouth utensils. No plotter will ever replace the hand of an artist.

FIG. 35.   Hachured map in the scale of 1:10 000

POSSIBLE APPLICATIONS OF HACHURING FOR THE GRAPHIC DEPICTION AND
ANALYSIS OF MORPHOMETRIC TOPICS. (AREA-SLOPE DTM'S)

As described in the previous chapters, the program computes for each hachure the
spatial coordinates XA, VA, HA and XB, VB, HB of its starting and end point respectively. Besides sending these values to a subroutine named 'SHADHAC' ('SHADe
HAChures!') for the computation of its width; they can also be stored on an external
file. In addition to this, the area of the hachure facet and its slope can also be registered on the file at the same time. This file which can be saved for future use may be
called an 'Area-Slope DTM'. In comparison to the original DTM which contained
heights only- although in matrix form-the 'Area-Slope DTM' is a much more detailed

source of terrain data. Not only are the hachure coordinates in such a file a very dense height registration of the terrain, but the entirety of the XA, YA and HA coordinates alone are also a digital registration in distance mode of the contour system used for the hachuring.

In addition to this, the terrain is registered in the form of little area patches together with their steepest slope. These terrain-facets approximate the true earth's surface far better than the original DTM. The information was, of course, contained implicitly in the initial DTM, but now it can be registered explicitly as a result of the data processing executed by program 'HACHURE'. The quality of the approximation depends primarily on the density and the precision of the surveyed data of the initial DTM, and on the vertical interval of the hachured belts and the proximity of the hachures.

The most straightforward application is, naturally, the drawing of the width of the hachures as a function of the slope angle, assuming a vertical light illumination. These are the 'Slope hachures' mentioned several times previously. Figure 36 shows such a depiction.

For this drawing, no height exaggeration should be applied as it would falsify the slopes. A different graduation scale for the widths of the hachures must therefore be applied as shown in the legend of Figure 36 (Compare with table 1, showing the graduation for oblique illumination).

Another way to present slopes is to draw in the place of the hachure centres little discs whose radii are proportional to the slope angle (see Figure 37). This is, in fact, a computer executed method suggested by M. Eckert[8] in 1921 and in principle by A. H. Robinson[9] who also suggested the use of dots, but of equal radii. The number

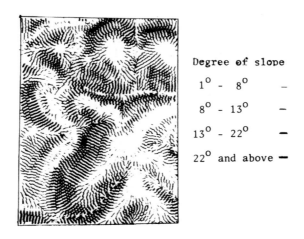

Degree of slope

1° - 8° —

8° - 13° —

13° - 22° ▬

22° and above ▬

FIG. 36. 'Slope hachures'

8. Eckert, M. (1921) **Die Kartenwissenschaft**, Vol. 1, Walter de Gruyter, Berlin.

9. Robinson, A. H. (1948) 'A Method for Producing Shaded Relief from Areal Slope Data', **Surveying and Mapping**, Vol. 8, pp.157–60, Washington.

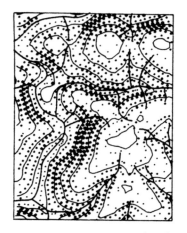

Degrees of slope

1^{O} - 8° .

8° - 13° .

13° - 22° •

22° and above •

FIG. 37. The 'dot method'

of dots per area unit were to be determined according to his method by the average degree of slope of an area unit.

A further possibility might be the drawing of 'selective slope maps'. In the three Figures 38a, 38b and 38c, three distinct slope groups are shown. The short program 'DRASLO' (DRAw SLOpes'!) which drew these figures has, among others, two parameters named SLOPMIN and SLOPMAX. Reading the slopes of the hachures stored on the 'Area-Slope DTM' file, only those slopes steeper than SLOPMIN or gentler than SLOPMAX, are hachured on the background of a contour map of the area.

Not only slope maps of various kind but also diagrammatic presentations, can be programmed based on the input of the stored file. Figure 39 shows a slope frequency diagram drawn by a short program named 'SLOFREQ', based on the 5310 hachures of Figure 25c.

FIG. 38. Selective slope maps

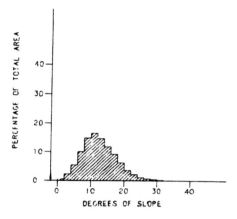

FIG. 39.   Slope frequency diagram

A similar diagram showing the area-height relation and others could also be programmed.

Another intriguing application might prove to be the possibility to draw '*flow line profiles*'. Starting from a hilltop in a certain direction, hachure might be added to hachure, by searching on the file for that hachure whose starting point A is nearest to the end point of the preceding hachure, until no further continuation can be found.

BIBLIOGRAPHY

Eckert, M. (1921) **Die Kartenwissenschaft**, Vol. 1, Walter de Gruyter, Berlin.

Imhof, E. (1982) **Cartographic relief presentation**, Walter de Gruyter, Berlin-New York.

Lehmann, J. G. (1799) **Darstellung einer neuen Theorie tier Bergzeichnung der schiefen Fliichen im Grundriss oder der Situationszeichnung der Berge**, Leipzig.

Monkhouse, F. J. and Wilkinson, H. R. (1971) **Maps and Diagrams**, Methuen & Co. Ltd., London.

Robinson, A. H. (1948) 'A Method for Producing Shaded Relief from Areal Slope Data', **Surveying and Mapping**, Vol. 8, pp.157–60, Washington.

Strahler, A. N. (1956) 'Quantitative Slope Analysis', **Bulletin of the Geological Society of America**, Vol. 67, pp. 551–96, New York.

Yoeli, P. (1982) **Cartographic drawing with computers, Computer applications**, special issue, Vol. 8, Department of Geography, University of Nottingham.

Yoeli, P. (1984) 'Cartographic Contouring with Computer and Plotter', **The American Cartographer**, Vol. 11, No.2, pp.139–155.

Yoeli, P. (1984), 'Computer-Assisted Determination of the Valley and Ridge Lines of Digital Terrain Models', **International Yearbook of Cartography**, Vol. 24, Kirschbaum Verlag-Bonn-Bad Godesberg, pp. 197–206.

Young, A. (1972) **Slopes**, Oliver & Boyd, Edinburgh.

Reflections on 'Topographic Relief Depiction by Hachures with Computer and Plotter'

KAREL KRIZ

University of Vienna

Professor Pinhas Yoeli can be seen today as one of the mentors of computer-assisted cartography of the early 1980s. Besides his abundance of published papers in manifold international cartographic journals he set a cartographic benchmark in 1982 with his book, 'Cartographic Drawing with Computers', the first of its kind, published by the University of Nottingham (Yoeli, 1982). The tenor of this unique publication was to encourage cartographers to use computers and to draw attention towards the automation of producing maps which are at least of the same high quality as those accomplished by conventional means — an ambitious goal that even three decades later is still a challenge, especially when dealing with large-scale topographic maps that embody numerous facets of relief depiction such as hillshading, rock and scree representation. Focusing his interest precisely in these areas, based on the principles of Eduard Imhof (Imhof, 1965), Yoeli hit the spot with his Henry Johns award-winning article 'Topographic Relief Depiction by Hachures with Computer and Plotter' that describes the revision of a method introduced centuries ago by J. G. Lehmann (Lehmann, 1799). The use of hachures is extended to not only depict topographic relief but also to be utilized as a source for the computer-assisted treatment of morphometric topics.

Topographic relief depiction has always played an important role in topographic mapping, particularly in mountainous regions. Over the centuries, cartography has searched for adequate methods to represent the surface of the Earth in a desirable way. The result of intensive engagement in this field is documented in a broad spectrum of cartographic products that display the diversity as well as the beauty of large-scale topographic mapping (Kriz, 2011). Utilizing computer technology, as Pinhas Yoeli describes in his article, was one of several pioneer attempts that reflected the modernization of the cartographic discipline and its impact in visualizing topographic features.

Relief depiction in maps has a long and extensive tradition with a wide range of outstanding artistic analogue examples. Nowadays, these methods of depiction remain in use through their enhancement in a digital context and can still be admired in current large-scale topographic maps, such as the well-known Swiss 1:25,000 topographic map series (swisstopo, 2013; Gilgen, 1998). Therefore, it is no wonder that the modern depiction of cartographic terrain in large-scale maps is currently focusing on reproducing this successful strategy that utilizes optimization and automation in a digital and above all time-saving environment (Jenny *et al.*, 2013). This leads to the fact that a special emphasis is given to design and aesthetic aspects in connection with hillshading as well as rock and scree representation. Even though these elements possess no cartometric reference, they play an important role in the overall impression of terrain in maps. Yoeli was well aware of this as he experimented with methods to optimize relief depiction using computers.

Hillshading in topographic maps is often responsible for a better perspective view of the Earth's surface, giving the observer a three-dimensional impression of the terrain. Imhof (1965) explained how hillshading can be integrated effectively in topographic maps. Although Imhof's book is nearly half a century old, it is still an essential reference and a 'must-have' for every cartographer dealing with terrain representation. Opponents of integrated hillshading in topographic maps, however, state that essential relief information is obscured when using this variant; it is important to show only authentic evaluated topographic information, without any artistic distortions or manipulations. Leonhard Brandstätter (1983) postulates that hillshading distorts the view of the essential topographic information and has only the effect of misleading the user. Certain areas become emphasized, others get suppressed, and can therefore convey false topographic information.

Similar to hillshading in topographic maps, rock and scree representation has mainly been performed by the hands of artists. Although there are many ways of depicting rock in topographic maps it is important to understand the technique and intentions behind the method applied. Present efforts in rock depiction on the one hand are concentrating in automated methods in order to achieve the quality standards of the past (Dahiden, 2008). On the other hand, experimenting with various methods allows a flexible approach with interesting solutions (Jenny *et al.*, 2010; reliefshading, 2013; screepainter, 2013). These are currently being addressed and vigorously discussed internationally in dedicated commissions within the International Cartographic Association, such as the Commission on Mountain Cartography (CMC, 2013).

Whether large-scale topographic maps with or without hillshading, or rock and scree representation are preferable or not depends not only on the accessibility of talented cartographic personnel and on the tools available but also on soft skills and how the user is cartographically socialized. The question therefore arises whether we are at all able to read maps the way in which they are intended to be read.

Yoeli presents a pragmatic answer to this question and clearly emphasizes that comprehensive collaboration in neighbouring fields unearths new perspectives and views that enhance scientific insight as well as public awareness. It is clearly a fact that through such contributions today's developments in terrain representation are transforming cartography into an interactive, modern, and scientific — as well as profoundly applied — discipline.

REFERENCES

Brandstätter, L. (1983). 'Gebirgskartographie', in **Enzyklopädie, Die Kartographie und ihre Randgebiete,** Band II, ÖAW, Wien

CMC (2013). ICA Commission on Mountain Cartography, http://www.mountaincartography.org/ (accessed 31 July 2013).

Dahiden, T. (2008). 'Methoden und Beurteilungskriterien für die Analytische Felsdarstellung in Topografischen Karten', Dissertation ETH-Zürich Nr.17674.

Gilgen, J. (1998). 'Felsdarstellung in den Landeskarten der Schweiz', in **Hochgebirgskartographie**, ed. by Kriz, K., (Wiener Schriften z. Geogr. u. Kartogr., 11), pp. 11–21, Silvretta.

Imhof, E. (1965). **Kartographische Geländedarstellung**, Walter de Gruyter & Co, Berlin.

Jenny, B., Gilgen, J., Hutzler, E. and Hurni, L. (2010). 'Automatische Gerölldarstellung für topographische Karten', **Kartographische Nachrichten**, 60, 188–193.

Jenny, B., Gilgen, J., Geisthövel, R., Marston, B. E. and Hurni, L. (2013). 'Design principles for Swiss-style rock drawing', The Cartographic Journal, in press.

Kriz, K. (2011). 'Topographische und Hochgebirgskartographie', in **50 Jahre Österreichische Kartographisch Kommission**, ed. by Kainz, W., Kriz, K. and Riedl, A., (Wiener Schriften z. Geogr. u. Kartogr., 20), pp. 105–116.

Lehmann, J. G. (1799). Darstellung einer neuen Theorie der Bezeichnung der schiefen Flächen im Grundriß, oder Situationszeichnung der Berge, Leipzig.

reliefshading (2013). http://www.reliefshading.com/ (accessed 31 July 2013).

screepainter (2013). http://www.screepainter.com/ (accessed 31 July 2013).

swisstopo (2013). http:// www.swisstopo.ch/ (accessed 31 July 2013).

Yoeli, P. (1982). **Cartographic Drawing with Computers**, University of Nottingham, Nottingham.

A Knowledge Based System for Cartographic Symbol Design

J. C. MÜLLER AND WANG ZESHEN

Originally published in *The Cartographic Journal* (1990) 27, pp. 24–30.

INTRODUCTION

As micro-computer and graphic devices are becoming widespread, a great number of thematic mapping packages are being developed and marketed. Many organizations and private firms involved in geo-scientific activities use these packages to produce quick maps for analysis and display. In the meantime, poor quality map products can be observed in the scientific literature, magazine, newspaper and other media, because the cartographic packages are easily accessed by a large community of users who are cartographically untrained. Hence, there is a need to guide the user of micro-computer based mapping packages towards the production of cartographically acceptable products. The system discussed in this paper provides such guidance and is a first step in the direction of a comprehensive tutorial system for the design of thematic maps. Only the choice of cartographic symbols based on cartographic principles and graphic semiology are considered.

MAP SYMBOLS AND COMMUNICATION

The main function of a map is communication.[1,2,3,4,5] See Figure 1. The communication process starts with real world data. These data are collected, analysed and portrayed in the form of a map. The analysis and the perception of a map generate in the user's mind a view of reality which usually deviates from the real world. This explains why the boxes I and If in Figure 1 are only partly coincidental. The extent to which mental representation departs from reality depends heavily on the accuracy and the effectiveness of the cartographic representation, however. The classic example of

1. Board, C. (1967) 'Maps as models', in **Models in geography**, Chorley, R. and Haggett, P., Methuen, London.
2. Kolacny, A. (1970) 'Kartographische Information-ein Grundbegriff und ein Grundterminus der modernen Kartographie', **International Yearbook of Cartography**, 10: 186–193.
3. Ratajski, L. (1973) 'The research structure of theoretical cartography', **International Yearbook of Cartography**, 13: 217–228.
4. Robinson, A. H., Sale, R. D., Morrison J. L. and Muehrcke, P. C. (1984) **Elements of Cartography**, John Wiley & Son Inc.
5. Ormeling, F. J. and Kraak, M. J. (1987) **Kartogra/ie ontwerp, production en gebruik van kaarten**, Delftse Universitaire Pers, Delft.

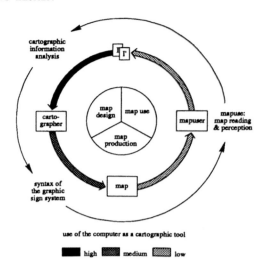

FIG. 1.   Outline of the cartographic communication model.[7] Note the partial overlapping between box I and box I'. Map reading leads to a view of reality (I') which only partially coincides with the real world (I)

Bertin[6] can be cited here as a point in case. Suppose we want to map the population of settlements. Two maps employing different symbologies, one using varied forms and another using sizes, are shown in Figure 2. In one map, the user can read a particular symbol by referring it to the legend. But nobody can see the distributional pattern of large cities versus small cities, even when a legend is available. Instead, the other map gives the map users the overall information (regional pattern) about population distribution readily, because people associate size with different quantities automatically. Hence, a map can provide more or less information depending on the choice of the symbology. In order to convey optimally accurate information, the map maker must be aware of perceptual and mental responses to particular symbologies, otherwise ineffective maps may be produced. Other factors that affect the symbol design should be also taken into account, such as map use requirements and conventional associations.

MAP MAKING WITH COMPUTER INTELLIGENCE

More and more non-cartographers are mapping with the help of cartographic packages. Although they may produce deceptive maps, we cannot expect a mapping package to be used only by cartographers. A more realistic solution is to include

6. Bertin, J. (1983) **Semiology of Graphics, Diagrams, Networks, Maps,** The University of Wisconsin Press, Madison.
7. Ormeling and Kraak, 1987.

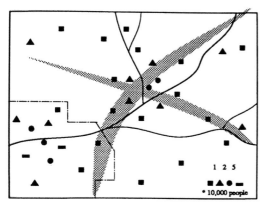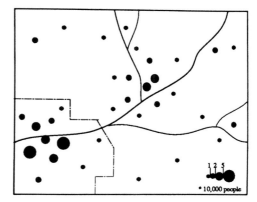

FIG. 2.   Quantitative information represented by two different symbologies

elements of the expertize of the human cartographer into the package so that major cartographic blunders may be avoided.[8]

Expert systems are computer programs that use inference techniques — involving formal reasoning and utilising the expertize of a human expert to solve problems in a specific domain. At the present time, most expert systems should be more accurately called knowledge systems because they can not perform at the level of a human expert. The study of expert systems is a very active subject in the domain of artificial intelligence (AI). From known facts the human being infers an unknown fact. In conventional computer languages, we can use the conditional statement 'IF ... THEN ...' to simulate the inference. But the 'facts' inside are just described as simple strings. For example, the fact 'John likes smoking' and 'Sue likes salads' are recognized as completely different facts in these languages. However, facts may also be symbolized in commonly used AI languages, such as Prolog:

likes('John', smoking).
likes('Sue', salads).

Here 'John', 'smoking', 'Sue' and 'salads' are objects, 'likes' is the relationship between the objects. These two facts can be recognized in the computer as having the same relationship but different objects. Only when the facts are symbolized can formal reasoning be carried out. Based on the above facts, some rules can be created to infer new facts. For example, the rule:

bad_habit(Person) if
likes(Person,smoking).

means that a person has a bad habit if he/she likes smoking. When this rule is invoked, the new fact, John has a bad habit, will be proved by the relevant facts.

8. Müller, J. C., Johnson, R. D. and Vanzella, L. R. (1986) 'A Knowledge based Approach for Developing Cartographic Expertise', **Proceedings of the Second International Symposium on Data Handling**, Seattle.

There is now a substantial body of cartographic research which has dealt with the principles of cartographic symbology and design. Although there is no perfect congruence in the results emerging from these studies, one can claim that there is a basic foundation in the most obvious principles which could be translated into rules and implemented in AI languages.

CARTOGRAPHIC SYMBOL DESIGN IN THE EXPERT'S VIEW

For a well designed map the box I and I' should have a large overlap part (Figure 1), which, to a great extent, depends on the adequacy of the relationship between data characteristics and cartographic symbology. The relationships between data characteristics and cartographic symbologies as well as the issue of data presentation accuracy, graphic semiology and mapping conventions will be quickly reviewed in the following sections.

Characteristics of spatial information

There are at least three essential characteristics affecting cartographic symbol design:

- measurement level,
- dimensional characteristic, and
- organizational structure.

Measurement level
There are four measurement levels listed below in increasing order of sophistication:

- Nominal scaling. The data are simply grouped by type;
- Ordinal scaling. The data are ranked, but not yet quantified;
- Interval scaling and ratio scaling. Although these two levels correspond to different measurement scales, they may be, for all practical purposes, treated equally under the heading quantitative mapping.

Dimensional characteristic
Geo-information may be transcribed in terms of its topological dimensions. On a paper map, those dimensions appear in the form of a point (zero dimension), a line (one dimension) or an area (two dimension). Time (third dimension) may be added under the form of map series or animated displays. The corresponding symbols usually have the same dimensional characteristics. But in quantitative mapping, the symbol dimension does not necessarily match the information dimension. Proportional symbols, for instance, can be used to represent absolute information for both point and area.

Organizational structure

When the information is hierarchically structured into qualitatively different sets and subsets, the symbol system must be able to reflect such a structure. For example, vegetation must be differentiated from water bodies. Within the vegetation set, there may be two kinds of woods and three kinds of bushes to be mapped. The map reader must be able to recognize class membership as well as class differences immediately.

Graphic semiology

Symbol design is based on two kinds of knowledge: the characteristics of spatial information on the one hand and the visual properties of graphic symbols on the other. The second type of knowledge is provided by graphic semiology, i.e. the study of the relationships between graphic symbols and their perceptual properties. The universe of cartographic symbols may be partitioned according to six different visual variables: size, value, texture, colour, orientation and form. Each of the variables has a perceptual property. The associative, ordered and quantitative perceptions correspond to three measurement levels, i.e. qualitative (nominal), ordinal and quantitative information. In addition to these three, one perception property called selective property is necessary for showing all the data types. Selection allows the eye to isolate one set of graphic marks from other sets. The pattern of this isolated set can then be used to answer the question: where is what? Without such a property, the map becomes a 'map to be read instead of a map to be seen'.[9] Table 1 is a summary of the relationship between visual variables and perception properties.

Other factors

Besides the map content, data characteristics and graphic semiology, several factors influence map symbology.

- Map use requirements, including the circumstances under which a map may be used, the type of map usage and map user. An example is that pictorial symbols are preferred when the map is made for children.
- Natural association between the information and its representation. When using the association properly, the communication process will be carried quicker and more accurately.
- International and national conventions such as those agreed upon in geological mapping and aeronautical charts.
- Psycho-physical laws. It is not unusual that the perception of a human being is quite different from the real value of the symbols, for instance, in the underestimation of larger proportional point symbol sizes when compared to smaller ones.
- Production and cost aspects.

Table 1.   Perception properties of visual variables (after Bos, 1984).

Visual Variable Perception property	Form	Orientation	Colour	Texture	Value	Size
associative	+	+	+	o	–	–
selective	–	o	++	+	+	+
ordered	–	–	–	o	++	+
quantitative	–	–	–	–	–	+

++ = very good; + = good; o = moderate; – = poor/no.

9. Bertin, J., 1983.

KNOWLEDGE REPRESENTATION

The purpose of expert systems is to help the decision making process on the basis of information and knowledge. The amount of cartographic knowledge available in the scientific literature and in the brain of the human expert is far beyond the computer capacity. On the other hand, many workers on expert systems suggest that expert system development must start with a small domain.[10,11] The reason to do so partly originates from the nature of AI language, since the order of clauses inside an AI program does not affect the operation sequences. Hence we can start with a small program until its value is proven, then add new rules and facts.

A system for symbol selection was developed according to this philosophy. General information is given, followed by the goal and domain knowledge definition.

General description of the system

The specific purpose of the work described in this paper was to develop an 'intelligent' thematic mapping package which could be used for graphic output in the already existing geographic information system called IL WIS (Integrated Land and Water Shed Management System).[12] With such a system, the users of ILWIS should be able to reach adequate decisions for the symbolic representation of statistical information based on physical or administrative units.

The basic hardware and software configuration for the implementation of the system is listed as follows:

- IBM AT or compatible, with MA TROX 960 GP graphics board;
- A table digitizer, with a digitizing program, named DIGITIZE, developed by ILWIS;
- A line printer and Tektronix colour plotter series or an IBM colour ink-jet printer for output of hard copy.

Turbo Prolog and Turbo C were selected as programming language for symbol design and map generation, respectively.

Goals and domain knowledge definition

The system is intended to portray statistical information at various measurement levels. Hence, six map types classified in terms of information characteristics were recognized: qualitative point and area mapping, ordered point and area mapping, and absolute quantitative and ratio (i.e. choropleth) quantitative mapping. As multiple quantitative mapping is also quite widely used, we have seven map types in total. It should be noted that each map type has several possibilities, e.g. for qualitative point mapping, the user can employ either the variable colour or form, and the choropleth map has many different classification methods. These different solutions for the same map type are not exclusive, some solutions may be combined to produce

10. Townsend, C. (1986) **Mastering Expert Systems with Turbo Prolog**, Howard W. Same & Company.
11. Hu, David (1987) 'Programmer's Reference Guide to Expert Systems', **Compiler**, 3 (4): 5–8.
12. Meijerink, A. M. J., Valenzuela, C. R. and Stewart, A. (1988) **ILWIS Integrated Land and Watershed Management Information System**, ITC Publication Number 7, ITC, Enschede.

other solutions in a particular situation. Before the map is generated by the system, a series of values, whether quantitative or qualitative, must be attached to each symbol. This step is called actual symbol design. For the prototype system, only four visual variables are included: form, colour, value and size. Three steps in symbol design, including map type determination, optimal solution selection and actual symbol design, will be discussed in depth as follows.

Map type determination

Although proportional symbol maps and choropleth maps are most frequently used, there are many other forms of representations for quantitative mapping, including the dot and the isopleth versions. To simplify the situation, we limited the correspondence between the data type and the map type to a one-to-one relationship. According to this rule, the proportional symbol map is selected as a unique representation of absolute information, whereas the choropleth map is used for ratio data. A model illustrating the relationship between data characteristics and map type is shown in Figure 3.

The computer language Prolog is inherently suitable to a backward chaining system which uses inductive inference to pursue a goal. The conclusion/hypothesis is set up first, and the inference engine tries to find its antecedents. For example, a simplified rule for pursuit of the map type may be:

map_type(choropleth_map)if measurement(quantitative),quantity(ratio).

The rule should allow interactive query. One can see that a rule has two parts: the head, i.e. hypothesis, and the body consisting of the conditions. When all the conditions in the body are met, the hypothesis is proved.

The rules listed above are from a rule based system, a most common type of expert system. When the system grows and becomes complicated, the hierarchical structure of knowledge is lost (Townsend, 1986). Furthermore the integration of a good user

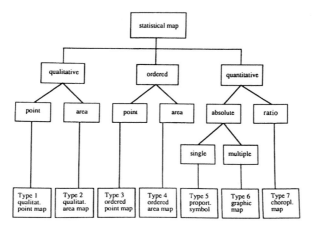

FIG. 3. Simplified Classification System for Map Types. See next section for an explanation of terminology for the different map types

interface is difficult in such a system. Therefore, we discarded the rule based approach and adopted a frame based strategy that is potentially more useful for development of large systems and interface design. A simplified program is listed below:

```
1    run if
2    frame(information,V),
3    put(Y).
4    put(Y)if
5    write(Y),
6    readln(C),
7    link(C).
8    link(C)if
9    frame(C,Y),
10   put(Y).
11   frame(information,[qualitative,ordered,quantitative]).
12   frame(qualitative,[quali_point_map,quali_area_map]).
13   frame(ordered,[ordered_point_map,ordered_area_map)).
14   frame(quantitative,[proper_symbol_map,choropleth_map).
```

Here the rules (line 1 to 10) and the taxonomy data base (line 11 to 14) are separated, so the hierarchical structure is very clear. The program can be initiated by typing 'run', and the recursive inference will take place between line 4 and 10.

Optimal Solution Selection

In the map taxonomy model defined in this system, each map type has several potential solutions, due to the facts that:

- several visual variables are relevant for the same map type, e.g. both the form and the colour variable are applicable for qualitative point symbol mapping;
- several formulae are available for data processing, e.g. different classification methods are available for choropleth mapping;
- several subgroups for some map types, e.g. proportional symbol mapping, can employ one, two or three dimensional symbols.

All the potential combinatorial solutions available in this prototype system for seven map types are listed below (refer to Figure 3):

TYPE 1 qualitative point map
  solution 1 symbols with the same colour and different forms;
  solution 2 symbols with the same form and different colours;
  solution 3 symbols with different forms and colours;
  solution 4 black symbols with conventional geometric forms;
  solution 5 black symbols with non-conventional highly contrasted forms
TYPE 2 qualitative area map
  solution 1 filling area with different colours.
TYPE 3 ordered point map
  solution 1 symbols with different values;
  solution 2 symbols with different sizes.
TYPE 4 ordered area map
  solution 1 filling areas with different values.

TYPE 5 quantitative point map (proportional symbol map)
    solution 1 Id., non-linearly scaled bars;
    solution 2 Id., proportional bars;
    solution 3 2d., proportional symbols;
    solution 4 2d., ratio reduced symbols;
    solution 5 3d., cubes.
TYPE 6 quantitative multiple point map (graph map)
    solution 1 pie charts;
    solution 2 bar graphs.
TYPE 7 quantitative area map (choropleth map)
    solution 1 equal interval classification;
    solution 2 equal ratio (geometric) classification.

As mentioned earlier, the choice of a solution depends on several factors, including map use requirements, numerical analysis and conventional associations. Two examples are given here:

Example 1. Qualitative point map (i.e. TYPE 1)
    Two subgroups can be recognized. One includes solution 1, 2, and 3, forming the coloured output group. The second group, consisting of solutions 4 and 5, is monochrome. The system will give a menu to separate these two groups, i.e.: OUTPUT: coloured or monochrome. In the first group, we know that the colour variable possesses a strong selective perception property, so solutions 2 or 3 should be the optimal solution.

Example 2. Graph map. (Le. TYPE 6)
    Bar graphs and pie charts are two widely used representations for multiple quantitative mapping. The ratio between components is easy to perceive in bar graph maps, whereas the differences between the total amounts are evident in pie chart maps. A query about user interest, i.e. INTEREST: total amount or each component, is available for selection of the proper graph map.

In the examples given above, the optimal solution depends on one factor; when more factors are involved, uncertainty may arise and more sophisticated methods should be applied. Three different approaches are widely used to handle uncertainty, including probability, fuzzy logic and certainty factor. The certainty factor approach was used in order to identify optimal solutions.

Three terms of certainty factor may be defined[13]:

CF[h:e] certainty of hypothesis h given e;
MB[h:e] measure of belief in h given e;
MD[h:e] measure of disbelief in h given e.

The certainty factor is simply the difference of the other two components:

CF[h:e] = MB[h:e] — MD[h:e].

Measures of belief (MB) and measures of disbelief (MD) can range from 0 to 1. Certainty factor (CF) can range from — 1 to 1. If two conditions support the conclusion, the measure of belief will be: MB[h] = MB[h1] + MB[h2] (1 — MB[h1]).

13. Shortliffe, E., Davis, R. and Buchanan, B. (1977) 'Production Rules as a Representation for a knowledge-Based Construction Program', **Artificial Intelligence**, 8 (1).

Actual symbol design

As the map type is determined and the optimal solution is selected, the system comes to the final step of symbol design. In this step, the actual values have to be assigned to each symbol, so a definition of the value domain is a prerequisite. As mentioned before, there are only four visual variables used in this package: form, colour, value and size. We can use list, which is defined as an ordered sequence of terms in Prolog, to record the available forms and colours:

var(form,[circle,square,triangle,rectangle,rhomb,trapez,plus,half_circle]).
var(colour,[red,green,blue,yellow,cyan,magenta,brown,orange,'b/w']).

On the other hand, value and size, the two other variables, can be only defined by formulae. However, coverage percentage, defined as the ratio of symbol-covered area over total mapping area, is still necessary for overall control and design purpose. A range of five to ten percent was suggested by Bertin.[14] Our own experience shows that 5% to 13% is reasonable.

An example is given here for demonstration. Suppose there are three kinds of building, say bank, hotel and restaurant, to be mapped, and qualitative point map (TYPE 1) with solution 2 (same form and different colours) is selected through the system. The solution is defined as follows:

recommend (type('qp'), num(2), note ('SYMBOLS with DIFF. COLOUR'), item ([form]), default ([circle]), loop ([colour]).

Here 'qp' indicates the qualitative point map. '(2)' means solution 2. The note 'SYMBOLS with DIFF. COLOURS' will appear on the top of the recommendation window. 'item([form])' and 'default([circle])' indicate that the circle is selected through the system as a default form. 'loop([colour])' means that different colours should be assigned to different features by loop. Before performing the assignment, the system will check the conventional knowledge base first and give priority to available associations. For example, suppose forest is a variable to be mapped. If there is a fact, CONVT (forest,colour,green), in the conventional knowledge base, the colour green will be assigned to forest symbols.

When the symbol design is performed and the solution is exhibited in the recommendation window, the system will ask for the user response by presenting the following menu: satisfied, next_recommendation or modify. The first option will initiate the map generation program and produce the map. The second option, next_recommendation calls the next solution which has lower certainty factor. The user's own design solutions will be invoked by the third option 'modify'. Any specification defined by the system can be revised by the user.

INTERFACE AND EXPLANATION

Menu selection, a well established technique for human-computer interaction, was adopted as the interaction style to implement the user interface of the system. Four criteria are traditionally used to evaluate an interface design: a short learning time, high performance speed, low error rate, and user satisfaction.[15] In order to reach these goals, the following constructs were implemented in the system:

14. Bertin, J., 1983.
15. Shneiderlman, B. (1986) **Designing the User Interface, Strategies for Effective Human-Computer Interaction**, Addison-Wesley Publishing Company.

1. Only three items should be kept in user's memory, so that operations remain simple and consistent:
   - Type the first letter to select option, and the option will be highlighted as a response;
   - Hit question mark to get more explanation;
   - Press Esc key to return to the last menu.
2. When the user types in a map type, or types in some typical map title/content, such as 'population density', the corresponding option in each menu will be selected and highlighted automatically.
3. Error preventing. Two safeguards have been embedded:
   - Typing wrong key does not cause any error;
   - The Esc key enables the user to go back to the previous menu when he/she finds that the wrong option has been selected in the last menu.

Since the user of expert system is assumed to be non-expert, an explanation is necessary in order to help the user in the menu selection process. In this package two levels of explanation have been implemented:

> level 1. Explaining the concepts, terminologies and operational instructions appearing on the menus, e.g. 'ordered', and 'modify' etc.;
> level 2. Explaining the inference procedure, i.e. why does the system present such a menu or reach such a conclusion instead of another.

The explanations at level 1 are on-line explanations which always follow the user's actions, whereas the level 2 explanation can only be retrieved by hitting the question mark.

STRUCTURE OF THE SYSTEM

The symbol design program in Turbo Prolog and the map generation program in Turbo C form the kernel of the prototype system. Since Prolog can not be used for sophisticated numerical analysis, a preprocessing program written in Turbo C was added as a foregoing part.

Finally, one important function of expert systems is learning. Since the experienced user spends time in analysis of the data and design of the symbol colours and forms, the user's solution entered after selecting the 'modify' option should be stored in a permanent data file for later use. The learning function was not totally achieved, however. This requires, in order to avoid wasting of computer memory space, the automatic discrimination between a sophisticated and a primitive user, which proves to be difficult. Hence, the conventional knowledge base is still edited manually.

Four fundamental components together form the system:

- data analysis,
- symbol design,
- map generation,
- system learning.

A number of temporary files were created to pass parameters and facts between these programs.

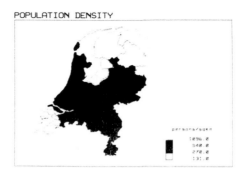

POPULATION DENSITY

persons/sqkm

1896.0
940.0
278.0
131.0

FIG. 4.   Dutch population density (reduced version of coloured original)

EXAMPLES

A command file, name WANG, is created for calling the four programs listed above. By typing WANG, the first program, data preprocessor, starts with questions about the location of the statistics to be mapped, i.e. file name and the column number. The system will read the data and calculate some parameters, such as maximum, minimum and the distribution type. Then the second program, map design process, will be invoked, and the user has to select a few menus about data characteristics and map use requirements. The numerical analysis given by the first program and conventional associations will be taken into account, and the optimal solution will be worked out by the system. If the user accepts it, the resulting map will be generated on the graphic screen.

Two examples are given here for demonstration.

Exercise 1. Dutch population density

QUESTION/MENU	ANSWER/SELECTION
statistical data file:	NED.POP
how many column:	1
MAP TITLE	Population Density
QUANTITY TYPE	derived
USER SELECTION	satisfied

When the user has answered two questions and selected three menus, the system will automatically chose a choropleth map as a representation, determine the number of classes, and adopt a classification method. The design proposed will be displayed on the recommendation window. If the user is satisfied, a map like in Figure 4 will result, otherwise the class number, colour, and classification methods may be specified by the user.

Exercise 2. Land use in The Netherlands
There is a file NED.LND storing land use for each province. We want to show two variables, say, the urban area and the forest area, and compare them to each other.

QUESTION/MENU	ANSWER/SELECTION
statistical data file:	NED.LND
how many columns:	2
MAP TITLE	Urban and Forests
QUANTITY	absolute
INTEREST	component
USER SELECTION	satisfied
remark on legend (yIn):	y
remark:	in hectares

Similar to the last example, the map type, colour (red for urban and green for forest), and the symbol size controlled by coverage percentage (7070) are defined by the system although any change can be made by the user if necessary.

CONCLUSION

A cartographic symbol design system, using a frame-like inference engine, was constructed. The system enables us to reach a decision about thematic map type selection, visual variable selection and representation based on generally accepted principles of map design. The knowledge based system presents the following characteristics:

- Non-professional user can learn and master cartographic operations quickly;
- Serious error in map type determination can be prevented;
- Optimal design solutions can be selected by the system;
- On-line explanations provide the user with necessary guidance during execution (the user does not need to remember the syntactic commands);

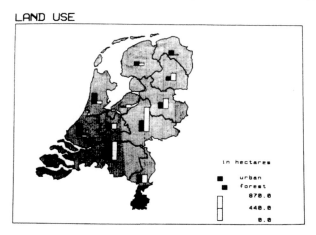

FIG. 5. Dutch land use per province (reduced version of coloured original, e.g. urban = red; forest = green)

- The map design process is 'shorter and simpler than in conventional packages, since design choices are made automatically;
- The user can learn cartographic practice from the explanations provided by the system.

The system is flexible and could easily be expanded into an advanced statistical mapping package using the present modular framework divided into map type determination, — optimal symbol design selection and actual symbol design.

One limitation which hampers further developments at present is the limited stack memory for recursive operation in Turbo Prolog version 1.5, since the frame system adopted consumes the stack memory very quickly. A new version of Prolog may provide larger space for recursive operation so that more menus and questions may be added.

The system learning program also causes some problems, because automatic execution of this program will make the dynamic knowledge base grow too quickly. How to record the valuable knowledge into the dynamic knowledge base, and in the meantime control the size of the knowledge base, is a serious problem -for most automatic learning systems.

Three aspects may be subject to further development, referring to the nature of the information, the data manipulation procedures and the representation.

- The system only allows information at the same measurement level. Further developments should be able to show combined information;
- The system can only handle information with two kinds of characteristics. Those are measurement level and dimensional characteristics. The organizational structure should be included by extending the value domain of colour and adding patterns. Adequate redundancy by combining visual variables should also be allowed;
- The classification procedure for choropleth mapping should include more options, with a rule based strategy for method selection;
- Other map types must be included as possible options for cartographic representations, including dot map and isopleth map.

BIBLIOGRAPHY

Bertin, J. (1977) **Graphic and Graphic Information Processing**, Walter de Gruyter & Co., Berlin.
Bos, E. S. (1984) 'Systematic Symbol Design in Cartographic Education', **ITC Journal**, 1: 20–28.
Jackson, P. (1986) **Introduction to Expert Systems**, Addison-Wesley Publishing Company.
Jenks, G. F. and Coulson, M. R. C. (1963) 'Class Intervals for Statistical Maps', **International Yearbook of Cartography**, 3: 119–134.
Jenks, G. F. and Caspall, F. C. (1971) 'Error of Choropleth Maps: Definition, Measurement, Redaction', **Annals of the Association of American Geographers**, 61: 217–244.
Kraak, M. J. (1988) **Computer-assisted Cartographical Three dimensional Imaging Techniques**, Delft University Press, Delft.
Wilkinson, G. G. (1987) 'The Search Problem in Automated Map Design', **Cartographic Journal**, 24.

Reflections on 'A Knowledge Based System for Cartographic Symbol Design'

DAVID FORREST

University of Glasgow

The early 1980s saw the emergence of an interest in cartographic circles in the relatively new research area in Computing Science of Expert Systems (ES). The general aim of ES is to interact with an end user to answer a question, or solve a problem. What makes them different from traditional algorithmic programmes is that they find solutions (sometimes multiple solutions) by using inference procedures to search a knowledge base. Ideally the interaction with the user is in the form of a conversation, much as one would have with a human expert; they should not follow a slavish set of questions, but should react to what information they already know and ask questions that will most likely take them towards a solution. Other characteristics include being able to explain how they have come to a conclusion, and to be able to learn from experience to develop their knowledge base. When I embarked on my PhD research in 1985, the University of Glasgow library had about 12 books on artificial intelligence and expert systems. Ten years later there was over a shelf of books on expert systems, plus more on other aspects of artificial intelligence, such as machine vision, robotics and natural language processing.

The inspiration for research into expert systems for cartographic design is clearly expressed by Müller and Wang: more and more non-cartographers were producing their own maps using an expanding range of desktop mapping packages. However, many of these users were unfamiliar with good cartographic practice and produced maps which broke basic rules of cartographic design. While many of the systems were capable of producing sensible maps, they offered little help to users and often the default settings were poor or inappropriate. Indeed, some systems even encouraged incorrect mapping choices. Figure 1 shows a typical example where absolute numbers are mapped using the choropleth method and the colour choices bear no systematic relation to the value range. It only took a few clicks to create a sensible map, but many users would unknowingly accept these erroneous offerings.

A very simple example of typical problem: a widely used statistical mapping package at the time would by default use five classes for choropleth maps; in doing so, five sensible shadings resulted. If the user increased the number of classes to seven, the first five classes got the five default shadings and the final two were blank. The user then had to manually change the shading and choose appropriate settings for seven classes. By building cartographic knowledge into the system, changing the number of classes would automatically select the appropriate shading set. This may seem a very trivial matter, but by building rules for several such situations, we move from a 'dumb'

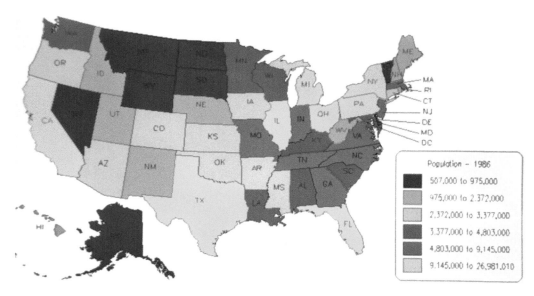

Fig. 1. An example of a map produced by a desktop mapping package using default values. Such maps, breaking basic cartographic principles, were often used in promotional brochures at the time

system to one which seems to exhibit some level of 'intelligence'. In effect, the default settings should react to other factors, not just be set once and for all. This, amongst others, is the kind of issue Müller and Wang tackled.

To a significant extent there was a negative view expressed by cartographers towards the idea of expert systems for map design. Perhaps some of this came from the term itself and idea that a computer could perform as well as an 'expert'. In fact most researchers in ES acknowledged that this was unrealistic, and that the term 'knowledge-based system', as used in the article by Müller and Wang, was more appropriate, the general aim being to create systems that could support decision making by users who has *some* relevant knowledge. Granted, some of the early attempts at expert systems were aimed at non-trained users, but most practical expert systems were designed for and used by people who were already knowledgeable in their field.

Cartography was not immune from some of the hype; a UK-based project known as Map-Aid aimed to produce a system that could design anything from a 1:500-scale detailed plan to a 1:30-million-scale general map for an atlas, hydrographic charts, geological maps, etc. (Robinson and Jackson, 1985). Clearly, particularly in the early stages of the development of expert systems, such an ambitious scheme was unrealistic. Experience quickly showed that successful knowledge-based systems were clearly targeted at solving a much more limited problem (i.e. limiting the problem domain to use computer science terminology). Success in the cartographic field was achieved by some very targeted systems on topics such as map projection choice, name arrangement, and some aspects of generalization. Most notable of these was the work by Chris Jones and various doctoral students (e.g. Cook and Jones, 1990; Jones *et al.*,

1991) which went on to become Maplex. This achieved commercial success and was bought by Esri (Environmental Systems Research Institute) and still forms part of ArcGIS. Useful reviews of progress with developments in knowledge-based systems in cartography around that time can be found in Buttenfield and Mark (1991) and Forrest (1993).

Design is classified as an unstructured problem and attempts at creating expert systems for design were generally less successful than those for classification, diagnosis and other types of problem. Some 'design' systems were effectively solving 'configuration with constraints' problems — such as kitchen design (what cabinets and appliances to order and where to put them) — that do not exhibit the combinatorial explosion of decisions involved in more complex design tasks like making a map. Arguably Maplex, with a focus solely on label placement, falls into this category.

Of the authors, Müller was one of first to write about a working ES for cartographic design (Müller *et al.*, 1986). This earlier paper described a system that reverse-engineered the design parameters of a range of maps and created a knowledge base from this. The system was then able to successfully regenerate specifications for similar maps, but it would have been impractical to extend the approach to a more flexible system.

In his PhD research, Wang developed more sophisticated structures for the knowledge base reported in the article and adopted a 'frame-based' approach, also adopted by others working on ES for map design (e.g. Forrest, 1999; Su *et al.*, 1993). While expert systems for map design are no longer a topic of conversation, this frame-based approach shares many characteristics with the object-oriented approach successfully used by some GIS. By encapsulating behaviours and methods with objects we can effectively incorporate cartographic knowledge in the system. For example, the Laser-Scan Gothic system knew that a contour object could not intersect itself or other contours. The concept of inheritance is also shared with object-oriented systems where characteristics of generic objects can be refined for more specific instances of the object. Therefore, the system can refine an existing map specification to meet new requirements, without the user having to specify all elements.

Although by the time of publication of Müller and Wang's paper there had been many papers on the application of expert systems to cartography, as far as can be determined, that by Müller and Wang is the first article in a refereed journal specifically on a working prototype ES for map design — other publications until then being in conference proceedings, mainly of the Autocarto and Spatial Data Handling series of conferences.

Although the approach they presented was a significant step forward in this field, it still falls very short of the requirements for a fully operational expert system, some of which are pointed out by the authors in their conclusion. If implemented today, such a system would have to interface with GIS, but in addition to cartographic design rules, more knowledge is also required about the nature of the data in the database than is typically included in geodatabases. The architecture of such a Cartographic Design Expert System is illustration in Fig. 2.

Wang went on to complete his PhD on Expert Systems for Cartographic Symbol Design in 1992 and published other articles with Müller on rule-based approaches to cartographic problems, mainly concerned with generalization (e.g. Müller and Wang, 1992). He later became a senior programmer with Esri in Redlands, USA. Müller

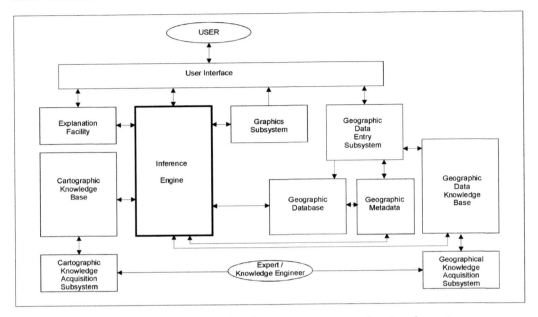

Fig. 2. The architecture of a cartographic design expert system showing the main components of such a system (after Forrest, 1999)

left ITC (now the Faculty of Geo-Information Science and Earth Observation, University of Twente, The Netherlands) and became Professor at Bochum University in Germany.

Müller and Wang showed that by asking a few simple questions, a knowledge-based system could select an appropriate type of map to suite the data and produce sensible symbols for it; despite many developments in GIS over the last 25 years and despite all their capabilities and sophistication in many aspects, GIS still cannot do this.

I have never regarded Expert Systems as a threat to cartographers; people will always make their own maps and technology has made this even easier. It also seems unlikely that ES would produce as well designed maps as even a modestly talented cartographer. However, what knowledge-based systems can do is prevent those lacking an appropriate background breaking the basic rules of cartographic design, and I think we would all welcome that!

REFERENCES

Buttenfield, B. P. and Mark, D. M. (1991). 'Expert systems in cartographic design', in Geographic Informa-
tion Systems: The Micro-Computer and Modern Cartography, ed. by Taylor, D. R. F., pp. 129–150,
Pergamon Press, Oxford.
Cook, A. C. and Jones, C. B. (1990). 'A Prolog rule-based system for cartographic name placement',
Computer Graphics Forum, 9, pp. 109–126.
Forrest, D. (1993). 'Expert systems and cartographic design', The Cartographic Journal, 30, pp. 142–148.

Forrest, D. (1999). 'Developing rules for map design: A functional specification for a cartographic design expert system', **Cartographica**, 36, pp. 31–52.

Jones, C. B., Cook, A. C. and McBride, J. E. 1991. 'Rule-based control of automated name placement'. In **Proceedings of the 15th International Cartographic Conference**, pp. 675–679, Bournemouth, 23 Sept–1 Oct.

Müller, J. C., Johnson, R. D. and Vanzella, L. R. (1986). 'A knowledge-based approach for developing cartographic expertise', in **Proceedings of the 2nd International Symposium on Spatial Data Handling**, pp. 557–571, Seattle, WA, Jul 5–10.

Müller, J. C. and Wang, Z. (1992). 'Area-patch generalisation: A competitive approach', **The Cartographic Journal**, 29, pp. 137–144.

Robinson, G. and Jackson, M. (1985). 'Expert systems in map design', in **Proceedings of Autocarto 7**, pp. 430–439, Washington, D. C., Mar pp. 11–14.

Su, B., Zhang, H., Li, H., Zhang, X., Zhang, Y., Zhu, X. and Li, J. 1993. 'A knowledge-based system based on GIS for thematic mapping', in **Proceedings of the 16th International Cartographic Conference**, pp. 486–477, Cologne, May 3–9.

The Role of the Ordnance Survey of Great Britain

D. W. RHIND

Originally published in *The Cartographic Journal* (1991) 28, pp. 188–199.

This paper is concerned with the future of the British national mapping agency in a society which is markedly different from that in which the Ordnance Survey Review Committee of 1979 worked. The objective is to ascertain which topographic information is needed, who should provide it, on what terms and through which mechanisms. Prior to making an attempt to answer these questions, the essential characteristics of Ordnance Survey (OS) are summarized as deduced from available documentary evidence[1]; the changing attitudes to information as a commodity, the growing competition in British mapping and the government's stringent requirements from the Survey are also outlined as just three of the many complexities which affect OS. Building upon Smith's classic 1979 paper and subsequent experience, the rationale for government involvement in mapping is examined. It is concluded that the Survey has a continuing vital role though there are a number of steps which the OS should take in order to adapt to changing circumstances.

OS IN A CHANGING WORLD

For Ordnance Survey (OS) to have survived 200 years without significant change of name or main purpose is an extraordinary achievement. But institutional longevity is no guarantee that the taxpayer's money is still being expended appropriately: long-established organisations such as universities, government departments and even major computer companies typically have strong in-built resistance to change. Unfortunately for such organisations, externally forced change in the social, economic and political fabric of life is now endemic and increasing in tempo. Part of this is due to the advent of computerisation. There are now many well-documented examples of how its introduction has changed radically the way in which organisations function and are best structured. In addition, events such as the tenfold increase in numbers of

---

1. The original of this paper was written in March 1991 and presented at a meeting in the Royal Geographical Society on May 23rd of that year. In August 1991, the author was appointed Director General and Chief Executive of the Ordnance Survey. The main aspects of the paper remain as originally presented though it has been reduced in length substantially on the advice of the referees. Thus it still represents the view of an individual external to OS at the time of writing and is based on publicly available evidence. In no sense therefore should any proposals herein be construed as necessarily forming part 'of OS or of British government policy.

private cars over the last 50 years have transformed the role of many official mapping agencies. The collapse of communism and restructuring of East European society in the last three years is already having comparable effects for some such agencies. Moreover, we now live in a time when certain major public services in the UK and elsewhere have been or are being commercialized in one form or another. In some cases this has taken the form of outright privatization (e.g. of the UK's regional electricity boards). In others, it has been manifested by demands for greater cost recovery or operations on a much slimmer budget.

In short, the entire context in which Ordnance Survey operates has been transformed in recent years. This is, of course, also true of other national mapping organizations; many of them — such as those in Australia, Canada, Germany, New Zealand and the USA (see, for instance, CAG[2]; NRC[3,4]; Robertson[5]) — have recently been investigated and, in some cases, amended substantially. This paper therefore attempts to examine whether the form in which OS still exists is appropriate to the needs, norms and fashions of the times. It is done by considering a set of key questions relating to what topographic information should be provided and on what terms.

THE KEY QUESTIONS TO BE ADDRESSED

The three key questions are as follows:

- what geodetic and topographically-related data and information products are needed?
- who should pay for them and how much will they pay?
- how can these products best be produced and how does this differ from the present situation?

Before discussing each of these in detail, however, we summarize the present characteristics of OS as defined by published documents and then review some of the problems and opportunities facing the Survey.

OS AT PRESENT

General-purpose versus special-purpose mapping

The single most important concept, underlying the case for OS is that one type of mapping product can be made which meets many different goals and can satisfy many different users. Thus the OS large scale maps have traditionally been designed as a compromise to meet the needs of local authorities, utilities, the land registration

2. CAG (1990) **Audit of the Department of Energy, Mines and Resources Surveys, Mapping and Remote Sensing Sector,** Canadian Auditor General, Ottowa.

3. NRC (1990) **Spatial Data Needs: the Future of the National Mapping Program,** Board on Earth Sciences and Resources, National Research Council, National Academy Press, Washington DC.

4. NRC (1991) **Research and Development in the National Mapping Division, USGS: Trends and Prospects,** Board on Earth Sciences and Resources, National Research Council, National Academy Press, Washington DC.

5. Robertson, W. (1990) 'Funding of National Mapping', paper given to **'Surveying 2000' Conference** at Royal Institution of Chartered Surveyors, London, September 1990.

authorities and others. In other countries, much of the need for a topographic frame-work for the utilities and local authorities is met by each of those organizations them-selves and, in some countries, is compiled to different specifications and up-dated on different criteria. Typically, mapping for land registration is produced by chartered surveyors either in a cadastral organization or in private practice.[6]

We have, at present, no real indication whether the mixture of features collected by OS is *ideal* for the largest volume consumers of OS products. Certainly the utilities' private trials in map-making found a simpler specification and fewer features than used by OS at that time to be acceptable for their immediate purposes: partially as a result, OS then simplified their own data gathering operations. Nor indeed do we have any costings of the relative merits of integrated, general-purpose mapping as opposed to special-purpose mapping to meet the users' needs. Any such analysis is confounded by the existence of the present OS topographic archive.

An important distinction can also be drawn between the traditional situation and that now pertaining: the great bulk of what was collected previously was also shown on the OS maps. However, because of the availability of a range of topographic information in computer form, the content of the product can now be tailored to the needs of individual clients. To that extent, OS maps need no longer be general purpose ones but what is collected is still determined by the perceived needs of a range of clients.

The context in which OS operates

No other change has affected OS (and most other organizations) as much as the decline in cost of computing power. In general terms, this has decreased by about a factor of ten every six years over the last thirty. Figure 1 illustrates this by reference to a number of important events in the Survey's history. At the time of the OS Review Committee in 1978/79, for instance, the raw cost of computing power was about 100 times greater than that in 1991. This change has transformed what is economically

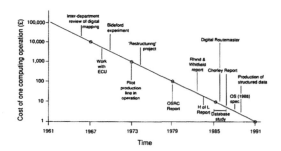

FIG. 1.   The decrease in computing costs since 1961 and key events in the evolution of OS digital database

6. Dale, P. F. (1991) 'Land Information Systems', in **Geographical Information Systems: Principles and Applications**, Maguire, D., Goodchild, M. F. and Rhind, D. W. (eds.), Vol. 2, 85–99, Longman, London.

feasible and equivalent changes seem inevitable at least in the next six year period. Associated developments, in particular the inevitability of widespread use of the Global Positioning System (GPS), provide additional challenges and opportunities for OS.

In the past, as typically supplied small scale artwork to other publishers under licensing arrangements. They then redrew this and published their own maps. Only Bartholomews were outside this arrangement. However competition is increasing at small scales and promises to appear even at large ones. For instance, OS' new 1:250,000 scale digital database is in competition with products from the Automobile Association and Bartholomews. The advent of the Bartholomews' 1:5,000 scale digital map for London and the publication of increasing numbers of street maps of cities, all based upon mapping claimed to be independent of OS, extends the competition into the larger scale domain. On several grounds therefore the context in which OS operates now differs greatly from that at the time of the OS Review Committee.

The anatomy of OS: the inputs

The situation may be summarized as follows. Ordnance Survey is now an Executive Agency of government, charged with the official topographic survey and mapping of Great Britain. It remains a separate government department, funded by vote, operates on a net running cost basis and is answerable to the Secretary of State for the Environment. The latter sets financial and other targets for OS. The long term objective is to become profitable (OS Information Paper 1990/1) which is unlikely to be possible until the major expenditure on digitising its maps is over. In the meantime, OS aims to maximize the recovery of its costs and to provide an efficient and effective service, based upon the concept that 'the national interest is best served by having accurate and readily available topographic data accessible to the public'.[7]

Figure 2 and Table 1 show the key financial and manpower details about the Survey and how these have changed since the time of the OS Review Committee. All money figures are converted into 1989 prices using the Implied GDP deflator with a 1989 base at market prices[8] and are expressed in millions of pounds. The original sources of all figures are OS Annual Reports or OS itself. Analysing the picture is complicated by various changes in the manner of reporting in the 1980s, notably within the Annual Reports, as a consequence largely of changes in government practice.

Over the period since the OS Review Committee reported in 1979, OS has shed some 30% of its then staff. Despite this, its expenditure has remained remarkably constant, varying between £60 million in 1983/84 and £70 million in 1986/87 but averaging £65 million in 1989 prices. At the same time, revenue has risen sharply to 177% of its 1979/80 value and this is reflected by the rise in the overall cost recovery from 37% to 66% over the same period. This latter figure is, of course, driven by the government's insistence that core activities should produce a minimum specified cost recovery and that all other services should be profitable (i.e. have a cost recovery of 100% or more). The increase in the level of copyright income over the period from £7.4 million to about £12 million in 1989/90 should also be noted.

7. OS (1990) **Ordnance Survey Corporate Plan 1991–4**, Ordnance Survey, Southampton.
8. CSO (1991) **Economic Trends Annual Supplement 1991 Edition**, Central Statistical Office.

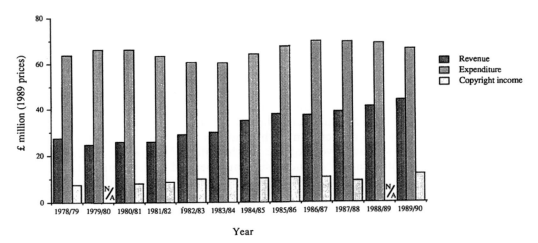

FIG. 2. Key financial statistics for OS for the period since the Report of the Ordnance Survey
Review Committee

Table 1. Key financial and manpower statistics for OS since the report of the Review Committee.

Date	Revenue in 1989 prices (£m)	Expenditure in 1989 prices (£m)	Copyright revenue in 1989 prices	% cost recovery overall	% cost recovery on core required by government	Total staff
1978/79	27.7	63.6	7.4	43		3,638
1979/80	24.8	66.3		37		3,475
1980/81	25.9	66.3	8.1	39		3,371
1981/82	26.1	63.4	8.8	41	25	3,151
1982/83	29.3	60.5	10.1	49	30	2,970
1983/84	30.1	60.1	10.0	50	30	2,814
1984/85	35.2	63.9	10.4	55	30	2,961
1985/86	38.0	67.4	10.7	57	35	2,981
1986/87	37.4	69.6	10.8	54	40	2,965
1987/88	38.9	69.2	9.2	55	40	2,757
1988/89	41.1	68.6		60	40	2,648
1989/90	44.0	66.2	12	66	40	2,525
1990/91			13.5§	64	55	2,457
1991/92*				70		
1992/93*					65	

* = target set by government.
§ = expressed in 1990 prices.

Table 2. Expenditure and income by category in OS; all figures are in £m and expressed in 1989 prices (see text for explanation).

Date	Core income	Core expenditure	Commercial income	Commercial expenditure	Repayment income	Repayment expenditure	Overseas income	Overseas expenditure
1985/86	19.8	50.2	3.7	3.5	8.4	8.3	6.1	5.5
1986/87	20.8	53.2	3.7	3.4	8.3	8.4	4.6	4.7
1987/88	24.1	56.9	4.0	3.3	7.3	6.7	3.5	3.2
1988/89	27.6	55.9	4.2	3.6	7.3	6.9	2.1	2.1
1989/90*	29.8	52.4	5.7	5.2	6.7	6.7	1.8	1.8

* = provisional figures.

Following the government's acceptance of the OS Review Committee's recommendations and various changes of policy in the 1980s, the organization's activities are now split up into four main categories. Paramount amongst these are the 'core' activities, covering the geodetic survey, the topographic survey and production of basic scale mapping and the production of the 1:50,000 scale mapping. Two thirds of the OS revenue is generated by these activities and four-fifths of the expenditure is consumed by them. Table 2 covers only those years for which the detailed figures are readily available. It shows that, over a five year period, core income grew 51% whilst expenditure grew only 4%. Commercial expenditure and income grew roughly in accord, with about a 10% surplus of the latter over the former; in total this accounted for only about 19% of core income and about 10% of expenditure. The volume of repayment services has declined in real terms by about 20% but income and expenditure are normally roughly in balance. The most disappointing picture is presented by the overseas activities, formerly carried out by the Directorate of Overseas Surveys before it became incorporated within OS in 1984: whilst income and expenditure are roughly in balance, both had shrunk by 1989/90 to 30% of the level at the start of the five year period. In part this reflects the decisions of the government's Overseas Development Administration that survey and mapping are now less important elements of the aid budget.

In analysing these figures, two particular factors need to be borne in mind. The first is that OS was engaged in a massive re-survey of the country recommended by the Davidson Committee until the early to mid-1980s. Additional funds were voted to speed this process in the 1960s and were generated by raising the cost recovery targets from 1966 onwards.[9] As a consequence, the then staffing was much higher than would be necessary in a steady state organization monitoring only change; in 1975, for instance, some 4,577 staff were in post. The second factor is the 'one off' nature of a digitisation process which requires the encoding into computer form of at least the great bulk of the 220,000 paper maps; the end of the mass digitising programme is now in sight, with estimates of the encoding of all the 1:1,250 scale maps by end-1991 and of those 1:2,500 scale maps required by customers in that form by 1995. Though mitigated to an unknown degree by the utilities' digitisation of OS mapping (which

9. OSRC (1979) **Report of the Ordnance Survey Review Committee**, Her Majesty's Stationery Office, London.

is then available for OS to sell), the expenditure involved is considerable. Once this investment is complete, annual running costs can reasonably be expected to fall significantly even though updating the database will be of paramount importance.

The anatomy of OS: the outputs

All of the above relates to inputs. How do we judge the outputs produced from this expenditure? Where the 'product' now largely consists of revisions to existing maps, this is difficult to measure in any readily interpretable way. The OS Corporate Plan for 1991–94 does give targets for the volume of survey of change (867,000 'house units' — which was handsomely exceeded in 1990–91) and for the number of maps to be converted to digital form (14,000) in 1991–92. Yet how do we know if these are easy or difficult (or even appropriate) targets? The first at least is much influenced by the economic climate: it is much easier to achieve house unit survey targets from entire new housing estates than from individual 'infill'. Clearly the build-up of and sales from the digital version of the topographic archive (Table 3 and Figure 3) are important indicators of success. In themselves, though, the improving sales figures shown in Table 3 may be most misleading: if customers only ever bought the data once, the effects would be catastrophic for OS over the next ten years or so. Recognising this, OS have now introduced a data leasing scheme which includes annual up-dates to the data, akin to the scheme used by the French national mapping organisation, Institute Géographique National or IGN.[10]

The anatomy of OS: the balance sheet

Because the significance (as opposed to the magnitude) of the outputs is difficult to quantify, it is difficult to prove that the country is getting good value for money in

Table 3. Large scale map sheets available in computer form.

Date	Available 1:1,250	Available 1:2,500	Σ digital maps	Sold 1:1,250	Sold 1:2,500	Sales as % of available	Map cost (1989 £)
1979/80			13,200				
1980/81			14,400				
1981/82			18,900				
1982/83			21,700				34
1983/84			23,000				35
1984/85	12,754	12,198	24,952	3,376	455	15.3	43
1985/86	14,864	12,696	27,560	5,667	3,022	31.5	40
1986/87	19,732	14,175	33,907	11,383	2,783	41.8	60
1987/88	27,745	16,537	44,282	13,787	6,249	45.2	97
1988/89	38,373	20,461	58,834	16,752	5,341	37.6	101 & 107
1989/90	49,736	31,326	81,062				110
April 1991			106,889				

N.B. the totals of large scale map sheets available in paper form have increased from 51,184 to 55,869 (1:1,250 scale) and from 158,967 to 163,061 (1:2,500 scale) in the nine year period from March 1981 to March 1990.

10. Rhind, D. W. (1991) 'Data charging, access and copyright: their implications for GIS', **Proc. European GIS Conference (EGIS '91)**, 929–45, Brussels 2–5 April 1991.

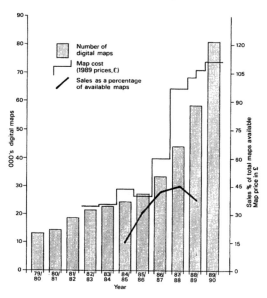

FIG. 3.   The number and cost of OS digital maps available at different times and sales of them as a proportion of those available

terms of the inputs consumed. Ultimately, informed value judgements are unavoidable since simple reliance on a battery of statistics is liable to be misleading. But it is important to note that the entire reporting of cost recovery includes almost none of the costs of assembling the topographic archive — estimated by the OS Review Committee in 1979 as being many hundreds of millions of pounds at that stage. These are not included as assets in the OS Annual Reports; in essence, the historical cost of survey and of assembling and digitising the archive has been written off. All cost recovery figures are compiled as if generated from activities solely in that year. It is, of course, non-trivial to make an actuarial assessment of the value of this information. Moreover, such an accounting practice is quite common in government. This contrasts with practice in some parts of the private sector: as an extreme example, some advertising agencies even count unsuccessful bids for contracts as assets on the basis that these can be quarried for later tenders.

OS history as tramlines

There are many considerations besides money and even government policy which influence the development of an organisation; one of these is the constraints imposed by history. Clearly, one of the penalties of longevity is the need to be consistent with what has occurred in the past. That organisations like INEGI (the Mexican national mapping — agency and government statistics organisation) have more sophisticated telecommunications links to their regional offices than does OS simply reflects the fact that technological jumps can now readily span several different generations of facilities. Indeed, if skilled personnel are available, it is often easier for out-of-date organisations to be brought to the international forefront more readily than other, more evolutionary ones. In this sense, as suffered greatly from initiating the world's first digital map production in 1973[11]; for almost fifteen years, they were 'stuck in the tramlines' of history, with little change to data specification or content but rapidly mounting numbers of maps which reflected the thinking and technical constraints of the early 1970s.

Indeed, the intrinsic nature of OS large scale mapping is an even more striking example: why is it that we publish almost no large scale photomaps even though these are commonly produced elsewhere in the world? In particular, orthophotomaps are being produced by some other agencies and the technology for producing them digitally is rapidly declining in price. It can be argued, of course, that user resistance to such new products is implacable or that the British countryside-differs from that elsewhere in the world. In reality, however, the investment in existing products and the whole structure of the existing organisation buttresses a view of the world described by points and lines and labels attached to them. Yet, despite the difficulties inherent in any change, OS needs to commission a study of the replacement cost of the entire existing archive by alternative forms, together with the maintenance costs under each option. We need to know the cost of ignoring history if rational decisions are to be made about the shape of the future.

INFORMATION AS A COMMODITY

We are already in the 'information age'. OS is in the market of selling information, increasingly stored in digital form. For this reason, it is essential to consider the nature of information as a commodity to be bought or sold.

Unhappily, it is not easy to define the distinctions between the commonly used terms 'data', 'information' and 'knowledge' except by example or in the most general fashion. Thus data is usually taken to be 'raw' or little-processed, such as population statistics for a given area; information is taken to imply direct relevance to one or more tasks or policies and generally involves some classification or other processing of the 'raw' data, resulting in (for example) the market size for a product in that area. Knowledge is generally used in a more abstract way but may also be used to denote a set of 'facts'. In principle, the data-to-information conversion somehow adds value by creating a new, derivative product without diminishing the value of the original; the converse process is generally impossible. Maps, however, are an especially difficult case because computer-readable data may be converted into other, noncomputer-readable data (a paper map, which is widely regarded as information) and this may then be converted back into (different) data by a digitising process. Hence there is a major terminological difficulty: what is data for one person and for one task may be information for others. Because of this, the terminology used below to cover all categories is 'information'.

The 'commodification' and the commercialisation of geographical information[12] seems to be a relatively recent phenomenon, at least so far as governments are

---

11. Sowton, M. (1991) 'The Development of Digital Mapping and Support for GIS in Ordnance Survey', in **Geographical Information Systems: Principles and Applications**, Maguire, D., Goodchild, M. F. and Rhind, D. W. (eds), Vol. 2, 23–38, Longman, London.

12. Openshaw, S. and Goddard, J. B. (1987) 'Some implications of the commodification of information and the emerging information economy for applied geographical analysis in the United Kingdom', **Environment and Planning A**, 19: 1423–39.

concerned. The prices of maps sold by Ordnance Survey until 1966, for instance, were based only upon the recovery of the costs of printing, of paper and of distribution although copyright charges for copying by users of published maps helped to defray total mapping costs. Prices of large scale maps were raised in 1966 to help meet the costs of speeding the Survey's work programme; seven years later, the organisation was ordered to maximize revenue on all products, subject to certain policy guide-lines (OSRC, 1979). As indicated in Table 1, OS must recover at least 70% of its total costs through supply of geographical goods and services for financial year 1991/92.[13] This is part of a general trend in UK government: it has laid particular stress on trying to identify, publicize and sell 'tradeable information' which it collects for its own purposes.[14] The aim has been to sell such information as a commodity — mostly through the private sector — in order to generate returns to the tax-payer.

Yet, as Openshaw and Goddard[15] and Hoogsteden[16] have shown, the whole concept of information as a commodity is also less than straightforward. It does not, for example, wear out through use though its value may diminish with time at a rate determined by the rate of change of the features it describes and the advent of competitors. It is also clear that different types of information display different kinds of characteristics: to the economist, much digital information could be considered a public good — it can be copied at near-zero cost and this may well limit the abilities of a vendor to secure acceptable profits. Set against this is that, in those situations where the currency of the data is critical (in military and a few civilian tasks), the market *can* (in theory at least) operate to generate profits.

However, the utility and value of digital spatial data — especially at the present time — are often highly related to the expertize, knowledge and imagination of the purchaser and exploiter. In these circumstances, the 'normal' relationships between supply, demand and pricing may well be distorted. In addition, all Geographical Information Systems (GIS) experience thus far strongly suggests that ultimate value is heavily dependent on the association of one data set with one or more others. Thus in the CORINE (and in perhaps every environmental) project, the bulk of the success and value came from linking data sets together.[17] Almost by definition, the spatial framework provided by topographic data is embedded in other data sets and/or these are plotted in relation to it; without this data linkage, almost no other geographical data could be analysed spatially or displayed. To that extent, OS information is liable to be central in all spatial data processing in Britain.

A final problem arising from a definition of information is how to measure a unit of it in order to set a proper charge for its use. Metrics such as total lengths of line, the number of bits used to store the data, the total numbers of map sheets (as normally used by OS) or the number of areas for which statistical values exist are all highly imperfect measures of value to be transferred to a customer. Indeed, use of such

13. OS, 1990.

14. DTI (1990) **Government-held Tradeable Information: Guide-lines for Government Departments in Dealing with the Private Sector,** Department of Trade and Industry, London.

15. Openshaw and Goddard, 1987.

16. Hoogsteden, C. (1989) 'The economics of mapping', unpublished Ph.D. thesis, University of Otago.

17. Mounsey, H. M. (1991) 'The Creation and Use of Multi-source, Multi-national Environmental GIS: Lessons Learned from CORINE', in **Geographical Information Systems: Principles and Applications,** Maguire, D., Goodchild, M. F. and Rhind, D. W. (eds), Vol. 2, 185–200, Longman, London.

metrics may well inhibit much use of information — unless the acquisition cost is low, organisations such as libraries (which generally have low modal access values for the bulk of items in their collections) will not wish to hold it. In practical terms, however, it is difficult to conceive better surrogates for value; however unsatisfactory, these may be the only *a priori* basis on which to price different products if rational, consistent and defensible pricing is a necessary requirement. The commercial alternative is simply to charge what the market will bear on all occasions. Converted into mapping terms, this would probably ensure higher prices for areas of high demand and/or where competition is weak.

Measuring national benefit

The nature of OS as an information gathering organisation is unusual within the UK government. Almost all other government departments collect information primarily to support the administrative or monitoring functions assigned to them. Ordnance Survey collects information but makes no policy-related use of it internally. It reformats information and sells it on in map form; OS is thus the major information wholesaler and retailer in government. As such, its unique character causes considerable difficulties through the lack of direct comparators except in the research councils (where the funding of the surveying role of the British Geological Survey has been a matter of dispute between government departments and the Natural Environment Research Council). Yet, as indicated earlier, the centralized provision of mapping services in Britain has been legitimized by the breadth of external usage of OS mapping. The argument is that a single set of products can be used for many different purposes and thus generate very substantial national benefits. Unhappily for the Survey and others, the mere assertion of 'national benefits' is not a popular yardstick at present. For a free-standing government agency, the acceptable means of demonstrating national benefits and hence cost-efficiency is by charging all customers — government as well as everyone else — at levels which are the maxima that different sectors of the market can pay.

Despite that, there are ways in which 'national benefit' from information supply can be quantified. One such example relates to the decennial Census of Population. In Britain, this is totally predicated upon the availability of up-to-date large scale maps on which all planning of Enumeration Districts is carried out (the ED is the areal unit for which one human being distributes and collects census forms, as described in Rhind[18]). The total cost of the 1991 Census in Britain is likely to have been about £135 million and, of this, the purchase cost of the 10,700 1:10,000 scale and 62,000 larger scale maps plus reproduction of them must have been about £2 million; the manual and clerical operations involved in drawing the ED boundaries on the maps involved must have cost a like sum. Compare this with the situation in the USA where, for the 1990 Census, a whole new digital map base was generated by the US Geological Survey and the Bureau of Census based upon new 1:200,000 scale mapping. The USA has something like 260 million people compared to about 56 million in the UK; this figure is, however, misleading because about 65% of the households in the USA receive and send back their census form by mail. Moreover, economies of scale should be attainable in dealing with larger operations. Despite all this, setting

18. Rhind, D. W. (ed.) (1983) **A Census User's Handbook**, Methuen, London.

up the TIGER system to provide the maps and other required census geography products cost something like $330million or $1.33 per head, some eight or nine times the apparently comparable figure in the UK. This comparison is far from ideal because the British procedures are paper-based, because the US data base can also be used for other purposes (notably because it has been embraced by the private sector as a result of the lack of copyright protection) and because of other factors. Nonetheless, it strongly suggests that the real value to OPCS — and to the country as a whole since the Census results are ubiquitously used — is much greater than can be measured through the sale at near-standard prices of OS maps. There may well be other examples of total dependency on OS mapping and the Survey would be well-advised to define them and the benefits involved in some detail.

ANSWERING THE KEY QUESTIONS

We now turn to the most important part of this paper — an attempt to answer the key questions set out at the beginning of the text.

What topographic information do users need?

The obvious way to proceed here is to ask the users. That is a necessary requirement and OS presently meets it through its user liaison committees, through periodic market research (such as the 1983 User Survey for Large Scale Digital Data and that for Small Scale Digital Data a year later) and through informal contacts via sales and sector representatives. The liaison committees are essentially 'rainbow coalitions'; to judge from the minutes of their meetings, many seem to be concerned with a variety of detail rather than with fundamentals. This *may* reflect a view by all of them that the existing products are near-ideal. Guidance also comes however from the results of studies by others (see, for instance, Bromley and Coulson[19]). In the USA a more formal and structured annual approach is used, known as the A-16 procedure.[20] That process however has severe shortcomings: since customers do not have to pay the real costs of what they request, they inevitably overstate their needs and priorities.

The prime disadvantage of the as elicitation approach is that it is slow to anticipate (and may even inhibit) new potential requirements. It also is a mechanical, 'bottom-up' approach which may not match well with 'top-down' policy guidelines. A quite different approach was taken by Smith[21] who considered the needs of government and society on a much more philosophical basis. From this, he deduced the need for particular categories of information (Figure 4). Though now dated because of the changing nature of society since 1979, this is still perhaps the best 'from first principles' attempt anywhere in the world to consider the essential role of a national mapping organisation and a number of his arguments bear repeating.

19. Bromley, R. and Coulson, M. (1989) 'The value of corporate GIS to local authorities', **Mapping Awareness**, 3, 5, 32–5, Oxford.

20. Starr, L. and Anderson, E. A. (1991) 'A USGS Perspective on GIS', in **Geographical Information Systems: Principles and Applications**, Maguire, D., Goodchild, M. F. and Rhind, D. W. (eds), Vol. 2, 11–22, Longman, London.

21. Smith, W. (1979) 'National mapping: a case for government responsibility', **Conference of Commonwealth Surveyors 1979**, Paper M2, Cambridge.

Examples of geodetic and topographic information	'ORDER' : (ESSENTIAL)														'PROGRESS' : (OPTIONAL)																
	1. Security		2. Protection of Life & Property						3. Fiscal & Administration						4. Education & Cultural Progress				5. Economic Progress											6. Social Progress			
	Defence	Police	Flood	Earth movements	Fire	Danger areas	Ambulance	Property boundaries	Water resources	Planning	Rating	Admin, general	Pollution control	Transport	History/ archaeology	Ecology	Scientific	General education	Agriculture	Forestry	Transport	Mineral surveys	Mining	Aeronautics	Urban development	Population studies	Public utilities	Offshore	Postal	Broadcasting	Leisure/ sport	Tourism
Planimetric control (Primary)	✓	×	?	✓	×	×	×	×	×	×	×	×	×	×	×	×	✓	?	×	×	×	×	×	×	×	×	×	✓	×	×	×	×
(Other)	✓	×		✓	×	✓	×	✓	✓	✓	✓	✓	✓	×	×	×	✓	?	×	✓	✓	✓	✓	×	✓	✓	×	✓	×	?	×	×
National grid	✓	✓		✓	✓	✓	✓	✓	✓	✓	✓	✓	✓	✓	✓	✓	✓	✓	✓	✓	✓	✓	✓	?	✓	✓	✓	?	✓	✓	✓	✓
Height control (Primary levels)	✓	×		✓	×	×	×	×	✓	×	×	×	×	×	×	×	✓	?	×	×	×	✓	×	×	×	×	×	×	×	×	×	×
(Other)	?	×		✓	×	×	×	×	✓	✓	×	×	✓	×	?	?	✓	?	✓	✓	✓	✓	×	✓	×	✓	×	✓	×	✓	×	×
Contours 5m	✓	×		S/L S/L	×	×	×	×	S/L S/L	×	×	×	×	L	L	✓	?	?	?	?	S/L S/L	×	S/L	×	?	×	✓	S/L	S/L	×
15m	S	S	S	S	S	×	S	×	S	S	S	S	S	S	S	✓	✓	S/L S/L	S/L S/L	S/L	S/L	S/L	×	S	S	S	S
Roads	S/L S/L	S/L S/L	S/L S/L	S/L S/L	S/L S/L	S/L L	S/L S/L	S/L L	S/L S/L	S/L S/L	?	S/L	S/L S/L	S/L S/L	S/L S/L	S/L	L	S/L S/L	?	S/L L	S	S/L L	S
Railways	S/L S/L	S/L S/L	S/L ×	S/L L	S/L S/L	S/L S/L	S/L L	S/L S/L	S/L S/L	?	S/L	S/L S/L	S/L S/L	S/L S/L	S	L	S/L S/L	?	S/L	×	S/L S/L	L
Footpaths	S/L S/L	S/L S/L	?	S/L S/L	S/L L	S/L S/L	L	S/L S/L	S/L S/L	×	S/L S/L	?	?	S/L	S/L S/L	×	×	×	×	?	?	×	×	S/L ×	×	S/L S/L	L
Buildings	S/L S/L	S/L S/L	S/L S/L	S/L S/L	S/L L	?	S/L L	S/L S/L	S/L L	S/L	?	?	S/L	L	?	×	×	S/L S/L	L	S/L S/L	×	S/L ×	×	S/L S/L	L
Areas	×	×	L	L	×	×	×	L	L	L	L	×	×	?	?	?	?	L	✓	×	×	×	×	?	?	?	×	×	×	×	×
Watercourses	S/L S/L	S/L S/L	S/L S/L	S/L S/L	L	S/L S/L	L	S/L S/L	✓	S/L	S/L S/L	✓	S/L S/L	S	L	S	S/L	×	S	×	S/L S/L	L
Vegetation/woodland	S/L S/L	S/L ×	S	×	S	L	S/L S/L	L	S/L	?	?	S	S/L	?	S/L	S/L S/L	?	×	×	S/L	L	S	×	×	×	×	S/L S/L	L
Fences/hedges	L L	L	?	L	×	L	L	L	S/L	L	S/L	?	?	L	L	?	S/L	S/L S/L	×	×	×	L	?	L	×	×	×	S/L S/L	L
Power lines	S/L S/L	S/L S/L	S/L S/L	×	S/L	L	S/L	S/L	L	S/L	?	S/L	?	?	?	?	S/L S/L	×	×	S/L	S	L	×	S/L	×	?	S	S/L	×
Antiquities	×	?	×	×	?	×	×	L	S/L S/L	L	S/L	×	?	S/L	?	?	S/L	S/L S/L	?	S/L	×	×	L	?	?	×	×	×	S/L L
Tourist information	×	?	×	×	×	×	?	×	×	?	?	S	×	?	?	?	?	?	×	?	?	×	×	×	?	?	×	×	×	×	S/L S	S/L L
Public rights of way	×	S/L	×	×	×	S	?	L	S/L S/L	L	S/L	×	?	?	?	?	?	S/L L	S/L	×	×	×	?	×	S	×	×	×	S/L S/L
Administrative boundaries	S	S/L	S	×	S/L	×	S/L	L	S/L S/L	S/L	S/L S/L	S/L S	S/L S/L	L	?	S/L	S/L S/L	S/L S	L	×	L	S/L S/L	S	S/L S/L	L	S	S/L S/L
Place names	S/L S/L	S/L L	S/L S/L	S/L S/L	S/L S/L	S/L L	S/L S/L	S/L L	S/L S/L	S/L S/L	S/L	?	S/L	S/L S/L	S/L S/L	S/L S	S/L S/L	S/L S/L	S	S/L S/L	L	L	S/L S/L	L	S	S/L S/L
Public buildings	L L	L	×	×	L	×	L	L	×	L	L	L	×	L	?	?	L	?	?	L	×	?	×	L L	L	L	×	L	L	?	L
House names & numbers	×	L	×	×	L	×	L	L	×	L	L	L	×	L	×	×	×	×	×	×	×	×	L	L	×	L	L	?	?

FIG. 4.   Topographic features required by map users (from Smith 1979)

Key (right margin):
Information considered essential	Information considered desirable	Information of doubtful value	Information not needed for this activity	
✓	✓	?	×	All
S	S			Small scales only
L	L			Large scales only
S/L	S/L			Both large & small scales

Since (then Director General of Ordnance Survey) argued that government had two primary responsibilities. These are to ensure order and to encourage material and other progress (such as encouraging economic growth, raising health standards and conserving the environment). The first is a mandatory responsibility and the second is optional.

Ensuring order

This involves the state in safeguarding national security and protecting individual rights to life, property and liberty. The price paid for this by the citizenry is the surrender of some individual rights and resources in the common interest. Smith (1979) argued that security purposes demanded three criteria be met by the topographical information collected: that it is up-to-date, is homogeneous (i.e. to a common reference system and common specification) and is continuous in spatial extent (i.e. exhausts the national territory). He pointed out that, whilst much of the information might be collected by the private sector, the specification of the product and the quality control constituted an irreducible minimum state responsibility. He also pointed out the dangers involved in developing responsibility for funding and agreement on the product to subsidiary bodies such as local governments: the parlous state of large scale mappings in many areas of the USA and the varied levels of county-by-county funding of school-level education in the UK testify to the force of this argument.

The protection of individual rights certainly requires the collection of much geodetic and topographic information. Monitoring crustal movement for flood defence studies and recording land parcel ownership details are two obvious and important examples. Yet even here the private sector may act as the government's agent. For instance, the government of Ontario has entered into partnership with a commercial enterprise to assume the responsibility for the land registration system. In Britain, the cadastral process is substantially divorced from the collection of topographic information: HM Land registry regards OS largely as a contractor. Unhappily for OS, many British statutes do not explicitly specify the use of OS information even where it exists; as one example, the assessment of local taxes in Britain (even after the abolition of the Community Charge) seems unlikely to involve measurement of the area of the land owned though such information may be created as a by-product from OS digital maps.

Thus far in the discussion, then, the case of having topographic information is clear. Though more than one method exists for collecting and holding the information, some type of state control seems essential.

Encouraging progress

So far as this is concerned, Smith advanced three reasons why government '. . . should in any case be involved in the collection of topographic information'. These he summarized as:

* when it is impractical for individuals or particular sections of society to provide it for themselves [and, by implications, when they need (c.f. desire) it];

- when it is less expensive for the nation as a whole if it is provided centrally (permitting economies of scale);
- when the pooling of many needs for a variety of information makes it cheaper to be met this way than to be met separately (permitting economies of aggregation).

Clear merit exists in each of these but — in principle — none *necessitate* that government alone must be involved. Crudely stated, the views of the political party in power since 1979 are that the government apparatus is best suited as an enabling mechanism and a regulatory one, with the private sector creating wealth (and tax payments) within this framework. Despite this, there are at least six other reasons why government might need to play a more activist role in the creation of geodetic and topographical information. These are:

- some future requirements are unpredictable. In practice, therefore, some information is collected on a 'just in case' basis. Parallels to this include curiosity-driven research in universities and government's application of the same precautionary principle in reducing CO_2 emissions even though the scientific evidence for global warming is not conclusive. Use of the large scale OS maps in the aftermath of the Lockerbie disaster is an example where such precautions paid off though it is difficult to put a cash value on such information. Arguably, only governments would be a customer for such information which may never be needed. Hoogsteden[22] terms this the option value of geographical information and suggests that indirect financial support via taxation is the logical mechanism;
- on purely profit-driven motives, largely rural areas would never be mapped except when a major project was planned (and then only small areas would need to be described in great detail); whilst OS large scale maps are important to many individual farmers, the size of the total market in this category is relatively small and the number of maps involved is large. In a wholly commercial activity, resources would be focussed on areas of high population density and of rapid change: some 40% of the OS 1km grid squares in mainland Britain, for instance, contain no people whatever.[23] It is no surprise therefore that the most spatially extensive utility only holds about 130,000 of the 210,000 large scale OS maps. In essence, then, the justification thus far for comprehensive coverage at 1:10,000 or larger scales is essentially either a social one (akin to the preservation of public telephone services in remote areas), a 'just in case' or a military one;
- with the advent of copying machines and computers, paper maps and digital information respectively have acquired many of the characteristics of a public good — and hence one from which it is difficult to make a profit. As indicated above, the information may be transferred at near-zero cost, the producer need lose none of the original material by disposing of it and the product does not wear out (though its value may well diminish differentially as its currency decreases; see Rhind[24]). In such circumstances, commercial enterprise might well

22. Hoogsteden, C. (1991) 'The benefits of national mapping', **Land and Hydrographic Survey**, 2: 5–7, Royal Institution of Chartered Surveyors, London.
23. CRU/OPCS/GRO(S) (1980) **People in Britain**, Her Majesty's Stationery Office, London.
24. Rhind, 1991.

be wary of providing such information. The failure of the commercialisation of remote sensing information in the USA is a relevant example;

- access to land to 'field complete' mapping is presently guaranteed to OS staff by statute but this may not be possible for private sector mappers;
- there is a statutory requirement on a few users (notably HMLR) to record their information on OS maps;
- since the activities of a national mapping agency underpin many other vital operations of the state or of basic service providers, it may well constitute an unacceptable risk to have this function outside of direct government control;
- government has an obligation on behalf of its citizenry to maintain a historical record of change in the natural and cultural landscape. OS has provided this for 200 years and, through the agency of the copyright libraries, the public has had access to its past.

The types of product needed

Two main types of information should be supplied. The first is that which is already being supplied to and used by customers. The second is that which might be valued by customers if only it were available and known to them. In practice, no hard and fast distinction can be made between these since the latter might well include re-formatted or reorganized versions of the former; moreover, any success of the latter might be at the expense of the former (but could be of greater benefit to the customers, might generate greater income for OS and might well 'lock out' competitors).

At present, we know something of what existing customers want by way of static topographic information — or at least what they will pay for. Some customers certainly want full, topologically structured and edge-matched (or 'seamless') vector databases whilst others will apparently be happy with raster-scanned versions of the existing paper maps (even if they are somewhat out-of-date) to act as a backcloth to other information. Thus the same geometric product is needed in at least two forms. But we know relatively little about what users really *need* by way of the frequency of topographic up-dates and what value they place upon increasing frequency of up-dates. This is clearly critical for OS manpower planning and, indeed, for the totality of its activities. The indications are that the needs vary between users. The ideal form of the new data is also unknown: do users want up-dates to their existing files (which may be operationally difficult for a number of reasons) or would they simply prefer complete new versions of the entire mapping? There is reason to believe that some users are only interested in current information rather than the change of it through time, thereby favouring the latter approach. But this is certainly not true of users such as lawyers and engineers concerned with land ownership or site investigation who typically need access to historical information provided via the copyright libraries.

Adding value in GIS is normally achieved by linking data sets together; with essentially only one data set of their own — topography — to offer at present, it might be thought that OS success in the value-added business is likely to be small. Yet that one data set is critical for almost all the others and should enable many partner-ships to be forged. One example of how this can be done is provided by IGN Paris who invited bids from 1600 companies to develop software and applications for IGN data (for which seven free data sets and technical advice were given). From the 150 organisations answering, some 42 were selected and some 65 products were produced and marketed with the IGN 'seal of approval'. Thus it would indeed be surprising if OS was not much more involved in 'down stream' applications of their data in years to come, the majority of these being carried out in partnership with the private sector.

The problem in any marketing lies in detecting latent need, in making the latent demand real and in meeting it. The indications are that OS presently has an inadequate (or at least an unstructured) 'anticipation of need' role compared with some other mapping organisations, is rather conservative in testing the market and slow in bringing products to the market. Digital Elevation Models are, for instance, by far the largest selling files so far as US Geological Survey and IGN Paris are concerned and they exist for much of Britain thanks to the Ministry of Defence: yet OS initially seemed extremely reluctant to market them, sales are said to be very small and customers complain of delays in supply. It may however be that the real problem is a quite different one: the development of new products needs considerable risk capital. Typically, most successful commercial geographical information products have cost substantially more to market than to create.

A common factor in all of the above is the nature and role of the staff. In contemporary circumstances, all of OS staff must be part of the marketing operation. What is being marketed is not floppy disks or tapes of data: it is a service so 'this is our data in a standard format and now it is your problem' cannot be the approach used. This implies the need for a close awareness of what the customers require and of how they wish to handle the data: it seems essential, for instance, that either OS or its agents should run in-house all the main software packages run by their users and ensure that the OS data sets work on these before distribution. Building on this, a possible development is that of OS as a skills centre which would provide advice to clients on a paying basis. This scheme seems applicable very generally, including overseas. It is disappointing that an organisation which had prided itself on being at the leading edge in high technology mapping for the last twenty years has not developed overseas programmes exploiting such skills.

Who should pay for the products?

What role should the state play in making available information collected on behalf of its citizenry? The answer is, of course, one for each state itself to answer but it is appropriate here to summarize some of the issues. Much of this section draws upon publications of the Australian Land Information Council (e.g. ALIC[25,26,27,28]). It also

25. ALIC (1990:1) **Data Custodianship/Trusteeship**, Australian Land Information Council Paper number 1, Canberra.

26. ALIC (1990:2) **A General Guide to Copyright, Royalties and Data Use Agreements**, Australian Land Information Council Paper number 2, Canberra.

27. ALIC (1990:3) **Charging for Land Information**, Australian Land Information Council Paper number 3, Canberra.

28. ALIC (1990:4) **Access to Governmental Land Information: Commercialisation or Public Benefit?** Australian Land Information Council Paper number 4, Canberra.

draws upon the paper by Maffini[29] who, largely from a commercial sector viewpoint, considered five factors to influence the distribution of government databases: political, legal, fiscal/administrative, economic and moral.

A traditional view is that governments collect information to support statutory purposes or to achieve specific public benefits for which they were elected. Since members of the public have already paid for its collection through taxes, the information already used for these purposes should — it is argued — be made generally available to them at no more than the cost of reproduction. It is also often argued that low-cost dissemination maximizes the breadth of use, helps foster industry and thus facilitates the creation of taxable wealth and jobs. The existence of a thriving industry in the USA is usually attributed in no small part to the pursuance of this policy by the federal government.

Set against these by proponents of charging are five arguments. The first is that only a small number of citizens may benefit from the free availability of data which has been paid for by all and hence this is unfair. The second argument partially follows from the first: any legal method of reducing taxes through recouping of expenditure is generally welcomed by the citizenry. The third is based on the experience that considerable efforts need to be devoted to packaging, documentation, promotion and dissemination of information to maximize use of products and that these activities are invariably expensive. Most government agencies are not well-suited to such activities, rarely having adequate promotion budgets, etc. Since nothing is for free, commercial firms will only do all this if substantial profits are to be made. The fourth argument in favour of non-marginal pricing is that putting such a charge on information inevitably leads to more efficient operations and forces consumers to specify *exactly* what they require. Finally, there are good examples where proposals for cost-sharing in the creation of data has been vitiated by the knowledge that 'staying out' of the consortium ensures that the data can be obtained more cheaply afterwards; thus the decision to make information widely available at minimal cost may actually delay its availability.

All of the above points contain some element of truth. Which conclusion on the proper level of pricing is reached will be influenced by one's views on the role of the state as much as by evidence of economic advantage. However, the fact that there are so many possible arguments means that organisations like OS are liable to suffer significant changes of policy on change of government and this could have massive effects upon their activities.

The cost recovery mechanism

Organisations which recoup money typically do so by two means — sale of their information or (increasingly) by leasing it. The former only works over long periods when up-dating is critical for certain classes of users or if 'planned obsolescence' is built-in, as was once common in the motor industry. Thus sales of CD-ROMs of road networks for use in car guidance systems are predicted to continue in perpetuity — not because the road network changes fundamentally every year but because different or additional hotels and other value-added information is integrated with each new

29. Maffini, G. (1990) 'The role of public domain databases in the growth and development of GIS', **Mapping Awareness**, 4(1): 49–54.

release of the 'core' data. The price of the CD-ROM may well be subsidized for road users through advertising paid by these hotels. In general terms, however, data leasing is a more attractive option for many organisations supplying information, especially those with a preponderance of large volume customers: it smooths cash flow but is sensitive to the effect of new competitive products becoming available in the market place.

The mechanics of recovering money either from sales or from data leasing ensures that it is essential for information not to leak to users who have made no financial contribution. How this is enforced (e.g. through copyright laws) varies in detail from country to country and can not be covered here; the Commission of the European Community has published a Green paper on 'Copyright and the challenge of technologies' and CERCO (Comité Européen des Responsables de la Cartographie Officielle) and other organisations have been discussing this with a view to protecting the rights of owners and custodians of geographical databases. Most countries have up-dated their local equivalent of the Berne Convention on copyright (in Britain, for instance, the Copyright, Designs and Patents Act of 1988 is supposed to fulfil this function). In practice, many items may require to be resolved by case law where a dispute occurs.

Cost recovery is particularly difficult when the use of information is indirect. Consider, for instance, the most widely used geographical information at present — statistics pertaining to administrative or other areas. Most of these are derived by aggregation from bureaucratically-compiled records pertaining to human individuals. As one example, Rhind[30] has shown how population information is sometimes assembled in this manner. In certain countries (notably — but not only — Britain), privacy legislation constrains the uses to which the 'raw information' may be put when it is held in computer form — even though defining privacy is not an easy matter.[31] Yet commercial exploitation is commonplace: there are reckoned to be 4,500 machine-readable files in Britain of names and addresses derived from customer records or from encoding the annually revised electoral roll. In many cases, the information owners enhance the value of their file by inferring commercially relevant information from Population Census details for small areas to which these individuals may be linked through knowledge of their home location. Other possible applications of data related to human individuals are legion and burgeoning: aggregation by postcode area of residence of credit card spending, of electricity consumption or other such routinely (and increasingly automatically) collected information would, for instance, facilitate lifestyle deductions and hence have commercial value in target marketing. Such aggregation to 'sensible areas' and linkage of data sets can most easily be done either by use of the OS National Grid or through the hierarchical structure of the postcode system. But portrayal of the results in map form and many types of analysis can only be done readily through the medium of the National Grid. The current financial return to Ordnance Survey in all this is negligible even though the Survey provides the basic infrastructure without which the operations cited all

30. Rhind, D. W. (1991) 'Counting the People: the Role of GIS', in Geographical Information Systems: Principles and Applications, Maguire, D., Goodchild, M. F. and Rhind, D. W. (eds), Vol. 2, 127–137, Longman, London.

31. Danby, B. (1991) 'In search of privacy', GIS World, 4(1): 108–111.

would be very difficult. Given government policy, some way of 'taxing the benefit' is desirable.

Finally, perhaps the most contentious question of all is how the market segmentation is viewed. In the UK, for instance, local government has traditionally been treated by information-providing agencies of the state in a fashion different to other branches of central government — even though they are often involved in the data collection process by form-filling, etc. The accident of how functions are distributed across central government departments may also affect the ease and cost with which data may be obtained. Other state-funded organisations such as Higher Education (HE) are normally treated as being outside the realms of government and charged accordingly. There is surely something wrong when HE finds it easier and cheaper to purchase information from the private sector than from OS!

Policing the use of information

Policing the distribution of their information has lately become an obsession for OS — for understandable reasons. 'Leakage' of it to unauthorized users deprives them of sales income or leasing revenue when government expects progressively higher levels of cost recovery. Unfortunately, policing becomes progressively more difficult and expensive as they move from dealing with a relatively small number of 'respectable' quasi-public sector bodies (the Land Registries, the utilities and local governments) to what could well turn out to be a mass-market. Three main considerations exist in such circumstances:

- is the cost of policing disproportionate to the cost of the data?
- can policing ever be successful given the ease of copying the data (at least when its use becomes as widespread as computer games)?
- does some data leakage (e.g. to young experimenters) actually help to develop new applications and hence the market?

No simple answers exist to these questions and the matter is, indeed, one for government as a whole rather than merely one for OS. It is, however, in the Survey's interests at least to get official recognition of the cost of policing.

Public access to geographical information

Until now, we have considered only the British situation in detail though the American federal insistence on supply of information at reproduction cost has been noted. The right of the citizenry to have access to information compiled by the state is a matter of greater importance in some other countries than it currently is in the UK, notably in the USA as enshrined in the Freedom of Information Act. But the nation state — at least in Europe — is no longer the sole arbiter of what information can be made available or even of its pricing. An EC Directive of April 1990 requires that all official agencies of all member states must make available all their environmental holdings of information to the general public at 'reasonable cost'. It is not clear what constitutes environmental information but topographical information might be a necessary part. Nor is it clear what constitutes reasonable cost except in so far as no one organisation should expect to price its products in this category much more highly than others. In parallel with EUROSTAT moves to harmonize the characteristics of area-aggregate statistics on the environment and its provision of more geographically-detailed demographic and socio-economic statistics for Europe, these developments may well eventually affect the independence of official information-compiling and -disseminating organisations within the countries of the Community.

How can these products best be produced?

If we do not know in detail which products are needed in future, if our government sets a goal of full cost recovery and yet detailed mapping of all the nation is still required, what is the optimum way to proceed? In principle, we could achieve our desired ends by:

- a wholly state-run facility generating all topographic information. This is unlikely since commercial sector players are already active in the market and are unlikely to withdraw in the short term;
- a wholly private sector scheme. This would be unprecedented world-wide and might well represent a high-risk strategy (see below). Moreover, since OS exists, it could only really be brought about by the transfer of the Survey to the private sector;
- some mixture of private and public sector, with an identical or some different balance to that of the present.

The most compelling reasons for continuing to vest primary responsibility for topographic mappings and geodetic control in a government-owned Ordnance Survey are:

- the fact that some of the uses for OS large scale map data are essential to the functioning of the modern state (chiefly in relation to land registration, to emergency service provision and to the maintenance of law and order). Entrusting the whole task to one or more commercial enterprises seems very risky indeed. Even commercial enterprises as large as News Corporation get into deep financial trouble; as a consequence, that enterprise has sought to divest itself very quickly of at least some of its geographical information gathering subsidiaries to whoever will provide ready cash. At least some of the work of those organisations has been severely disrupted as a result. This is not a sane basis on which to maintain the infrastructure of a modern state.
- some of the roles that OS plays — notably the provision of geodetic information and the provision of 'just in case of disaster' mapping — are hardly likely to be commercially attractive;
- the 'as of right' access to land for survey purposes provided by parliament would be more difficult to grant to a commercial sector body;
- OS has a measure of statutory protection (e.g. the Land Registry are compelled to use OS maps in certain of their operations) and this would probably necessitate change of legislation;
- the spatial framework provided by OS currently provides a structure which is used by many other organisations and considerable costs are involved in any change, both for them and for the Survey;

- OS already has a long-established infrastructure in place to maintain the mapping;
- multiple alternative suppliers of information are unlikely to maintain the coherence of the existing map coverage indefinitely, even with a state-funded regulatory agency.

We should, however recognize that a near-monopoly by the state in certain markets often has real disadvantages for the customer. This can be true even where those providing the service are as totally committed to giving the public good service as OS, the water companies and other organisations in recent years. But the dangers of providing 'Rolls-Royce' solutions, risk aversion, setting monopoly prices at a higher level than would occur in competitive situations, overstaffing and preservation of the *status quo* are great in such circumstances. How then are such dangers best avoided whilst capitalising upon the merits of continuing with OS as the primary mechanism for obtaining and supplying consistent, high quality topographic and geodetic information? How can the optimal way forward be decided given that some parts at least of the private sector have a vested interest in reducing the magnitude of OS' activities?

The only solution to this is to ensure that OS is reviewed on a regular basis. Part of this is already done every year through the published financial reports and through investigations by the National Audit Office. But what is required is a formal professional audit, rather than a financial one. Taking into account the time required to build national databases, the appropriate interval between such reviews would be somewhere between five and eight years. Those involved in the review would be drawn from OS customers and from nomination by the responsible government minister; it would also be logical to have OS represented in some fashion. In between these reviews, it would be the responsibility of the Director General to ensure that the organisation was effective and efficient, to report annually on how s/he had met customer complaints and demands and to demonstrate how new as well as existing customers had benefited from the activities of the national mapping agency.

There are many other matters which the Survey should be considering. These include the possibility of diversification into different but related products (many of which are not currently available from other suppliers), the carrying out and publication of cost/benefit studies of the value of topographic data to user organisations, devising enhanced methods of anticipating user need and market stimulation. Much of it can only be done by an improved understanding of the users' business and tools. Standing out above all of this, however, is the need to extend the range of applications which are made using even the existing as topographic and related information. Given the levels of past and continuing investment by the taxpayer, together with the general desirability of widespread access to public information in a democracy, much greater use should be made of collaboration with the private sector in dissemination of OS information and the addition of value through data integration and sale of the new products.

CONCLUSIONS

This paper has documented the rapid technological, economic and social change in which OS is enmeshed. It has — to the author at least — demonstrated the continuing role of OS as the primary provider of detailed spatial information though the Survey clearly needs to address some key issues to maximize its utility to customers. The management of the change which is inevitable within the Survey will be difficult and will require contributions from staff at all levels. It will also require foresight and intelligence on the part of government to maximize the value of the substantial investment already made in as. Finally (and however trite it seems), customers, government, the citizenry, the private sector and the Ordnance Survey all have a vested interest in working together to keep each other efficient, honest and effective.

ACKNOWLEDGEMENTS

David Toft, head of OS Marketing Branch in early 1991, kindly provided some statistics used in the text. Tina Scally drew the diagrams on an Apple Mac. Drew Clarke, Peter Dale, Chris Denham, Hugh O'Donnell, Jean Phillipe Grelot, Tony Hart, Robert LaMacchia and others provided relevant background information on other organisations. Walter Smith contributed some powerful insights on OS and government whilst the two anonymous referees employed by the *Cartographic Journal* gave much sound advice. Naturally, however, the responsibility for any errors herein rests with the author.

BIBLIOGRAPHY

ALIC (1990:1) **Data Custodianship/Trusteeship**, Australian Land Information Council Paper number 1, Canberra.

ALIC (1990:2) **A General Guide to Copyright, Royalties and Data Use Agreements**, Australian Land Information Council Paper number 2, Canberra.

ALIC (1990:3) **Charging for Land Information**, Australian Land Information Council Paper number 3, Canberra.

ALIC (1990:4) **Access to Governmental Land Information: Commercialisation or Public Benefit?** Australian Land Information Council Paper number 4, Canberra.

Bromley, R. and Coulson, M. (1989) 'The value of corporate GIS to local authorities', **Mapping Awareness**, 3, 5, 32–5, Oxford.

CAG (1990) **Audit of the Department of Energy, Mines and Resources Surveys, Mapping and Remote Sensing Sector**, Canadian Auditor General, Ottowa.

CRU/OPCS/GRO(S) (1980) **People in Britain**, Her Majesty's Stationery Office, London.

CSO (1991) **Economic Trends Annual Supplement 1991 Edition**, Central Statistical Office.

Dale, P. F. (1991) 'Land Information Systems', in **Geographical Information Systems: Principles and Applications**, Maguire, D., Goodchild, M. F. and Rhind, D. W. (eds.), Vol. 2, 85–99, Longman, London.

Dale, P. F. (1991b) 'The challenges ahead', **Proc. Mapping Awareness Conference**, 333–339, Blenheim Online, London.

Dale, P. F. and McLaughlin, J. D. (1988) **Land Information Management**, Oxford University Press, Oxford.

Danby, B. (1991) 'In search of privacy', **GIS World**, 4(1): 108–111.

Danko, D. M. (1990) 'The digital chart of the world project', **Proc. GIS/LIS '90**, Vol. 1, 392–401, ACSM, ASPRS, AAG, URISA and AM/FM, Washington DC.

Department of the Environment (1987) **The Report of the Government. Committee of Enquiry on the Handling of Geographic Information**, Her Majesty's Stationery Office, London.

DTI (1990) **Government-held Tradeable Information: Guide-lines for Government Departments in Dealing with the Private Sector**, Department of Trade and Industry, London.

Hoogsteden, C. (1989) 'The economics of mapping', unpublished Ph.D. thesis, University of Otago.

Hoogsteden, C. (1991) 'The benefits of national mapping', **Land and Hydrographic Survey**, 2: 5–7, Royal Institution of Chartered Surveyors, London.

Maffini, G. (1990) 'The role of public domain databases in the growth and development of GIS', **Mapping Awareness**, 4(1): 49–54.

Mounsey, H. M. (1991) 'The Creation and Use of Multi-source, Multi-national Environmental GIS: Lessons Learned from CORINE', in **Geographical Information Systems: Principles and Applications**, Maguire, D., Goodchild, M. F. and Rhind, D. W. (eds), Vol. 2, 185–200, Longman, London.

NRC (1990) **Spatial Data Needs: the Future of the National Mapping Program**, Board on Earth Sciences and Resources, National Research Council, National Academy Press, Washington DC.

NRC (1991) **Research and Development in the National Mapping Division, USGS: Trends and Prospects**, Board on Earth Sciences and Resources, National Research Council, National Academy Press, Washington DC.

Openshaw, S. and Goddard, J. B. (1987) 'Some implications of the commodification of information and the emerging information economy for applied geographical analysis in the United Kingdom', **Environment and Planning A**, 19: 1423–39.

OS (1990) **Ordnance Survey Corporate Plan 1991–4**, Ordnance Survey, Southampton.

OSRC (1979) **Report of the Ordnance Survey Review Committee**, Her Majesty's Stationery Office, London.

Rhind, D. W. (1990) 'Topographic Databases Derived from Small Scale Maps and the Future of Ordnance Survey', in **The Association for Geographic Information Yearbook 1990**, Foster, M. J. and Shand, P. J. (eds), 87–96, Taylor and Francis, London.

Rhind, D. W. (1991) 'Counting the People: the Role of GIS', in **Geographical Information Systems: Principles and Applications**, Maguire, D., Goodchild, M. F. and Rhind, D. W. (eds), Vol. 2, 127–137, Longman, London.

Rhind, D. W. (1991) 'Data charging, access and copyright: their implications for GIS', **Proc. European GIS Conference (EGIS '91)**, 929–45, Brussels 2–5 April 1991.

Rhind, D. W. (ed.) (1983) **A Census User's Handbook**, Methuen, London.

Robertson, W. (1990) 'Funding of National Mapping', paper given to 'Surveying 2000' Conference at **Royal Institution of Chartered Surveyors**, London, September 1990.

Smith, W. (1979) 'National mapping: a case for government responsibility', **Conference of Commonwealth Surveyors 1979**, Paper M2, Cambridge.

Sowton, M. (1991) 'The Development of Digital Mapping and Support for GIS in Ordnance Survey', in **Geographical Information Systems: Principles and Applications**, Maguire, D., Goodchild, M. F. and Rhind, D. W. (eds), Vol. 2, 23–38, Longman, London.

Starr, L. and Anderson, E. A. (1991) 'A USGS Perspective on GIS', in **Geographical Information Systems: Principles and Applications**, Maguire, D., Goodchild, M. F. and Rhind, D. W. (eds), Vol. 2, 11–22, Longman, London.

Reflections on 'The Role of the Ordnance Survey of Great Britain'

DAVID FAIRBAIRN

Newcastle University

The development of cartographic techniques, products and applications has closely paralleled advances in human activities such as navigation, exploration, education, recreation, legislation and administration, throughout history. The nation state, as an instrument of societal organization, has influenced, and often directed, all such activity and has therefore impacted significantly on mapping. The epitome of such impact has been the emergence, maturity and success of national mapping agencies (NMAs), creating products of benefit in all these fields (along with more destructive activities, such as confrontation and conflict).

Contemporary NMAs can trace their origins to the seventeenth century Sweden and France, polities whose newly centralized and managerial governments led to the perceived need for systematic nation-wide mapping of topography, military landscapes and land ownership. For over three centuries, the major scientific advances in cartography, the largest volumes of map production, and the most significant employment of cartographers, took place in such mapping organizations.

The nature and perceived importance of their activities usually meant that such agencies were controlled by military personnel, and despite the resultant inherent secrecy, their *modus operandi* and outputs appeared to be almost uniform: there was little need or desire to publicize working practices, innovations, or types of output. Standardization of approach was evident world-wide, and even in war-time, one side could well surmise the nature of the other side's map holdings.

It was not until the twentieth century that any form of co-operation between NMAs could be explicitly seen. From a British perspective, the first meeting of the Conference of Empire Survey Officers in 1928 started a trend of information interchange which continued, quadrennially, into the twenty-first century. Later renamed the Conference of Commonwealth Survey Officers, and eventually the Cambridge Conference (after the usual location of its meeting), it now takes place every two years, and is attended by NMA leaders from around the world. This invitation-only event is a valuable networking, and informal innovation- and awareness-raising, event which also results in an excellent published (but not widely disseminated) record: its Proceedings of papers addressing concerns, trends and experiences in NMAs worldwide.

Amongst the most important of these papers in the Proceedings is a strikingly original and important paper (Smith, 1979) delivered by the then Director-General of Ordnance Survey (of Great Britain), Walter Smith, appointed in 1977 as the first-ever civilian leader of the Ordnance Survey. Smith's background was in commercial land survey, but his presentation to the Cambridge Conference in 1979 took a decidedly philosophical look at the *raison d'être* of a national mapping agency, aligning its activities to overall concepts of the purpose of government, following a 'Jeffersonian' model in which the state exists to ensure order and encourage progress. Despite the long history of NMAs, Smith's was one of the first attempts to justify and validate the existence and operation of such organizations from a political and an economic perspective.

One of the benefits of the paper by David Rhind, 'The role of the Ordnance Survey of Great Britain' (discussed here), is that it gave exposure to Smith's views in *The Cartographic Journal*, a wider arena than the rarefied atmosphere of the Cambridge Conference. Written in early 1991, this paper accepts the basic premise of Smith's idealistic approach, and despite viewing the Ordnance Survey from an external position (Rhind was an academic at the time of writing it), uses readily available statistics (from Annual Reports, etc.) and policy documents to paint an accurate picture of the state of Ordnance Survey at the start of the 1990s, along with a fuller exploration, using the Smith template, of its purpose and tasks.

Rhind starts by posing three key questions, relevant to Ordnance Survey (OS) at the time, but also fundamental to the organization of mapping at a state level throughout history. Firstly, what products and datasets are required from a NMA; secondly, how should such an agency be administered, especially from a financial perspective; and thirdly, how can production be improved and made more efficient in the future? Clearly each of these questions embraces a host of subsidiary enquiries: how are the products and datasets appraised by the end users? What is the nature of the information being handled, and how does its commodification affect NMA policy? How is a NMA aligned to similar governmental organizations in terms of financial model? What are the requirements in terms of manpower planning? What current and future government duties and responsibilities should a NMA be called upon to support? How can contemporary technology be best applied to the regular work of an NMA and its future methods? There are even idiosyncratic (yet engaging) questions directly posed, such as 'why has OS never published any form of image mapping?'

In a detailed way, Rhind's paper addresses such issues in an OS context, but it also considers more generic issues including the comparison between organizational models for delivery of governmental data in the US and the UK, the quantification of the national benefit of the type of spatial data handling typically undertaken by NMAs, and, most notably, Smith's philosophy of what a government's responsibilities are and the extent to which society relies on the work of national mapping agencies.

During 1991, OS was celebrating two centuries of world-leading surveying and mapping activity, but also facing an uncertain future direction. In fact, OS had been subject to uncertainty from the moment of its inception: in contrast to the earlier French and Swedish examples, its creation was the result of a disparate set of demands and recommendations from a range of military and political personalities; at many times its direction was guided by short termism and disconnected initiatives; it has been subject to a significant number of influential, official, external governmental reviews (along with many regular, often contradictory, internal departmental appraisals) which have determined policy in both re-active and pro-active ways; it has been able to successfully embrace opportunities in a visionary way; and, often concurrently, has weathered a series of externalities which have occasionally threatened its very existence.

That OS was regarded as a world-leading example of an effective national mapping agency, able to adapt to the contemporary world at the end of the twentieth century is testament to the skills and nature of both its leadership and its staff up to that time. It is testament to the abilities of David Rhind that he was able, against significant internal competition, to succeed to the position of Director-General later in that two-hundredth anniversary year. His article was a masterful summary of the environments — financial, political, business and social — experienced by an organization which was relentlessly moving its focus from map production to maintenance of a national spatial data infrastructure. Originally read as a paper at a Royal Geographical Society meeting celebrating the bi-centenary, the paper could be regarded as a comprehensive position paper indicating the opinions of someone strongly interested in the soon-to-be vacant role of Director-General.

The article also set a trend for more direct discussion of issues related to the function and operation of NMAs. By the time of the 1995 Cambridge Conference, David Rhind (as Director-General of OS, and, therefore, customary Chair of the conference) was leading the production of a valuable book, *Framework for the World* (Rhind, 1997a), which developed ideas from the conference in an even more public manner and with valuable commentary on common practices and challenges for NMAs around the world. If Rhind's 1991 article was the foundation of his tenure as Director-General of Ordnance Survey, perhaps his final chapter in that book, with its provocative title (Rhind, 1997b), was the coda. This increased propensity, from the 1990s onward, to investigate, analyse, and make constructive suggestions about both broad strategy and detailed operations, resulted in increasingly confident and open approaches to documentation and commentary on NMAs (e.g. the extraordinarily comprehensive and economically insightful report by Hoogsteden and Hannah (1999) which considered the impact of perhaps the most radical adjustment to NMA practices happening anywhere at the time: the changes to New Zealand's national mapping organization).

The environments within which NMAs now operate incorporate many influences and inputs barely foreseen in 1991: a host of radically different factors impinge upon their daily work, including citizen science and user generated geographical data (Rhind's 1991 paper did, in fact, predict the enormous impact of civilian use of GPS); supra-national spatial data infrastructures and attempts at international standardization; new models of service oriented mapping, along with radically different methods of product delivery using web services; the external and global impact of commercially-sourced geobrowsers, and democratization of map creation and data accessibility; the paradigm of governmental economic retreat from public service provision, along with global recession, leading to radically different financial and administrative pressures. These all affect any contemporary overview of NMAs, such that recent published strategies and reviews, from Canada (NMSWG, 2010) to Ireland (Greenway, 2007), from France (Lubek *et al.*, 2005) to South Africa (Denner and Oosthuizen, 2008), and in many other nations, have continued to develop the type of enquiry initiated by Rhind's original thought-provoking and influential article.

REFERENCES

Denner, M. and Oosthuizen, H. (2008). 'The strategic positioning and configuration of national mapping organisations as enablers of economic and social growth in South Africa', **South African Journal of Business Management**, 39, 41–55.

Greenway, I. (2007). 'National mapping — funding the public good', Strategic Integration of Surveying Services, in **Proceedings of FIG Working Week**, Hong Kong SAR, China, May 13–17. Paper TS 3E.

Hoogsteden, C. and Hannah, J. (1999). 'Of capital importance: Assessing the value of the New Zealand geodetic system', University of Otago, Dunedin, http://www.linz.govt.nz/geodetic/standards-publications/published-paper-presentations.

Lubek, P., Cannard, P., Cousquer, Y. and Champagne, V. (2005). 'Rapport d'enquête sur le référentiel à grande échelle de l'institut géographique national', English summary of recommendations, http://epsiplatform.eu/content/rapport-d-enquete-sur-le-r-f-rentiel-grande-chelle-de-l-institut-g-ographique-national.

NMSWG (National Mapping Strategy Working Group) (2010). 'Canada's national mapping strategy', Canada Council on Geomatics, Ottawa, http://www.ccog-cocg.ca/strategy-strategie_e.pdf.

Rhind, D. (Ed.) (1997a). **Framework for the World**, GeoInformation International, Cambridge.

Rhind, D. (1997b). 'Redesigning and rebuilding Ordnance Survey', Chapter 21, in **Framework for the World**, GeoInformation International, Cambridge.

Smith, W. (1979). 'National mapping: A case for government responsibility', Paper M2 in **Proceedings of Conference for Commonwealth Surveyors**, Cambridge, UK, July 23–1 August.

Line Generalization by Repeated Elimination of Points

M. VISVALINGAM AND J. D. WHYATT

Originally published in *The Cartographic Journal* (1993) 30, pp. 46–51.

This paper presents a new approach to line generalization which uses the concept of 'effective area' for progressive simplification of a line by point elimination. Two coastlines are used to compare the performance of this, with that of the widely used Douglas-Peucker, algorithm. The results from the area-based algorithm compare favourably with manual generalization of the same lines. It is capable of achieving both imperceptible minimal simplifications and caricatural generalizations. By careful selection of cutoff values, it is possible to use the same algorithm for scale-dependent and scale-independent generalizations. More importantly, it offers scope for modelling cartographic lines as consisting of features within features so that their geometric manipulation may be modified by application- and/or user-defined rules and weights. The paper examines the merits and limitations of the algorithm and the opportunities it offers for further research and progress in the field of line generalization.

INTRODUCTION

The identification of the salient character or caricatures of a line is central to the process of line generalization. The caricatures produced by the Douglas-Peucker algorithm[1] are believed to be most successful. The algorithm is reputed to select the critical, shape-preserving points on a line. However, several researchers have expressed reservations about this and other point-based methods.[2,3] These methods are believed to be suitable only for minimal simplification, and not for generalization, especially of complex lines. Caricatural generalizations preserve only the major distinctive features and omit smaller, less significant ones in their entirety. Results obtained using an area-based algorithm (specified by Visvalingam and first tested by Whyatt[4]) suggest that it can be used to filter out features within features progressively and thus achieve both minimal· simplification and caricatural generalization.

1. Douglas, D. H. and Peucker, T. K. (1973) 'Algorithms for the Reduction of the Number of Points Required to Represent a Digitised Line or its Caricature', **The Canadian Cartographer**, 10(2), 112–122.

2. Visvalingam, M. and Whyatt, J. D. (1990) 'The Douglas-Peucker Algorithm for Line Simplification: Re-evaluation through Visualisation', **Computer Graphics Forum**, 9(3), 213–228.

3. Visvalingam, M. and Whyatt, J. D. (1991a) 'Cartographic Algorithms: Problems of Implementation and Evaluation and the Impact of Digitising Errors', **Computer Graphics Forum**, 10(3), 225–235.

4. Whyatt, J. D. (1991) **Visualisation and Re-evaluation of Line Simplification Algorithms**, Unpublished Ph. D. Thesis, University of Hull, pp. 259.

BACKGROUND

Attneave[5] used a caricature of a sleeping cat to illustrate the presence of 'information loaded' critical points. He proposed that people perceived these points of high curvature along lines as perceptually important and high in information content. His concept of critical points has influenced research on line simplification in digital cartography. White[6] observed that the widely used Douglas-Peucker algorithm identified more of these critical points than others she studied. McMaster repeatedly asserted that this algorithm was 'mathematically and perceptually superior' to other algorithms he evaluated because it picked out more of these critical points and produced least displacement from the original line (see for instance McMaster[7]).

Waugh[8] and Wade (as reported in Whyatt and Wade)[9] proposed extensions to the algorithm; tag values were associated with each point as an indicator of its significance. These tag values may be used to rank points into a hierarchy of critical points. The concept of a fixed rank order of critical points is convenient· since a tolerance parameter may be used to filter out the required points at run time. Visvalingam and Whyatt[10,11] used these tag values to study the assumptions underpinning the algorithm's wide use, other properties of the method and their implications. They concluded that this algorithm does not necessarily select points which cartographers select from complex lines and reviewed other criticisms of the method. Even using relatively simple test data, White[6] only found a 45% agreement between points selected by the algorithm and by respondents in her study. Furthermore, Ramer[12] working in the field of Pattern Recognition had already noted that the algorithm selects some redundant points. Visvalingam and Whyatt[13,14] pointed out that shape distortions can occur as a result of selecting the furthest point from the anchor-floater point, since these extreme points can fall on spikes and minor insignificant features or be the outcome of tiny rounding and digitizing errors. It is well known that the Douglas-Peucker method is only suitable for minimal simplification of lines in Jenk's[15] class of imperceptible generalization. Consequently, many believe that point-based methods are incapable of even approximating the cartographer's art of producing caricatural generalizations. This paper describes a point-based algorithm which is capable of achieving both minimal simplifications and caricatural generalizations.

5. Attneave, F. (1954) 'Some Informational Aspects of Visual Perception', **Psychological Review**, 61(3), 183–193.

6. White, E. R. (1983) **Perceptual Evaluation of Line Generalisation Algorithms**, Unpublished Masters Thesis, University of Oklahoma.

7. McMaster, R. B. (1987) 'Automated Line Generalisation', **Cartographica**, 24(2), 74–111.

8. Waugh, T. C. and McCalden, J. (1983) **GIMMS Reference Manual**, GIMMS Ltd, Edinburgh.

9. Whyatt, J. D. and Wade, P. R. (1988) 'The Douglas-Peucker Line Simplification Algorithm', **Bulletin of The Society of University Cartographers**, 22(1), 17–25.

10. Visvalingam and Whyatt, 1990.

11. Visvalingam and Whyatt, 1991a.

12. Ramer, U. (1972) 'An Iterative Procedure for the Polygonal Approximation of Plane Curves', **Computer Graphics and Image Processing**, 1, 244–256.

13. Visvalingam and Whyatt, 1991a.

14. Visvalingam, M. and Whyatt, J. D. (1991b) 'The Importance of Detailed Specification, Consistent Implementation and Rigorous Testing of Cartographic Software', in Rybaczuk, K. and Blakemore, M. (eds.), **Mapping The Nations vol 2, 15th Conference of Int. Cartographic Assoc.**, Bournemouth, 856–859.

15. Jenks, G. F. (1989) 'Geographic Logic in Line Generalisation', **Cartographica**, 26(1), 27–42.

THE AREA-BASED METHOD

Although this algorithm produces very encouraging results, we continue to view it as just one step towards a more intelligent system for line generalization. The basic idea underpinning this algorithm is to iteratively drop the point which results in least areal displacement from the current part-simplified line. This results in the progressive elimination of geometric features, from the smallest to the largest. Area was chosen because other metrics, such as shape, only start to have an impact when the size of a feature exceeds a perceptual limit. This initial study set out to evaluate the impact of geometric size on line generalization.

Elimination rather than selection

The Douglas-Peucker algorithm drops all intermediate points if they fall within a tolerance band of the straight line, called the anchor-floater line, connecting the first and last points. This is reasonable.[16] However, the assumption that the further point from this arbitrary anchor-floater line must be a critical point is questionable. The area-based method is based on the assumption that it is easier to filter points on lines by a process of elimination rather than selection.

Effective area

Size, as measured by area, sets a perceptual limit on the significance or otherwise of other perceptual indicators. It is also the most reliable metric for elimination since it simultaneously considers distance between points and angular measures. Area measures are widely used in traditional cartography[17] and have been used for eliminating polygonal features by others working in digital cartography.[18] This concept may be extended to the simplification of linear features. The area-based algorithm associates with each non-terminal point its effective area, which is the area of the triangle formed by the point and its two neighbours. This is the areal displacement which would occur if that point alone was omitted from the current line (Figure 1a).

McMaster[7] used least areal displacement from the original line to compare the performance of line simplification algorithms. In the area-based algorithm the areal displacement is measured relative to the current, part simplified line, and not the original source line.

The algorithm

The algorithm is as follows:

- Compute the effective area of each point (see previous section);
- Delete all points with zero area and store them in a separate list;
- REPEAT

16. Visvalingam and Whyatt, 1991b.

17. Robinson, A. H., Sale, R. D., Morrison, J. L. and Muehrcke, P. C. (1984) **Elemenls of Cartography**, Wiley & Sons, New York.

18. Deveau, T. J. (1985) 'Reducing the Number of Points in a Plane Curve Representation', **Proc. Auto Carto 7**, Washington DC, 152–160.

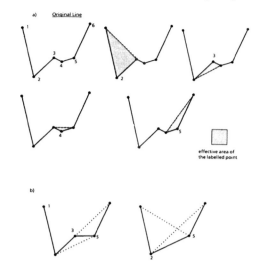

FIG. 1. Area-based line generalization: a) effective area of points, b) generalization by repeated elimination of the smallest area

    ○   Find the point with the least effective area and call it the current point. If its calculated area is less than that of the last point to be eliminated, use the latter's area instead. (This ensures that the current point cannot be eliminated without eliminating previously eliminated points.) Delete the current point from the original list and add this to the new list together with its associated area so that the line may be filtered at run time;

    ○   Recompute the effective area of the two adjoining points (see Figure 1b);

 • UNTIL

    ○   The original line consists of only 2 points, namely the start and end points.

The ability to rank points in some manner is useful in scale-free mapping since it expedites the filtering of points at run time. However, Visvalingam and Whyatt[19] demonstrated why a fixed rank-order of critical points, no matter how they are devised, limits the scope for producing appropriate scale-related displays. Here, at this early stage of experimentation with a relatively simplistic metric, we were more concerned with assessing whether area could be used for progressive filtering out of features within features.

When expressing the above algorithm as a computer program, attention must be paid to the precision and range of co-ordinates to avoid problems of overflow and underflow. Visvalingam and Whyatt[20] pointed out that the implementation of an algorithm as a computer program requires a consideration of special geometric cases, numeric problems and digitizing errors. For example, different implementations of the Douglas-Peucker algorithm produce different results.

19. Visvalingam and Whyatt, 1990.
20. Visvalingam and Whyatt, 1991a.

Locally derived metric but holistic view of line

The processes of selection by the Douglas-Peucker method and elimination using effective area differ in two distinct ways. The Douglas-Peucker method requires the calculation of the distance of each point from an anchor-floater line. Therefore, all points are considered initially. Once the point of maximum offset is selected, the two parts of the line can be processed independently of each other. Thus, contrary to the claims made by others, the algorithm ceases to be global and holistic after the selection of the first point.

Even though the effective area is calculated using only three neighbouring points, in the area-based method, the process does require a holistic view of the line whilst progressively eliminating detail. It involves a comparison of all remaining points along the line when seeking to eliminate a point.

Selection of cut-off values

The algorithm for ranking points is objective, even if somewhat simplistic and arbitrary. In theory, it therefore should be possible to use the smallest detectable or resolvable area as the threshold for filtering features for a target scale. However, this is not possible at present since the rank order of points is arbitrary — it takes no account of shape nor the context of the triangle. The eye perceives the area of features rather than of individual triangles. Thus, the selection of cut-off values remains subjective, particularly at higher, caricatural levels of generalization. At these levels, the number of points to be retained has to be determined by the shape of geomorphic features; Topfer's law[21] may be inappropriate. Since caricatures consist of a minimal set of points, the inclusion/omission of even one point can alter the shape of the feature.

DATA, RESULTS AND DISCUSSION

Data

Two complex stretches of coastline, around Carmarthen Bay and Humberside (Figure 2), were selected since they describe features within features. The data were extracted from the files containing the boundaries of British Administrative Areas (digitized from 1: 50,000 source maps by the Department of Environment and Scottish Development Department; which were held at the South West Universities Regional Computer Centre).[22]

Results

The results are encouraging and indicate that it may be possible to model cartographic lines as consisting of features within features. The Carmarthen Bay test data shows a progressive elimination of size-related features (Figures 3 and 4). During the initial stages, the most noticeable changes occur in the boxed region in Figure 2 which

21. Topfer, R. and Pillewizer, W. (1966) 'The Principles of Selection', The Cartographic Journal, 3(1), 10–16.

22. Wise, S. (1988) Using the DoE and SDD Digitised Boundary Data at SWURCC, User Guide, Bath University Computing Services, Bath, Avon.

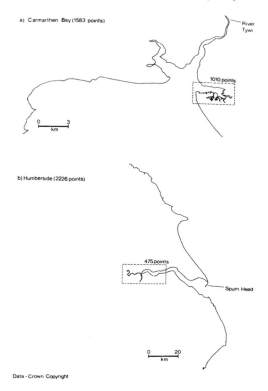

a) Carmarthen Bay (1583 points)

River
Tywi

1010 points

0   3
km

b) Humberside (2226 points)

475 points

Spurn Head

0   20
km

Data - Crown Copyright

FIG. 2.   Test data

contains many creeks. At all levels of generalization indicated in Figure 3, the area-based method eliminates more points from the creek region than from the rest of the coastline; the number of points retained in the latter are provided with the figure heading. Figure 3a shows that 77% of the points in the creek region may be eliminated without any significant departure from the original section. In Figure 3b, further elimination of points results in a shortening of the creeks (marked A) and a noticeable straightening of the coast (along B). After this, in Figure 3c, shorter creeks (marked C) and some headlands (marked D) are eliminated in their entirety. The result is a noticeable simplification of the coastline and creeks. On further generalization, the presence of creeks is only intimated by the retention of a couple of creeks, which have become noticeably widened and shortened (Figure 3d). The method first produces minimal simplification, in the realms of imperceptible generalization as Jenks[23] put it, then perceptible simplification followed by typification.

Figure 4 shows, at the 1:1 million scale, the effect of further elimination. Figures 4a-c indicate that the generalization and removal of the creeks precede the noticeable straightening of the coast at this scale and the elimination of the feature marked X on

23. Jenks, 1989.

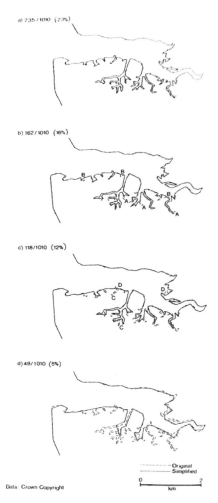

a) 235 / 1010 (23%)

b) 162 / 1010 (16%)

c) 118 / 1010 (12%)

d) 49 / 1010 (5%)

- - - - - Original
———— Simplified

0   2
km

Data Crown Copyright

FIG. 3.   Minimal simplification and typification. Corresponding figures for all of Carmarthen
Bay: (a) 570/1583 (36%);(b) 416/1583 (26%); (c) 332/1583 (21%); (d) 174/1583 (11%)

Figure 4c (which is actually an error introducing during digitizing). There is then a progressive shortening of the three river estuaries (Figures 4d-f) into the distinctive three-pronged caricature seen on 1: 1m and smaller scale atlas maps. The four-point representation of the bay would not be an inappropriate generalization if the stretch of coastline occurred within a more extensive geographic area shown on very small scale, 1: 10m, maps.

Figure 5 shows the results of the Douglas-Peucker algorithm using equivalent numbers of points. None of the maps are satisfactory and there is gross distortion of shape when less than 4% of points are used. Figure 6a shows the coastline depicted by the 200 points digitized by the Ordnance Survey from the 1: 625,000 Route Planner

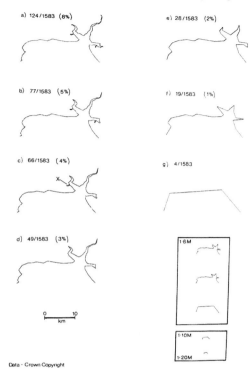

FIG. 4.   Caricatural generalizations

map. Figure 6b shows that comparable results may be produced using just 77 points, i.e. 5% of points. In comparison, the Douglas-Peucker method cannot achieve appropriate simplifications with 200, let alone 77, points (Figures 6c and d). When compared with simplifications produced by the Douglas-Peucker algorithm, using similar numbers of points, the area-based algorithm produces better balanced, aesthetically more pleasing and cartographically more appropriate simplifications. Whyatt[4] also found that the method produces better results than the algorithms proposed by Roberge[24] and Dettori and Falcidieno.[25] However, as with other point-based methods, the area-based method has limitations.

Discussion

The method is attractive for the following reasons. It is very simple in concept and builds on existing ideas within cartography. It confirms the importance of size in

24. Roberge, J. (1985) 'A Data Reduction Algorithm for Planar Curves, Computer Vision', **Graphics and Image Processing**, 29, 168–195.

25. Dettori, G. and Falcidieno, B. (1982) 'An Algorithm for Selecting the Main Points on a Line', **Computers and Geosciences**, 8(1), 3–10.

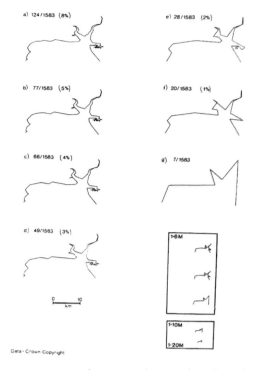

FIG. 5. Output from Douglas-Peucker algorithm

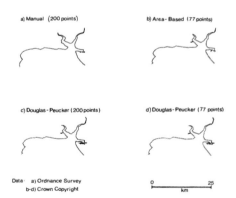

FIG. 6. Comparison of (a) manual simplification with output of (b) area-based algorithm and (c & d) Douglas-Peucker algorithm

geometric simplification. Minor features, such as spikes, are rapidly eliminated since they enclose very small areas. It is achieving both minimal simplifications and caricatural generalization. The impression is one of entire sub-features being eliminated before features at the next higher level are noticeably modified. Since the test data demonstrates that the method performs adequately with more than three levels of features, it is reasonable to assume that the method will scale up to national and global levels. This needs to be tested. Attention needs to be paid to the implementation of the algorithm which must adjust the precision of co-ordinates during calculations to avoid overflow.

It appears that cartographic lines may be modelled as consisting of overlapping hierarchies of nested features. The explicit representation of features will enable the user to associate weights to these to modify the normal geometric processing. The algorithm may be used for achieving both scale-dependent and scale-independent generalization.[26] Except when eliminating the larger features, the most noticeable changes occur when size-related features are dropped in their entirety. Unlike the Whirlpool program, based on Perkal's algorithm, the area-based method does not leave lakes at the head of the estuary.[27] In between the removal of features at different levels of the hierarchy, the appearance of the map is not substantially affected by the elimination of further points. Thus, features at each level of the hierarchy can be represented by a minimal or a larger set of points. A minimal set appears to be adequate for reduced smaller-scale displays while a larger set may be more appropriate for scale-independent generalization. The algorithm is also capable of suggesting typifications. For example, in Figure 3d, the presence of an irregular coastline of creeks is only indicated.

Like all simple point-based algorithms, the area-based method has limitations. These outstanding problems raise some interesting issues. At present the algorithm considers only the size or area of triangles. Manual generalization considers the size, shape and the geometric and geographic context of features consisting of several points. We were therefore surprised that this germ of an algorithm performed as well as it did. However, the cut-off values, yielding the characteristic caricatural shapes, had to be found by trial-and-error using interactive software as explained earlier. The selection of cut-off values remains subjective; it is directed by purpose and influenced by the cartographer's wider knowledge of geography. However, even as it stands, the algorithm can be used to derive an approximation of the required shape. The cartographer can then adjust the shape if necessary. None of the maps included in this paper have been manually modified.

As with many other point-based methods this algorithm can also cause complex lines to cross each other. Figure 7a shows that the banks of the River Ouse can stand some minimal simplification before they start to cross (Figure 7b). Similar crossings were observed on other narrow features, such as Spurn Head at the mouth of the Humber estuary (Figure 2b). In the past this has been regarded as a test of an algorithm's performance, mainly because there was limited scope for resolving the problem. However, in the context of the area-based method, crossing lines flag the need for more intelligent generalization. Figures 7a–d point to different considerations

26. Robinson, 1984.
27. Beard, K. M. (1991) 'The Theory of the Cartographic Line Revisited', **Cartographica**, 28(4), 32–58.

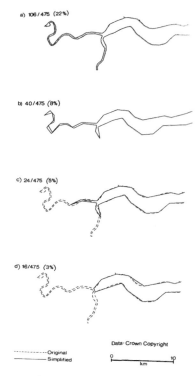

a) 106/475 (22%)

b) 40/475 (8%)

c) 24/475 (5%)

d) 16/475 (3%)

Data: Crown Copyright

------- Original
——— Simplified

0   10
|———————————|
km

FIG. 7. Progressive simplification of rivers at the head of the Humber Estuary

requiring different approaches to generalization. Figure 7a is sufficient if all that is required is minimal simplification. At grosser levels of simplification, it may be sufficient to show only the head of the Humber estuary (Figure 7d). At intermediate levels different strategies may be adopted, depending on the purpose of the map. If there is no reason for showing the rivers, only their mouths need to be indicated as shown in Figure 7e. If it is necessary to show the rivers, they can either be exaggerated by displacing the banks or be shown by an abstracted line. A similar approach can be adopted with other special features like Spurn Head. Line crossings indicate a need for drastic modification but this has not been routinely undertaken in line simplification in the past since it is computationally very demanding. Using the area-based method it is possible to isolate features, such as rivers and spits; even after minimal simplification, a number of points which share the same rank (seventeen on the Ouse) are eliminated together. ·There is scope for more efficient and intelligent detection of line crossings. Remedial action can then be based on traditional cartographic principles. Thus, in the context of the area-based method, line crossings are not defects but rather clues for prompting intelligent generalization.

The area-based method also provides scope for achieving slightly different shapes using the knowledge that combinations of positive and negative areas indicate different types of features. Thus, navigational charts may choose to weight spits and headlands as more important than same-sized landward intrusions.

CONCLUSION

People have little difficulty in abstracting the features described by the geometry of lines. The automatic segmentation of a line into the constituent features remains an outstanding challenge. Computer simplification of lines is still largely driven by geometric rules of thumb. Much effort is being expended now on codification into computer-readable form of existing knowledge.[28] It is not enough to code already existing specifications on the 'what' and 'when' of generalization. Intelligent generalization requires a deeper understanding of 'why' and 'how to'. The automatic generalization of lines awaits algorithms for structuring the point samples into overlapping hierarchies of features. The features may then be used to link in geographic labels and functional information to ignore or guide the underlying geometric processing.

This paper described a novel area-based method for line generalization. The most important aspect of this algorithm is Visvalingam's scheme for generalizing lines by repeated elimination of points using a suitable metric. The effective area of triangles was chosen initially to test the hypothesis that line generalization is significantly influenced by size. We regard the work reported in this paper as only a step towards the evolution of a more intelligent system for line generalization. It indicates that there is some scope for modelling cartographic lines as consisting of a geometric hierarchy of features within features. Being a point-based algorithm, it has some inherent limitations. However, even in its present primitive form, it is already capable of producing acceptable simplifications and caricatural generalizations. It can be used to achieve both scale-dependent and scale-independent generalization of features at a given level of the hierarchy by using either their minimal or a fuller description. It is also capable of producing typifications. Furthermore, this approach provides sufficient clues for isolating distinctive features for omission, exaggeration or skeletal representation. It opens up opportunities for further research in digital cartography.

ACKNOWLEDGEMENTS

We are grateful to the UK Science and Engineering Research Council for the award of a studentship to Duncan Whyatt, to the South West Universities Regional Computer Centre (SWURCC) for providing access to the DoE/SDD boundary files and to the Ordnance Survey for permission to use the 1: 625,000 digital data. These data remain the property and copyright of the Crown.

28. Buttenfield, B. P. and McMaster, R. B. (eds.) (1991) **Map Generalization: making rules for knowledge representation**, Longman, UK.

Reflections on 'Line Generalization by Repeated Elimination of Points'

ALEXANDER J. KENT

Canterbury Christ Church University

Generalization is common to every map and central to the theory and practice of cartography. What may be termed the 'problem' of generalization, i.e. how to best preserve essential characteristics in the modelling of space, is therefore an important and recurring theme of research published in *The Cartographic Journal*. If the first 50 volumes show how advances in technology have shaped the discipline, few topics serve to highlight this so clearly; from the evolution of manual to automated methods of generalization, and from geometrically isolated techniques to object-oriented and multi-scalar approaches.

In 1985, Mahes Visvalingam founded the multidisciplinary Cartographic Information Systems Research Group (CISRG) at the University of Hull, with a research agenda largely inspired by the work of Thomas Peucker and his colleagues David Douglas, Nick Chrisman and David Mark. The early acquisition of an ICL PERQ computer exploited new developments in workstation technology, enabling the CISRG to demonstrate how window management systems based on UNIX could allow the interactive visualization of data and processes, especially for exploring population census and topographic data. Duncan Whyatt joined the CISRG in 1987 to begin his PhD in Computer Science under the supervision of Mahes Visvalingam, researching the computer-assisted visualization of line generalization algorithms.

The line is probably the most important graphical symbol in the construction of cartographic language. Far from being just an intuitive method with which to represent 'linear' features such as rivers, coastlines, roads and boundaries, the division or unity of areas and their complexity are indicated, implied and imposed according to lines. Choosing the appropriate level of detail for a given scale therefore raises questions of accuracy and conformity that fundamentally affect the character of linear symbology and the quality of their representation and also the file size of the digital dataset.

The Douglas–Peucker (DP) algorithm for reducing the number of points in a line was published in 1973, and over the next 20 years, had gained general acceptance as one of the most important and widely used line-simplification algorithms. By the 1990s, it had already been adopted by many GIS and digital mapping packages (João, 1998) and at the time of publishing their research, Visvalingam and Whyatt (1993, p. 46) believed the DP algorithm to be the most successful, but not unsurpassable, as Visvalingam (2014, pers. comm.) recounts:

The main merit of the DP algorithm was that it attempted to take a holistic view of the line, i.e. it searched the whole section of the line at each iteration. It took a lot of doodling on contrived test lines before I began to pay attention to the obvious — cartographers did not select points, they eliminated the smaller features in their entirety. However, feature recognition was still a research problem in pattern recognition, let alone cartography. So, I decided to focus initially on eliminating triangles formed by a sequence of three points. Use of the DP offset distance to eliminate points did not provide satisfactory results. At each iteration of the DP algorithm, the base line remains the same, so the offset distance gives the same ranking as the area of the triangles formed by points with their base line. But, when it comes to elimination, this is not the case. Cartographers often used the give-and-take rule when dealing with inflections. More hand doodling with the area metric was promising and identified how one should deal with special cases, such as inflections, and the data structure that was needed for filtering points in the correct order so that entire features, such as inflections and minor features, could be eliminated together. I then produced a formal specification for my algorithm so that Duncan could include it in his comparison of a selection of line filtering/generalization algorithms. His choice of coastlines made the results especially interesting. Other undergraduates, MSc and PhD students within the CISRG tested the algorithm with a variety of line types and DTMs to assess its wider applicability, not just to large-scale topographic data but also to identifying important breaks of slope on terrain, for terrain segmentation, and for producing sketches of terrain models.

While the DP algorithm omits all intermediate points if they fall within a tolerance band of a straight line connecting the first and last points ('the anchor-floater line'), the basic idea underpinning Visvalingam's algorithm is 'to iteratively drop the point which results in least areal displacement from the current part-simplified line' (Visvalingam and Whyatt, 1993, p. 47). Working through the DP algorithm by hand, Visvalingham had realized that it was most suitable for the minimal simplification of lines and not for caricatural generalization, which preserves only the major distinctive features. This inspired the creation of the new algorithm and was echoed by Shi and Cheung (2006) in their evaluation of various methods of line generalization some thirteen years later.

One of the advantages of Visvalingam's algorithm is its intuitiveness, making it straightforward to implement (van der Poorten and Jones, 2002). The algorithm can also be used to detect errors in the compilation of large-scale data, as it identifies spurious features attached to roads. Visvalingam (2014, pers. comm.) demonstrated this to Ordnance Survey in 2002, along with a Java program that used data pre-processed by her algorithm to automatically segment large-scale road data into in-line geometric features, such as roundabouts, lay-bys and side roads — offering distinct prospects for the automatic labelling of such features through pattern recognition and potentially enabling the feature- and object-based generalization of maps. Before its closure in 2003, the CISRG was twice invited to exhibit its research at The Royal Institution, where the program for segmenting OS road data was also demonstrated.

A modified version of Visvalingam's algorithm was later developed by Zhou and Jones (2004), where the triangles are weighted according to their flatness, skewness and convexity in combination with their area. The performance of this modification, together with that of the original and the DP algorithms, can be visualized (and tested with user-generated data) using the web-based interactive tool MapShaper (Bloch and Harrower, 2006). Advances in computer technology have led to the development of a plethora of automated methods and the optimization of techniques for generalizing vector graphics (particularly in deriving smaller from larger-scale topographic mapping) is an ongoing challenge. As Visvalingam (2014, pers. comm.) remarks, 'Computers churn out results faster than people can see or think — and tedious hand working and step-by-step computer-assisted visualization provide the opportunity for the birth of new insights'. Along with many other key developments in this field, the full utility of the Visvalingam's line generalization algorithm is yet to be realized.

REFERENCES

Bloch, M. and Harrower, M. (2006). 'MapShaper.org: a map generalization web service', in **AUTOCARTO 2006**, Vancouver, WA, Jun 25–28.

Douglas, D. and Peucker, T. (1973). 'Algorithms for the reduction of the number of points required to represent a digitized line or its caricature', **Canadian Cartographer**, 10, pp. 112–122.

João, E. M. (1998). **Causes and Consequences of Map Generalisation**, Taylor & Francis, London.

Shi, W. and Cheung, C. (2006). 'Performance evaluation of line simplification algorithms for vector generalization', **The Cartographic Journal**, 43, pp. 27–44.

van der Poorten, P. M. and Jones, C. B. (2002). 'Characterisation and generalisation of cartographic lines using Delaunay triangulation', **International Journal of Geographical Information Systems**, 16, pp. 773–794.

Visvalingam, M. (2014). Personal communication.

Zhou, S. and Jones, C. B. (2004). 'Shape-aware line generalization with weighted effected area', in **Developments in Spatial Data Handling: Proceedings of the 11th Annual Symposium on Spatial Data Handling**, ed. by Fisher, P. F., pp. 369–380, Springer, Berlin.

Map Design for Census Mapping

D. DORLING

Originally published in *The Cartographic Journal* (1993) 30, pp. 167–183.

High resolution colour mapping using modern technology has allowed us to explore the breadth of census data available to the contemporary social scientist. An examination of methods used to create cartograms, which minimize visual bias involves considering how densities and area boundaries should be mapped. The Modifiable Area Unit Problem is discussed and the advantages of using three-colour or trivariate mapping are outlined. Further modifications to traditional census mapping such as the mapping of change and flow are addressed.

> *For maps of larger scale, an artistic objective might well lessen our insistence on a strict geometric framework for maps and make room for the greater use of mental constructs of social, cultural, and economic space ... Such maps might well be considered the cartographic equivalent of 'mild' surrealistic art.*
>
> Arthur Robinson[1]

INTRODUCTION

This paper aims to give a flavour of new cartographic techniques which can be used to study census data. Techniques used to study differences between many small places which often require colour are the paper's main focus. What makes the census particularly valuable for social science is not the breadth or depth of the questions that it asks (as they are few and shallow), but the great spatial detail that is provided — showing how each neighbourhood, each block of streets, each hamlet, differs socially from its neighbours (for every place in the country simultaneously). The census also acts as the base for almost all other mapping of British social, economic, political, housing and medical statistics as it tells us where households, jobs, voters, dwellings and people are. Whilst we cannot have information on every individual, due to confidentially constraints, we are given many statistics for small groups of people aggregated into geographical areas. The question being addressed here is 'how can we map the detail this data contains but in its raw form conceals?'

The 1851 British census was the first to be mapped in any detail,[2] although American censuses were mapped before this and ancient maps can be found of the

---

1. Robinson, A. H. (1989) 'Cartography as an art', in **Cartography Past, Present and Future**, Rhind, D. and Taylor, D. R. F. (eds.), 91–102. p. 97.

2. Petermann, A. (1852) **England and Wales: distribution of the population**, Census of Great Britain, 1851, Population Tables I, Her Majesty's Stationery Office, London.

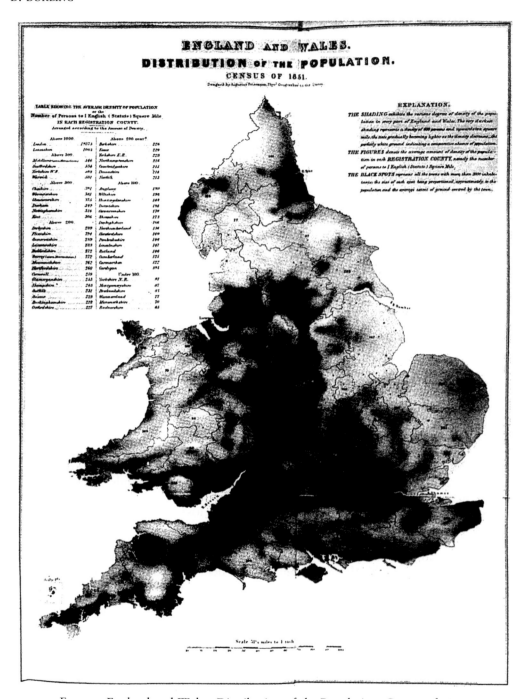

FIG. I. England and Wales, Distribution of the Population, Census of 1851

population in, for example, China, Figure 1 shows a dot map of England and Wales showing the distribution of the people in 1851. Graphical depiction of statistics was not in favour over much of the subsequent century.[3] The next British census to be mapped nationally and in detail was that of 1961 — by hand drawn choropleth maps using the local authority boundaries of the day.[4] Computers were first used to map the 1971 census. *People in Britain: a Census Atlas*[5] depicted characteristics of the population by colouring kilometre grid squares across the whole of Great Britain to show the detailed spatial structure of our society. In 130 years, however, the fundamental look of maps of the population had changed little. Since then, with two decades of development in computer graphics, a great deal has changed.

PLOTTING POINTS

The finest level of output of the 1991 census spatially is the 'enumeration district' in England and Wales and the 'output area' in Scotland. Each enumeration district contained, on average, four hundred people and consisted of just a few streets or a single block of flats in towns and cities, part of a village or a single hamlet in the countryside. Output areas are even smaller. There were over 131,000 enumeration districts in 1991 in England and Wales and 26,000 output areas in Scotland. We have been given information about very many small groups of people and all their geographical neighbours.

For each enumeration district a national grid reference to an accuracy of one hundred metres is given in what are called the *small area statistics*. This identifies the centre of population of the area. The points were chosen by hand and in the first release of the statistics several thousand were found to have been misplaced.[6] Figure 2 shows the corrected spatial distribution of these collecting units and through them the national population distribution. All the figures, apart from Figure 1, were produced on a home microcomputer using some very basic programming. The practicalities of census mapping are now a lot simpler than they were in 1851 (although disc space was not a problem then!); now it is the theory of how better to represent the population which is most challenging.

Along with the 1991 census data, digital boundaries of enumeration districts have been produced, in theory delimiting precisely where 'on the ground' people were counted (although the academic purchase of these are still due for release at the end of 1993). Enumeration districts range in size from less than one thousandth of a square kilometre (the base of a high rise block) to over one hundred square kilometres (encompassing empty moorland in the most remote areas). It would not be surprising to find that the largest is more than a million times the area of the smallest whilst still containing fewer people. Mapping with these boundaries will no doubt take place, but could produce some very misleading pictures because these areas encompass all

3. Beniger, J. R. and Robyn, D. L. (1978) 'Quantitative graphics in statistics: a brief history', **The American Statistician**, 32(1), 1–11.

4. Hunt, A. J. (1968) 'Problems of population mapping: an introduction', **Institute of British Geographers, Transactions**, 43.

5. Census Research Unit (1980) **People in Britain — a census atlas**, H.M.S.O., London.

6. Atkins, D. and Dorling, D. (1993) 'Connecting the 1981 and 1991 Censuses', paper presented at the **1991 Census Conference**, Newcastle, 13–15th September.

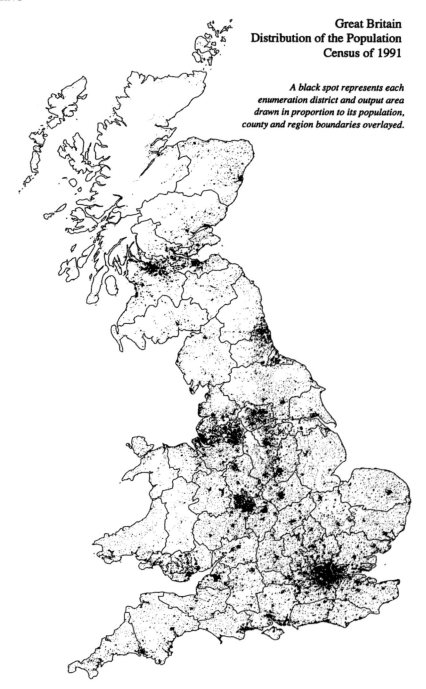

**Great Britain
Distribution of the Population
Census of 1991**

*A black spot represents each
enumeration district and output area
drawn in proportion to its population,
county and region boundaries overlayed.*

FIG. 2.　Great Britain, distribution of the population, census of 1991

the empty land as well as that occupied by people. Ian Bracken and David Martin have suggested some methods using Geographical Information Systems to avoid this problem.[7] Here I concentrate on alternative cartographic solutions.

Dasymetric mapping — shading only inhabited parts of a map — is seen as one possible way forward.[8] One solution is to take the dasymetric mapping of the 1851 census to its digital limits and attempt to draw a point to represent every household. Within enumeration districts we have a good idea of where the households are because a postcode is now linked to every census form and we know the rough location of the addresses allocated to each postcode. Figure 3 shows an inset of central London with a circle drawn at the location of each postcode, its area in proportion to the number of households living there. Unfortunately, as we portray physical reality more accurately, we are left with less and less space in which to show social reality. Even in the most densely populated part of Britain what dasymetric mapping shows most clearly are those areas which it leaves blank. The parks, rivers and roads at the centre of the capital are what stand out clearly on a dasymetric map of any variable for a city like London, while outside the cities the isolated places are most prominent, places which usually contain least people and for which averages and proportions are least meaningful.

How might we go about visualizing more fairly the characteristics of people in so many areas, which vary so much in physical size? Figure 4 shows a grid drawn over a map of Britain. The grid is not uniform, but is violently stretched and twisted. It is made up of almost two thousand squares each of which (on the mainland) contains 30,000 people. There are many such squares in the large cities, but few in the more rural areas. The picture is more complex than this, however; for in East Anglia, for example, the effect of Norwich can be seen, like a weight pulling the fabric of some giant net inwards. Each square contains the same number of people, yet some cover great swathes of land while others are barely visible. Perhaps it would be more sensible to pull all the lines straight — so as to form a rectangular grid — ensuring that each square would be of equal size so that their populations are more fairly represented than on a traditional map. This would create an *equal population cartogram*[9] which is discussed later.

DRAWING BOUNDARIES

For the 1981 census, digital boundaries were available only for wards, of which there were 9,289 in England and Wales and into which enumeration districts were nested. The electoral wards varied in population from 498 (Lower Swaledale) to 41,502 (Birmingham, Weoley), and in land area from 19 hectares (Skipton Central) to 44,789 hectares (Upper North Tyne). They exhibit a skewed distribution in which the majority are small in area where people huddle together on the land. On a conventional national

---

7. Martin, D. (1991) **Geographic information systems and their socioeconomic applications**, Routledge, New York.

8. Langford, M., Unwin, D. J. and Maguire, D. J. (1990) 'Generating improved population density maps in an integrated GIS', **Proceedings of the first European Conference on Geographical Information Systems**, EGIS Foundation, Utecht.

9. Tobler, W. R. (1973) 'A continuous transformation useful for districting', **Annals of the New York Academy of Sciences**, 219, 215–220.

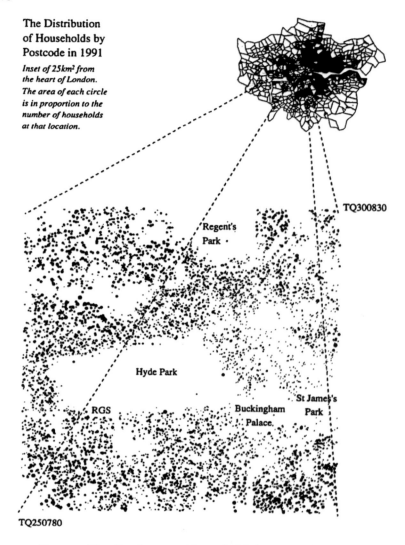

The Distribution
of Households by
Postcode in 1991

*Inset of 25km² from
the heart of London.
The area of each circle
is in proportion to the
number of households
at that location.*

TQ300830

Regent's
Park

Hyde Park

St James's
Park

RGS

Buckingham
Palace

TQ250780

FIG. 3.   The distribution of households by postcode in 1991

grid projection only a small proportion of the wards are visible. Figure 5a shows all wards' boundaries for the whole country (the inset in Figure 3 focuses in on London). At almost every level of magnification, below one thousand fold, there are some wards which cannot be seen and others, usually containing the least people, which dominate the image. Simply because the census provides geographical information using one projection does not mean that that projection must be' maintained, although almost all recent mapping does this. Similarly, we do not have to use the raw counts of the people that the census provides, but can transform these to use more meaningful scales.

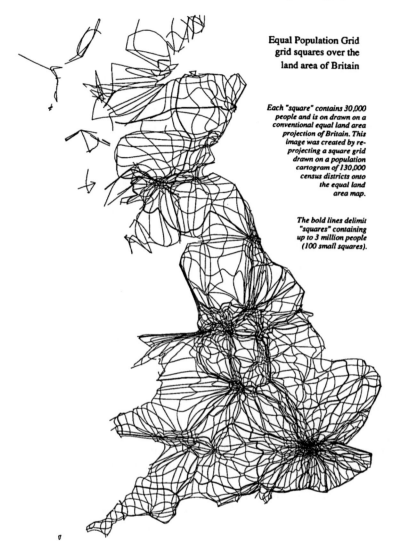

Equal Population Grid
grid squares over the
land area of Britain

*Each "square" contains 30,000
people and is on drawn on a
conventional equal land area
projection of Britain. This
image was created by re-
projecting a square grid
drawn on a population
cartogram of 130,000
census districts onto
the equal land
area map.*

*The bold lines delimit
"squares" containing
up to 3 million people
(100 small squares).*

FIG. 4.   Equal population grid squares over the land area of Britain

Chi-squared mapping[10] is a good example of how shading categories can be altered in an attempt to overcome the overemphasis traditional maps give to the characteristics of unpopulated areas. Traditional thematic mapping makes most use of ratios to characterize areas, for instance of the number of unemployed people divided by the total workforce. A signed chi-square statistic can be calculated in place of the more

10. Visvalingham, M. (1981) 'The signed chi-score measure for the classification and mapping of polychotomous data', **The Cartographic Journal**, 18, 1, 32–43.

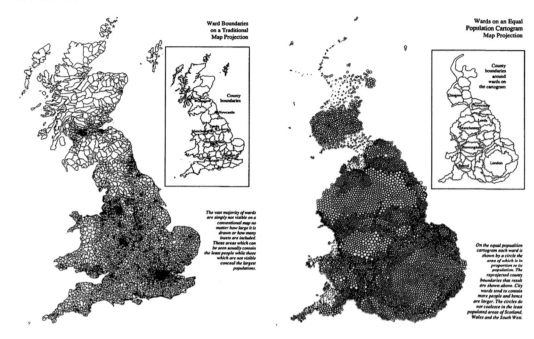

FIG. 5a.   Ward boundaries on a traditional
map projection

FIG. 5b.   Wards on an equal population
cartogram map projection

usual ratio for each area and areas are only shaded brightly when this statistic differs significantly from what would be expected. This avoids the basic problem with census statistics, that the most extreme ratios (and hence the areas which draw most attention on maps) are almost always found where there are the least people and hence where the statistics are most unreliable. Chi-squared mapping was used to map the 1971 census,[11] and is a good palliative if mapping with an equal land area projection is a high priority.

The information in the census concerns not land but people and households. In visualizing these, a primary aim can be that each person and each household is given equal representation in the image. Current computer versatility allows this, so a principal challenge of visualizing the 1991 census is to achieve this. Figure 5b shows an equal population cartogram of the ten thousand wards of Britain based on the 1981 census. In effect, the lines in Figure 4 have been straightened (for more details see Dorling[12]). On the cartogram each ward is represented by a circle, its area in proportion to its population and located as close to its original geographical neighbours as possible. The cartogram represents one level of abstraction beyond the equal area choropleth map — one level closer to the maps of social landscape we might aim to see if we wish to look both inside cities and across the nation, simultaneously.

11. Hunt, 1968.

12. Dorling, D. (1991) **The visualization of spatial social structure**, PhD thesis, University of Newcastle upon Tyne.

Many arguments can be made for why we should use cartograms to portray spatial distribution from the census. By trying to minimize visual bias, cartograms can be claimed to have advantages in census mapping, being more sensible statistically and more just socially. Their main disadvantage is that they are unfamiliar, but we do not learn from familiarity. Traditionally cartograms have been used in mapping election results and medical statistics.[13,14] This is because, theoretically and respectively, every elector's vote is equally important to the outcome of an election and we are concerned with the spread of disease among individuals, not over land. The British Labour party will never appear to have won on a traditional map of constituencies while incidents of disease always appear massively concentrated in cities on a pinmap. If you believe that every individual's circumstances are equally important in the spatial make-up of a society, then you should use area cartograms in mapping census data.

REPRESENTING PEOPLE

However carefully we define the value of a variable and assign a shading category to an area we will still be using arbitrary spatial boundaries for calculating that value in the first place. This dilemma is well known as the Modifiable Areal Unit Problem.[15] For census mapping the Modifiable Areal Unit Problem can be handled in a number of ways. The simplest is merely to illustrate it by using multiple boundaries, and with cartographic computer animation it is possible to redraw images instantly using different boundaries, to see the effects of these choices.[16] A more sophisticated approach to this problem is to think more carefully about what areas are meaningful to visualize. A set could be specifically defined (e.g. Housing Market Areas), or 'fuzzy boundaries' could be used — for instance by employing kernel mapping methods.[17]

Arbitrary boundaries, however, have their greatest influence on the impression gained in their use for portraying statistics, not in their use for calculating them. If the whole of a region is shaded dark grey because levels of unemployment are particularly high in one of its towns, is our image accurately reflecting reality? Just because proportions are calculated for one area does not mean they have to be shown using a scaled down replica of the boundary of that area (Le. by choropleth mapping). David Martin,[18] for instance, makes a strong case for almost never using choropleth mapping.

The severity of the Modifiable Area Unit Problem for conventional thematic mapping is due, as has already been touched on, to the fact that greatest emphasis is given to those places containing fewer people — where the arbitrary movement of

13. Hollingsworth, T. H. (1964) 'The Political colour of Britain by numbers of voters', **The Times**, October 19th, p. 18.

14. Howe, G. M. (Ed.) (1970) **National atlas of disease mortality in the United Kingdom,** Thomas Nelson and Sons, London.

15. Openshaw, S. (1982) **The modifiable areal unit problem,** Concepts and Techniques in Modern Geography No. 38, Gen books, Norwich, England.

16. Dorling, D. (1992) 'Stretching space and splicing time: from cartographic animation to interactive visualization', **Cartography and Geographical Information Systems**, 19(4), 215–227,267–270.

17. Brunsdon, C. (1991) 'Estimating probability surfaces in GIS: an adaptive technique', **EGIS '91 Proceedings**, pp. 155–164, Brussels, Belgium.

18. Atkins and Dorling, 1993.

boundaries can have the most influence on the values calculated. This is illustrated graphically below. In general, arbitrary boundaries can be seen to have much less influence on the impression gained from cartograms as opposed to traditional maps (Fotheringham[19] has claimed that visualization may provide a range of solutions to the problem). Our methods of visualization must be robust if one arbitrary choice or another is not to lead us to misinterpret what is important.

The 1991 census mislaid over a million people who are thought to live in Britain. This may appear disastrous, but our conventional mapping techniques are capable of hiding many more than this from our eyes. Conventional maps lose people because they are designed to show land and other features of traditional military significance such as hills, roads and woodland.[20] People are –concentrated in cities in such small areas of space that, when we map their distributions nationally and in detail, the characteristics of the majority are almost impossible to see. When we map at higher scales we convert millions of statistics into only a few dozen or a few hundred numbers and then draw I a picture of them — a great deal of information is lost. Pictures can contain much more information than this. The problem is to structure the information spatially in such a way as to maximize the content that can be seen (while still retaining the geographical topology) — to clarify reality through the portrayal of detail.[21]

In the cartograms shown here most wards have been placed adjacent to wards with which they share a common boundary, although occasionally this has not proved possible (thus they are 'noncontinuous area cartograms'[22]). Even when reproduced to the page size of this journal every ward is made visible in a single image. Here is a projection upon which it is possible to show the fortunes of every group of people at something akin to the neighbourhood level, nationally, without a particular bias against those groups who happen to live in the most densely populated areas.

Figure 6 shows two images which give an impression of where some of the people who were missed by the census might be. They show the proportion of residents who were 'imputed' to exist by the census authorities when nobody was found in a dwelling and it was thought by the enumerator not to be vacant. On the map, these people can be seen to be concentrated in cities; but the largest areas of underenumeration are on the west coast of Scotland. The cartogram shows that this is a false overall impression to gain of the spatial distribution of the location of imputed residents. Underlying the cartogram is the true distribution of the population so an indication of the actual relative scale of the problem in each area is given. More than a third of all those imputed nationally were in London. This can be estimated by comparing the sizes of the shaded areas which are in proportion to the populations.

MAPPING SOCIAL STR.UCTURE

The main subject of interest in mapping the census, however, is the characteristics of those people who did complete their census forms. These characteristics, when in turn

19. Fotheringham, A. S. (1989) 'Scale-independent spatial analysis', in **The Accuracy of Spatial Databases**, Goodchild, M. and Gopal, S. (eds.), Taylor & Francis, New York, Chapter 19. p. 223.

20. Harley, J. B. (1989) 'Deconstructing the map', **Cartographica**, 26(2), 1–20.

21. Tufte, E. R. (1990) **Envisioning Information**, Graphics Press, Cheshire, Connecticut. p. 37.

22. Olson, J. M. (1976) 'Noncontinuous area cartograms', **The Professional Geographer**, 28(4), 371–80.

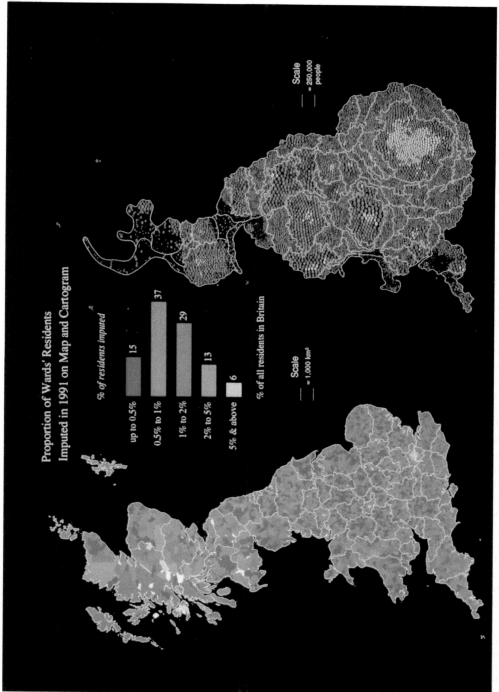

Fig. 6. Proportion of wards' residents imputed in 1991 on map and cartogram

used to characterize places, are often quite complex variables to map. For instance, although any single individual might be assigned to a single social class out of a three-fold division of occupations, the place in which that individual lives could be made distinct by the combination of the relative numbers of people in each of those three groups who are also living there. The census allows us to measure how much mixing of different social groups there is and how that is changing. There are statistical techniques to reduce such numbers down to a single index of, say, segregation, but information on the nature of social divisions is lost when these techniques are applied. The cartographic solution that census mapping has produced is to use different mixtures of colour to represent different mixtures of people. The US Bureau of the Census[23] pioneered automated bivariate colour mapping to show how two characteristics of the population were spatially related to one another. Colour, however, contains three primary components and so can, in theory, be used to show the relationships between three dependent variables.

Three-colour, or trivariate, mapping is the most sophisticated possibility when there is only a pinprick to shade. Three primary hues of varying intensity will produce a unique colour for every possible combination (while more than three colours do not). The three primary hues I choose to use are red, blue and yellow (mixing to purple, orange and green) as I find these most intuitive (as does Arnheim[24]). Sibert[25] prefers red, green and blue, while Bertin[26] recommends cyan, yellow and magenta. Whatever choices are made, when images use three-colour shading, the patterns formed are usually much more subtle than those found in univariate structures. It is often the case that the more complex a picture is, the more there is to learn from it. This process is not necessarily easy, but can be very rewarding in terms of gaining new insights.

Figure 7 uses a population cartogram of the 129,000 1981 census enumeration districts where each district is represented by a circle coloured one of sixty four possible shades to show the detailed spatial distribution of three occupational social classes in Britain. Clusters of enumeration districts form where the occupation of 'heads' of households is predominantly 'professional' (blue), 'intermediate' (yellow) or 'supervised' (red). Where there is a concentration of professional workers, for instance around Greater London, the area will appear bright blue. Where there are high proportions of people in both intermediate and supervised occupations (but not in professional occupations), for instance around central Birmingham, areas of orange emerge. Superficially, there is not a great deal of difference between this kind of image and a CAT (Computer-Aided Tomography) scan, with the latter showing a slice through the human brain and the former, a slice through society. One arguable difference is that a great deal of training is needed to interpret a map of the brain, with which most of us are unfamiliar, whereas a map of society can be much more easily, understood superficially (as we, ourselves, constitute society) and so can successfully show more complex structures.

23. US Bureau of the Census (1970) **Distribution of older Americans in 1970 related to year of maximum county population**, United States Maps, GE-70, No.2, US Department of Commerce, Washington DC.
24. Arnheim, R. (1970) **Visual Thinking**, Faber and Faber Ltd., London. p. 30.
25. Sibert, J. L. (1980) 'Continuous-colour choropleth maps', **Geo-Processing**, 1, 207–217. p. 214.
26. Bertin, J. (1981) **Graphics and graphic information processing**, Walter de Gruyter, Berlin. p. 163.

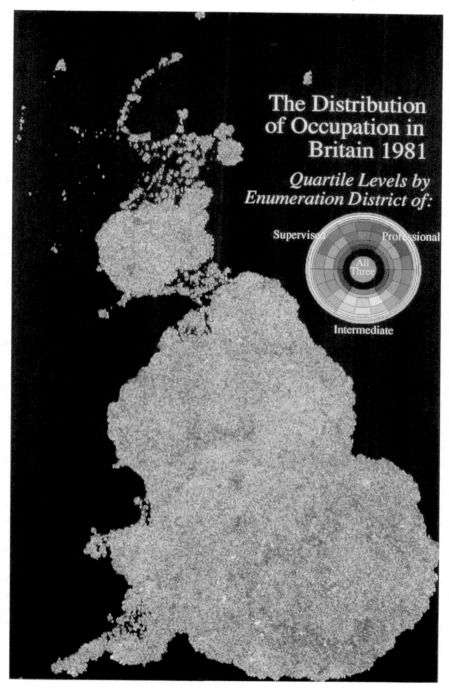

FIG. 7.   The distribution of occupation in Britain, 1981

Because the three distributions of occupations can be seen to spatially 'repel' each other (people in different classes of job tend not to live in the same streets), the primary colours chosen, of red, blue and yellow, dominate this image. Central London, Birmingham, Liverpool and Manchester have distinctly high proportions of people in 'supervised' occupations, although a 'snake' of professionals winds its way, north to south, through the centre of the capital. If the variables chosen had been birthplace (subdivided by, say, Irish, Asian and Afro-caribbean categorizations), then a great many places would have shown either high proportion of all three or low proportions, because these groups of people tend to cluster together in space, although many more subtle mixes of colours would also be visible.

The patterns in Figure 7 are quite difficult to discern because there is a 'bittyness' which reflects the reality that the social structure is textured. To make those patterns which form more apparent we can generalize the image by applying 'smoothing' operations of generalization (of the type used in remote sensing). Figure 8 shows the result of using ten passes of the most basic binomial filter (through which, in each dimension, each point is given a new value equal to half its old value and a quarter of the values of each its two closest neighbours[27]). In the figure the very complex distribution of occupational groups has been smoothed. Generalization very much simplifies the image and is particularly useful when reductions in size are required, but, as with all these decisions, the impression of the distribution changes. Watching the generalization take place as an animation can be very enlightening, the image eventually becoming a single blur of one muddy hue. As we generalize like this we are, in effect, using larger and larger (overlapping) areal units — exploiting the' modifiable areal unit problem to our advantage.

In Figure 8 the sharpest divide is confirmed to be in Inner London, where a white buffer appears between the blue epicentre of the capital and its red core. This is where the spatial generalization procedure failed to merge the colours because social divides at this boundary are too great. Moving outwards the red becomes orange, where more 'intermediate' workers are housed and then progressively yellow, green and blue as areas housing 'professional' workers are taken in. Purple is the least common colour, as would be expected, representing places where the lowest and highest categories of occupations mix — it is (in various shades), nevertheless, still present.

One final word of warning about detailed colour mapping: the apparent colour of an object is affected by the colours surrounding it. Isolated spots of unexpected colour stand out in otherwise uniform areas far more than they do where there is already great variation. The colour of each object you see is, in fact, the result of a mixture of colours on the page that is unique to every distribution shown.[18] This is not such a grave problem when visualizing geographic space as it is with other subjects as we may, anyway, choose to blur adjacent colours for theoretical reasons. We must just understand that our pictures can be as subjective as the conclusions we draw from them — and the decisions which led to their creation — despite the explicit consistency of their design. Each apparent weakness of a choice in map design can always be argued to be a strength!

27. Tobler, W. R. (1989) 'An update to numerical map generalization', in Numerical Generalization in Cartography, McMaster, R. B. (ed.), **Cartographics**, University of Toronto Press.

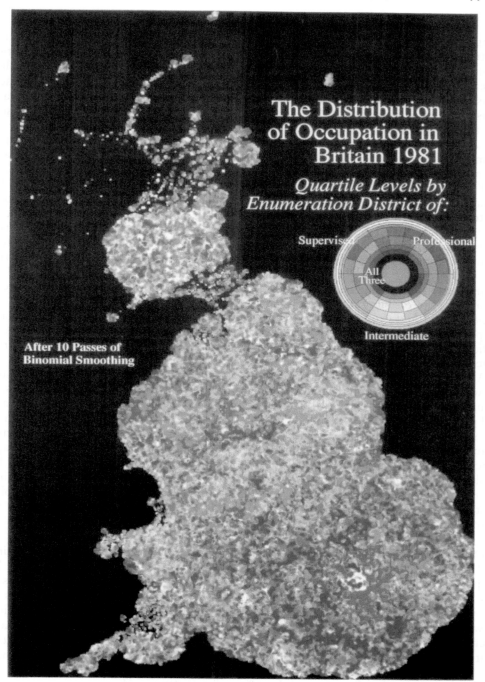

FIG. 8.   The distribution of occupation in Britain smoothed from EDs

MAPPING CHANGE AND FLOW

Mapping the census does not necessarily involve only the census, or only.one census. Pictures of change over time, and of the spatial relationships between characteristics found in the census and in other surveys often produce more unexpected results. Greater care is also needed when disparate datasets are related graphically. Each census is taken using a unique geographical base and so, at the finest level of detail, censuses cannot be related to one another spatially. Advantage as well as necessity for the use of less fine geographical units can also be argued for; as the sensitivity of the data increases once changes over time are calculated (slight errors in two large numbers result in larger errors once their difference is calculated — Cole[28] discusses other sources of error).

The enumeration districts (and output areas) of the 1991 census data can be amalgamated to fit, as well as possible, into 1981 wards (and Scottish part-postcode sectors — Atkins and Dorling[5]). Maps of change over time can then be produced using the 'frozen ward' as a common spatial unit. Figure 9 shows the change and the new static position in the relative distribution of those people working in 'professional' occupations using a simple five category classification of wards. As the statistics become more complex it is often worth making the map design simpler. The changes are subtle. The static picture of 'professionals' in 1991 at ward level is very similar to the distribution of 'blues' in the 1981 enumeration district cartogram of *Figure 8*. Marked geographical patterns are visible with the most, significant increase of professionals being in Inner London, and in other smaller areas of apparent gentrification elsewhere in Britain. Figure 6 showed that Inner London i~ also the area where the data is least reliable so some caution in interpretation is advisable.

Although the figures for' individual wards might be unreliable, where a cluster of wards have very similar values it is reasonable to assume that, in these places, a significant change has occurred. Central London has obviously experienced substantial gentrification over the decade, whereas other large British cities have not had quite the same experience. Figure 10 shows the same distribution as Figure 9 using a cartogram and a traditional land area map of Local Authority Districts. On these maps place names can be added, as has been done here, although few are likely to be legible when the figure is reduced for printing (the key advantage of desk-top mapping over paper publication is interactive panning and zooming). Here a much simpler spatial story is told, of a central core and declining hinterland to Britain: an easy to absorb tale, but how representative is it? This illustrates how important choices of design are in census mapping — in all senses they totally alter the picture.

Flow mapping is a classic example of the importance of design, and the commuting and migration flow matrices of the census are another set of related datasets to be mapped. These matrices contain information about those people who have crossed any of the boundaries separating over nine thousand areas in England and Wales in the week or year prior to the census date. Figure 11 shows all the significant migration flows drawn on a traditional map with the lines of flow coloured by the dominant occupational status of the ward of origin. The bright red centres of some cities show that the distances travelled by many people in low status areas are often short. When

28. Cole, K. (1993) **Sampling error and the 10% small area statistics**, Manchester Computer Centre Newsletter No. 22, University of Manchester.

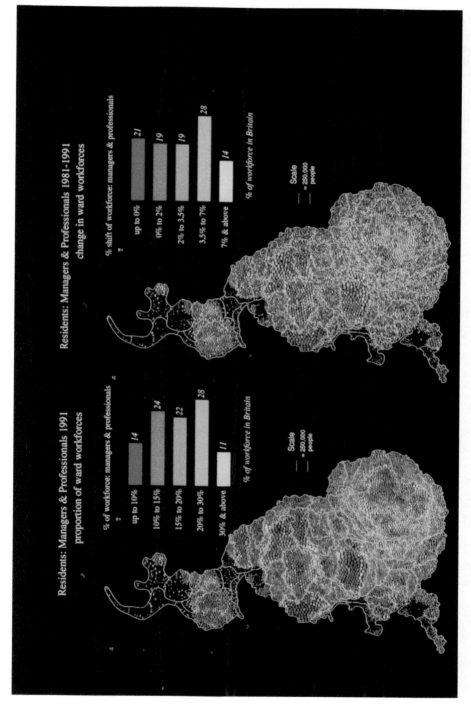

FIG. 9. Residents: managers and professionals, proportion and change at ward level

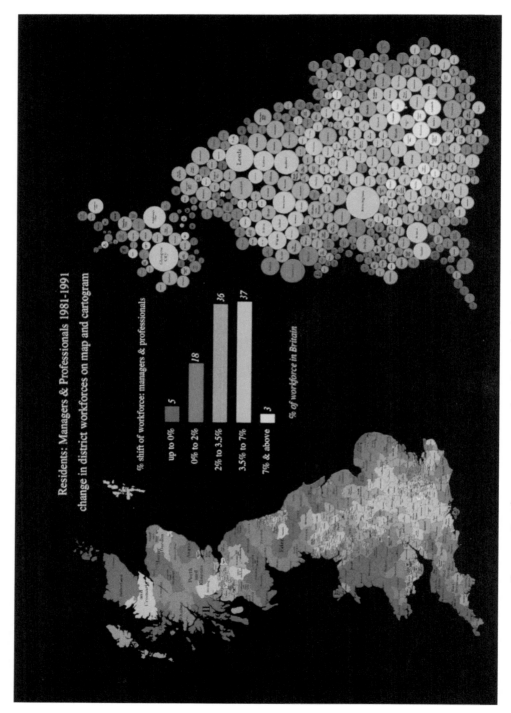

FIG. 10. Residents: managers and professionals change in district map and cartogram

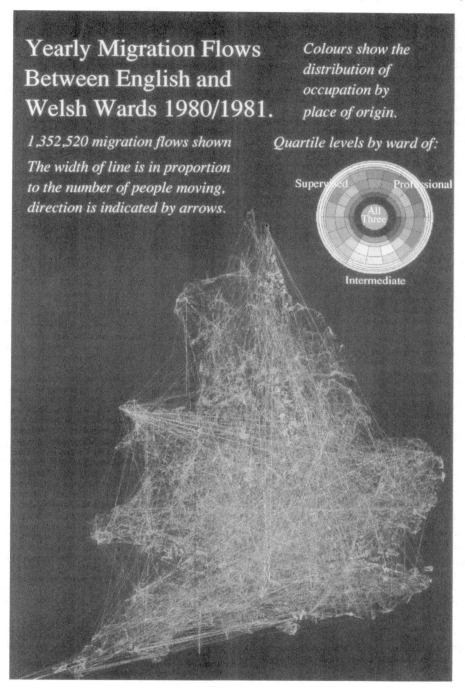

FIG. 11. Yearly migration flows between English and Welsh wards 1980/1981

reprojected on a cartogram the boundaries of Local Authority Districts can be discerned as barriers to people moving between council houses (although more intensive research would be needed to confirm this).

Long distance migration is, in general, the province of those leaving 'professional' or 'professional-intermediate' areas (blue or green/blue lines). Distinct concentrations of flows can be seen from London to the South Coast, and even to the wards of the Isles of Scilly. These may well be due to retirement and a map of flows for every age band could be drawn to discern this. Flow mapping, however, is still very much in its infancy. An area of considerable research challenge is the mapping of spatial change in flow matrices over time. Mapping in both time and space is an area where new technology promises more imaginative methods of visualization.

> The development of census cartography should be based first and foremost on the use of models of the dynamics ... Look upon the population and its various activities as part of a vertically-rising stream in space-time...
>
> Szego[29]

RELATING OTHER DATASETS

Politics and Health are two areas of concern to social scientists which are closely connected with census mapping. Figure 12 shows a cartogram of all the wards in Britain, used to visualize the results of the British local elections of 1987, 1988 and 1990 combined (all three years have to be included to get a complete coverage of the country). Each ward has been coloured according to 36 categories of election result, ranging from a Conservative marginal where Liberal came second and Labour third (cyan) to a ward where only a Labour candidate stood (blood red). The standard electoral triangle is used as a key in which the colours of all the possible categories are displayed to indicate the proportions of the vote they represent.[30] To explain adequately just this one diagram would take several pages of text. That is not the purpose here: what is of interest for this paper is to show that this detail is possible and to highlight its implications.

The lace-like patterns of local voting tally very closely with those of occupations which were shown in Figure 8 (as has been known for a long time, but can now be demonstrated visually). Features such as the concentrations of Liberal seats on the edge of Inner London, and the Liberal's general high propensity to appear where there is a greater degree of social mixing, might have been missed in a conventional analysis. The fact that everywhere there are small clusters of bright blue ('safe' Conservative seats), surrounded by purples (Conservative/Labour marginals) encroaching on reds with almost infinite repetition might well be missed without visual analysis — as too can be the stark message of left-wing support from most of the electorate in recent local elections.

Moving from politics to health, Figure 13 shows the spatial distribution of mortality by two of the most significant causes — cardiovascular diseases, and cancers. In mapping mortality data accurate census figures are needed to calculate the standardized mortality rates (so that the influence of concentrations of particular age groups and sexes does not dominate the patterns). Here data from both the 1981 and

29. Szego, J. (1987) **Human Cartography: mapping the world of man**, Swedish Council for Building Research, Stockholm. p. 200.

30. Upton, G. J. G. (1991) 'Displaying election results', **Political Geography Quarterly**, 10, 3, 200–220.

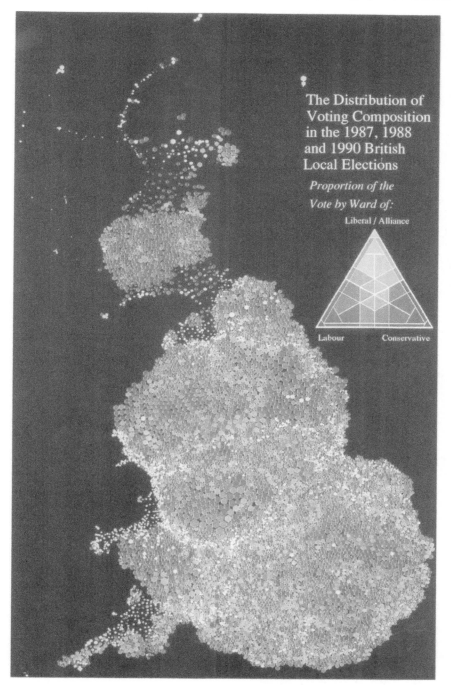

The Distribution of
Voting Composition
in the 1987, 1988
and 1990 British
Local Elections

*Proportion of the
Vote by Ward of:*

Liberal / Alliance

Labour   Conservative

FIG. 12.   The distribution of voting composition in the 1987, 1988 and 1990 elections

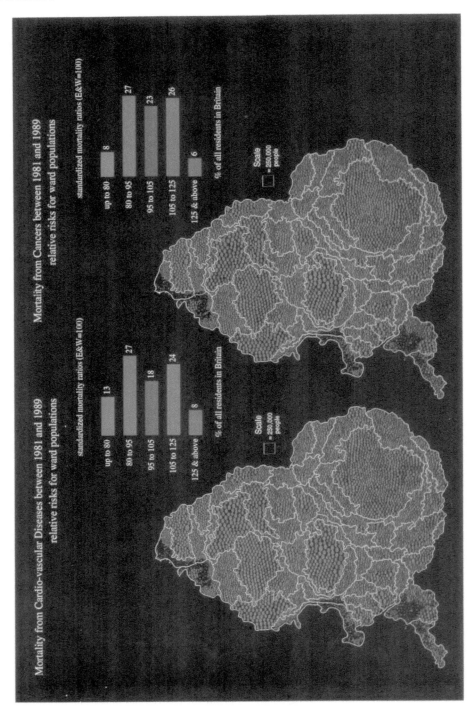

FIG. 13.   Mortality from cardiovascular diseases and cancers at ward level

1991 census have been used along with mortality records from nine of the intervening years. The North/South divide in Heart Diseases and the Inner/Outer City pattern to the spatial distribution of Cancers are both clearly evident. As are, most importantly, all the fallibilities of making such a simple statement about such complex spatial distributions.

CONCLUSION

This paper has only been able to take the briefest look at the new opportunities for map design which the modern census data and computer versatility provides. It has concentrated on high resolution colour mapping because the census is one of the few datasets to provide information sufficiently detailed for such designs. At lower resolutions (when mapping districts and counties for instance) many more graphical opportunities are opened up along with the space in which to implements them.[31]

The census was made for mapping. It contains more spatially referenced information than any other social survey and allows yet more (geographical) information to be meaningfully mapped by providing a base against which this can be done. If we are to do justice to the representation of the people we are claiming to show, then great care is needed in the design of these images. The census contains some seven thousand statistics for each of more than one hundred thousand areas. Combined with other censuses and related datasets access we quickly gain access to billions of numbers, which describe (admittedly, only in the briefest of terms) characteristics of the lives of millions of people. There will never be enough researchers with enough time to look at more than a fraction of this information, but if we can look at it efficiently, fairly and with a better idea of what it is we are looking for and how best to look, we might just learn something new about the society we constitute.

ACKNOWLEDGEMENTS

Thanks are due to Tony Champion, James Cornford, David Dorling and Nina Laurie who read and commented on earlier drafts of this paper, and for funding from the British Academy. Joseph Rowntree Foundation and The University of Newcastle upon Tyne which enabled the research this paper is based upon to take place.

The census figures and geographical boundaries used here remain, as always, Crown Copyright.

31. Dorling, D. (1993) 'Visualizing the 1991 census', in **A Census User's Handbook,** Openshaw, S. (ed.), Routledge, London.

Reflections on 'Map Design for Census Mapping'

CHRIS PERKINS

University of Manchester

Danny Dorling was awarded his PhD from the University of Newcastle-Upon-Tyne in 1991 and the article under discussion here is one of the first to emerge from this work, and from his subsequent Rowntree and British Academy-funded research. It is also one of the first in an incredibly prolific academic career: by 2013 Dorling had published 22 authored books, including eight atlases, and over 200 academic papers, with over £7.5 million pounds research income to the end of 2010. Most of his work has investigated how technology can be deployed to generate more effective social visualizations that might inform the policy process. He continues to be interested in social inequalities, and the mapping of census and other socio-demographic datasets. The central concerns of this paper are reflected many times in his subsequent work and many of his papers focus on the possibilities of the cartogram as a form. His recent appointment, from September 2013, as the Halford Mackinder Chair in Geography at the University of Oxford, is arguably one of the highest profile posts held by any author publishing in *The Cartographic Journal*.

Dorling's paper starts from the premise that new cartographic techniques can be deployed to study census data, and is published as part of a full-colour themed edition of the Journal, focusing upon Map Design. He establishes problems of mapping census data, charting the historical trajectory of attempts to map-out social distributions and reflects that since the advent of the first mapping of the census in 1851 remarkably little had changed. He focuses on the difficulties that stem from data collection and in particular considers the problems posed by the sampling frame. Whatever the values of a census variable, and however they are shaded, the arbitrary spatial boundaries framing data inevitably limit ways in which data may be represented. Places with fewer people are given undue emphasis in choropleth displays. Openshaw (1982) designated this issue the Modifiable Areal Unit Problem and the issue has occupied considerable research time in the years since. The solution presented in this paper was first explored on a home computer during Dorling's undergraduate dissertation and he develops these ideas and explains the significance of his solution in this paper. He redraws boundaries to reflect the number of households in each enumeration unit, and represents each unit as a circle, with an area proportional to its population, located as close as possible to its real geographical neighbours. No attempt is made to reflect the actual shape of the unit, or the precise relations to adjacent enumeration districts: every person and household is given equal value in his maps. Dorling cartograms, as they came to be called, minimize visual bias. They make it harder to recognize the individual geography of an area, but offer a fairer focus upon data distribution, which significantly improves the design qualities of the displays.

The second half of the paper explores application areas for the technique, discussing electoral and health mapping and evaluating some of the problems and potential of using cartograms to map the census in particular, and social structure in general. Dorling also considers the use of trivariate colour schemes, deploying red, blue and yellow hues to map out occupational difference. He also considers the mapping of change deploying frozen wards as a device to visualize difference between the 1981 and 1991 censuses. The potential of the technique to show detailed spatial patterns on a common scalar base is demonstrated in twelve figures, many of which are printed as full-page displays.

The Dorling cartogram has continued to be deployed to map social distributions in many of Danny's subsequent atlases, e.g. Dorling (1995), Dorling and Thomas (2004) and Shaw *et al.* (2008). The application described here has come to be superseded by the elaboration of the shape-preserving Gastner–Newman algorithm, which was operationalized a decade after this paper (Gastner and Newman, 2004). Dorling himself has moved on to apply the Gastner–Newman algorithm to the best available national global data sets in the *WorldMapper* project, and arguably has also shifted focus onto global as against British social inequalities (see Dorling *et al.*, 2006). But this paper was instrumental in focusing attention on novel visual ways of representing census data, at the very time these data were first becoming widely available in digital form. It is no coincidence that it deploys an unusual number of visualizations for *The Cartographic Journal* at this date. It was published relatively early in what was to become the commonplace application of computer processing to large amounts of spatial data, and has encouraged an ongoing focus on the possibilities of the cartogram as an alternative way of envisaging space. Cartograms are much more frequently now used to display thematic data than was the case before Dorling started his research, and this paper significantly re-energized interest in the design of this kind of thematic display.

REFERENCES

Dorling, D. (1995). **A New Social Atlas of Britain**, John Wiley and Sons, London.
Dorling, D., Barford, A. and Newman, M. (2006). 'Worldmapper: the world as you've never seen it before', **IEEE Transactions on Visualisation and Computer Graphics**, 12, pp. 757–764.
Dorling, D. and Thomas, B. (2004). **People and Places: A Census Atlas of the UK**, Policy Press, Bristol.
Gastner, M. T. and Newman, M. E. J. (2004). 'Diffusion-based method for producing density-equalizing maps', **Proceedings of the National Academy of Sciences of the United States of America**, 101, pp. 7499–7504.
Openshaw, S. (1982). **The Modifiable Areal Unit Problem**, Concepts and Techniques in Modern Geography No. 38, Geo Books, Norwich.
Shaw, M., Davey-Smith, G., Thomas, B. and Dorling, D. (2008). **The Grim Reaper's Roadmap: An Atlas of Mortality in Britain**, Policy Press, Bristol.

ColorBrewer.org: An Online Tool for Selecting Colour Schemes for Maps

MARK HARROWER AND CYNTHIA A. BREWER

Originally published in *The Cartographic Journal* (2003) 40, pp. 27–37.

Choosing effective colour schemes for thematic maps is surprisingly difficult. Color-Brewer is an online tool designed to take some of the guesswork out of this process by helping users select appropriate colour schemes for their specific mapping needs by considering: the number of data classes; the nature of their data (matched with sequential, diverging and qualitative schemes); and the end-use environment for the map (e.g., CRT, LCD, printed, projected, photocopied). ColorBrewer contains 'learn more' tutorials to help guide users, prompts them to test-drive colour schemes as both map and legend, and provides output in five colour specification systems.

INTRODUCTION

Colour plays a central role in thematic cartography. Despite this, using colour effectively on maps is surprisingly difficult and often exceeds the skill and understanding of novice mapmakers. On the one hand, a 'good' colour scheme needs to be attractive. On the other hand, the colour scheme must also support the message of the map and be appropriately matched to the nature of the data. Moreover, colour schemes that work for one map (e.g. a choropleth map of income) will not necessarily work for another (e.g. a map of dominant commercial sector by county). Relying on the same colour scheme for all thematic mapping needs is a mistake. Unfortunately, many novice map makers have become conditioned to use the default colour schemes built into commercial mapping and GIS packages and sometimes are even unaware that they can change those default colours. If colour schemes are not carefully constructed and applied to the data, the reader may become frustrated, confused, or worse, misled by the map.

An understanding of how to manipulate the three perceptual dimensions of colour (hue, saturation and lightness) is required to create attractive and logical colour sequences for thematic maps. These dimensions are three of the visual variables or basic 'building blocks' of graphics.[1] Nearly a century of colour theory development combined with perceptual testing has allowed us to better understand how individuals perceive and understand colour. From a semiotic perspective, colour works as a 'sign vehicle' when the map reader is able to understand that a specific colour on the map

---

1. Bertin, J. (1983). **Semiology of Graphics: Diagrams, Networks, Maps**, University of Wisconsin–Madison, Madison, WI.

represents or *stands for* something in the real world and that multiple occurrences of that colour stand for the same kind of thing.[2]

THE PROBLEM

Federal agencies such as the US Census Bureau, National Center for Health Statistics and National Cancer Institute produce numerous maps for both internal and external use (for example: Brewer and Suchan,[3] Pickle *et al.*,[4] 1996, Devesa *et al.*[5]). As collectors and distributors of geospatial data, these agencies make extensive use of thematic maps — especially choropleth maps — to communicate facts to the public and to facilitate exploratory data analysis by their own researchers. The individuals who produce these maps rarely have the time to worry about carefully crafting individual colour schemes every time they make a map. Although most GIS software incorporates default colour schemes for thematic maps, it provides no guidance on how to best use these default colour schemes, and many of these schemes are simply unattractive. More importantly, these default schemes are not appropriate for all mapping tasks.

Given the importance of maps in the activities of these agencies, and the central role colour plays on maps, poor use of colour is a concern. Agencies, and more generally novice map makers, would benefit from tools that could quickly guide them through the colour selection process and let them 'test drive' a colour scheme and see it on a map before committing to it. Such tools would let them work more *quickly* and *confidently* by reducing the chance of misapplied colour schemes on thematic maps.

THE SOLUTION: WWW.COLORBREWER.ORG

This paper outlines the development of ColorBrewer (Figure 1), an online tool that helps users identify appropriate colour schemes for maps. The reader is encouraged to try ColorBrewer online at www.ColorBrewer.org. The impetus for this work came from watching individuals that produce thematic maps at federal agencies struggle with selecting good colour schemes for maps. ColorBrewer is designed to take some of the guesswork out of this process. The system suggests possible colour schemes by prompting the user to identify how many data classes they have and what kind of colour scheme best suits their data and map message (i.e., sequential, diverging or qualitative; explained below).

ColorBrewer also provides guidance for varied display environments — CRT, laptop, colour laser print, photocopy and LCD projector — and warns where and when

2. MacEachren, A. M. (1995). **How Maps Work: Representation, Visualization, and Design**, Guilford, New York.

3. Brewer, C. A. and Suchan, T. A. (2001). **Mapping Census 2000: The Geography of U.S. Diversity**, Census Special Report, Series CENSR/01-1, US Government Printing Office, Washington DC, pp. 108.

4. Pickle, L. W., Mungiole, M., Jones, G. K. and White, A. A. (1996). **Atlas of United States Mortality**, National Center for Health Statistics, Hyattsville, MD.

5. Devesa, S. S., Grauman, D. J., Pennello, G. A., Hoover, R. N. and Fraumeni, J. F. (1999). **Atlas of Cancer Mortality in the United States, 1950–1994**, NIH Publication No. 99-4564367, National Cancer Institute, National Institutes of Health, Bethesda MD, pp. 367.

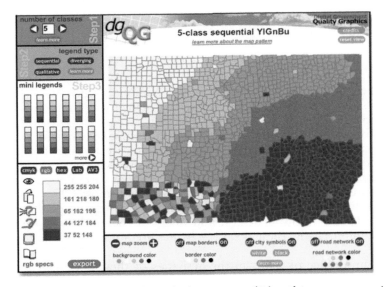

FIG. I. ColorBrewer walks users through the process of identifying an appropriate colour scheme and allows them to 'test drive' the colour scheme formatted as a thematic map and not simply as a colour legend. The Web-based tool was built in Flash 5

a colour scheme might fail. It is important to remind users that the success of a colour scheme is, in large part, a product of the display medium. For example, colours that are easily differentiated on a cathode-ray tube monitor (CRT) may look indistinguishable on a laptop LCD display. As anyone who has made colour prints knows, startling shifts in the appearance of colours can occur when moving from electronic displays to printed displays. Compounding this, many of the maps produced by federal agencies must work across multiple media (e.g. both as an electronic display and as a paper report) and finding truly robust colour schemes that work well in multiple environments is difficult.

In addition to simply suggesting colour schemes, Color-Brewer allows the user to 'test drive' each of the colour schemes to see how well they perform as both an ordered legend and as a choropleth map in order to give the user a better sense of how an individual scheme will function when it is applied to a real map. Differentiating colours within a complex distribution — especially when additional map information such as place names and linework is overlaid — is a harder perceptual task than seeing differences in simple and logically ordered data legends. Put another way, just because a map reader can see subtle differences between the individual colour patches in a legend does not mean they will be able to recognize those same differences on the map.[6]

6. Brewer, C. A. (1997). 'Evaluation of a Model for Predicting Simultaneous Contrast on Colour Maps', The Professional Geographer, 49, 280–94.

The appearance of colours can change with the presence or absence of enumeration unit borders, and the colour of those borders. To accommodate this, ColorBrewer allows users to turn enumeration borders on and off, as well as change their colour. Our goal was not to suggest what border colours to use, but to remind the user that this is yet another component to map design. Since thematic maps often contain additional base information, similar functionality has been built in with an overlay of roads and cities to demonstrate the effects of additional linework and point symbols on the appearance of the colours. The colour of these overlays can also be changed. Though the examples we have chosen are highways and cities, they should give users a good idea of how other linework, points, and text will function on their map. Some colour schemes make it easier to read base information than others, and this legibility may be more important for some map purposes than others.

Not all maps are displayed with a white background. Black backgrounds, for example, have become common in on-screen geovisualization environments. Since the appearance of a specific colour is influenced by the colours that surround it, ColorBrewer allows users to adjust the colour of the background display to determine how well different colours function against different backgrounds.

In short, the ColorBrewer system:

- asks the user to specify the number of classes and type of colour scheme;
- displays a selection of colour schemes for the user to choose among;
- presents the selected colour scheme on a map and in a legend display;
- displays guidance for each scheme on the potential for good results in different media;
- suggests whether a selected scheme will accommodate colour-blind people;
- allows the user to change colours of enumeration unit borders (or remove them);
- displays an overlay of sample line, point, and text base information;
- displays a variety of map background colours.

Beyond our description of system basics, a more in-depth discussion of the rationale behind and implementation of ColorBrewer follows in the next section.

KINDS OF COLOUR SCHEMES

A total of 35 colour scheme 'sets' are contained in Color-Brewer and they are divided into three groups: qualitative, sequential and diverging. These sets have been designed to produce attractive colour schemes of similar appearance for maps ranging from 3 to 12 classes. Not all colour schemes can be expanded to 12 classes, and many of the sequential scheme sets, for example, stop at 9 classes because further divisions within the scheme are perceptually unreliable (Figure 2).

Colour can be used to imply categorical differences (e.g. forest, city, marsh) or to imply ordered differences (e.g. population density rates). Figure 3 contains examples of sequential, diverging and qualitative colour schemes (also see Slocum,[7] Olson[8]). The kinds of data appropriately matched to these types of schemes are outlined below.

7. Slocum, T. A. (1999). **Thematic Cartography and Visualization**, Prentice-Hall, Upper Saddle River, NJ.

8. Olson, J. M. (1987). 'Colour and the Computer in Cartography', in **Colour and the Computer**, Durrett, H. J. (ed.), pp. 205–19, Academic Press, Boston.

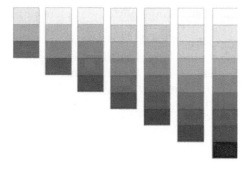

FIG. 2.   One sequential scheme that ranges from pink to blue-green is shown with 3 to 9 classes. Differences between adjacent colours become smaller as the number of classes increases

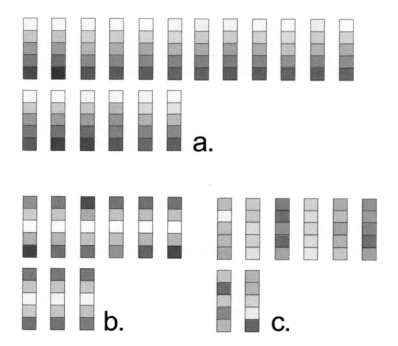

FIG. 3.   ColorBrewer contains 35 colour scheme sets that can be used to create legends for maps ranging from 3 to 12 classes. The sets are organized into three kinds of colour schemes: (a) sequential, (b) diverging and (c) qualitative

Sequential Colour Schemes

Sequential colour schemes (Figure 3a) imply order and are suited to representing data that ranges from low-to-high values either on an ordinal scale (e.g. cold, warm, hot) or on a numerical scale (e.g. age classes of 0–9, 10–19, 20–29, etc.). Lightness steps dominate the look of these schemes, usually with light colours for low data values and dark colours for high values. 'Dark equals more' is a standard cartographic convention. Sequential schemes can be either single hue (e.g. same blue, with different lightness and saturation levels) or multi-hued (e.g. light yellow through dark green). ColorBrewer includes 12 multi-hued sequential schemes and 6 single-hued schemes.

We included more perceptually-graded multi-hue sequential colour schemes in ColorBrewer for two reasons:

(1) they provide better colour contrast between classes and
(2) they are more difficult to create than single-hue schemes because all three dimensions of colour are changing simultaneously. Moreover, since the default sequential colour schemes in commercial mapping and GIS packages are usually single hued, we felt the inclusion of multi-hued schemes would help novice map makers who wished to use more sophisticated colour schemes.

Diverging Colour Schemes

Diverging colour schemes (Figure 3b) should be used when a critical data class or break point needs to be emphasized. The break or class in the middle of the sequence is emphasized by a hue and lightness change and should represent a critical value in the data such as the mean, median or zero. For example, a choropleth map of poverty rates might be designed to emphasize the national rate (midway through the range of rates shown on the map) so that places above and below the national rate are shown with different hues and thus have similar visual emphasis. Diverging schemes are always multi-hue sequences and, because of the way in which lightness is varied, do not make good black and white photocopies or prints (which only capture differences in lightness).

Although we have designed the diverging schemes to be symmetrical (e.g. equal number of colours on either side of the middle break point), designers may need to customize schemes by moving the critical break/class closer to one end of the sequence. For example, a map of population change might have two classes of population loss and five classes of growth, requiring a scheme with only two colours on one side of a zero-change break and five on the other (Figure 4b). To construct an asymmetrical scheme, the user should choose a ColorBrewer scheme with more colours than they need and omit colours (as needed) from one side of the scheme.

For this project both the sequential and diverging schemes were constructed from conceptual arcs across the outer shell of colour space. Colours were chosen that are well saturated and organized in orderly lightness sequences. The scheme was not designed using perceptual colour specifications, but knowledge of the relationships between CMYK colour mixture (cyan, magenta, yellow and black) and perceptually ordered colour spaces were used, such as Munsell,[9] to design the schemes. The

9. Brewer, C. A. (1989). 'The Development of Process-Printed Munsell Charts for Selecting Map Colours', **The American Cartographer**, 16, 269–78.

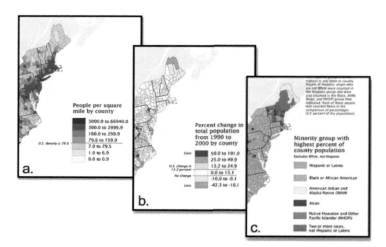

FIG. 4.   These examples from the *Mapping Census 2000 atlas* (Brewer and Suchan, 2001) illustrate appropriate uses of the three main types of colour schemes: (4a) sequential scheme for population density ordered from high to low; (4b) diverging scheme for population change with two hues representing loss and gain and the lightest colour representing the critical mid-range class of no change; and (4c) a qualitative scheme for prevalent minority groups.
These example maps are subsets of full maps (pages 11, 10, and 21 in Brewer and Suchan); they are also available online in PDF and Adobe Illustrator formats at
www.census.gov/population/www/cen2000/atlas.html

diverging schemes generally arc over the top of perceptual colour space (with white or light colours in the middle of the arc). The multi-hue sequential schemes include more hue change through the middle of the ranges and more lightness change at the ends of the schemes. ColorBrewer qualitative schemes generally maintain useful hue contrast with similar lightness and saturation for most colours (the exceptions are obviously the Paired and Accents schemes which intentionally include lightness and saturation differences).

Qualitative Colour Schemes

Qualitative colour schemes (Figure 3c) rely primarily on differences in hue to create a colour scheme that does not imply order, merely difference in kind. Since there is no conceptual ranking in nominal data it is inappropriate to imply order when depicting these data with colour (for example, by using a light-to-dark single-hue sequence). Qualitative schemes work best when hue is varied and saturation and lightness are kept or nearly constant. We do not recommend arbitrarily using strong 'neon' colours (i.e. high saturation) and pastel colours (i.e. light and low saturation) in the same qualitative colour scheme because these variations in saturation might imply order.

In addition to standard qualitative schemes, we offer two sub-categories that do not maintain consistent lightness: Paired and Accents (Figure 5). Accent schemes allow the

Qualitative Color Schemes

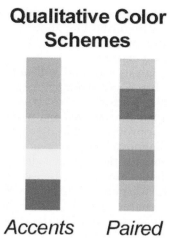

Accents   Paired

FIG. 5. Paired and Accents colour schemes are designed for specific qualitative data mapping situations in which the designer wishes to either emphasize some data classes or pair data classes together

designer to customize qualitative maps by accenting small areas or important classes with visually stronger colours. A small number of colours that are more saturated, darker or lighter than others in the scheme are offered as part of the Accent schemes. These accent colours should be used for classes that need emphasis for a particular map topic. Designers must beware of unintentionally emphasizing classes when using these qualitative schemes.

Paired schemes present a series of lightness pairs for each hue (such as light green and dark green). Often a qualitative map will include classes that should be visually related, though they are not explicitly ordered. For example, 'coniferous' and 'broad-leaf' woodland would be suitably represented with dark and light green land-cover classes. Although designers will probably not find use for an entire Paired scheme, these pairs can be combined with other qualitative schemes to build a custom scheme for a particular map. Qualitative schemes are, thus, more flexible than either diverging or sequential for which an implied order in the colour sequence is maintained.

NUMBER OF DATA CLASSES

Choosing the number of data classes is an important part of thematic map design. Although increasing the number of data classes on a thematic map will result in a more 'information rich' map by decreasing the amount of data generalization, if the cartographer uses too many data classes they may compromise map legibility — more classes require more colours that become increasingly difficult to tell apart. The maximum number of data classes ColorBrewer supports is 12, although only a few of the colour schemes can be divided into this many steps and remain differentiable. As a general rule of thumb, cartographers seldom use more than seven classes on a choropleth map. Isoline maps, or choropleth maps with very regular spatial patterns, can safely use more than seven data classes because similar colours are seen next to each other, making them easier to distinguish. The appearance of a map distribution will also be less varied among different classification algorithms (for example, quantile versus equal intervals) with more classes.[10]

ColorBrewer is not a data analysis tool and will not tell cartographers how many data classes they should use for a given mapping project (see MacEachren[11] and Slocum[12] for summaries of a method for objectively calculating a suitable number of classes). Rather, if a cartographer knows how many data classes they would like to use, ColorBrewer will suggest appropriate colour schemes. Because the main map in ColorBrewer is designed as a diagnostic tool for evaluating the robustness of different schemes in different display contexts, the system is designed to *dissuade* cartographers from attempting to use too many data classes (i.e. colours). Furthermore, a 10-class scheme that is reliable on a CRT display will most likely fail on a laptop LCD display because of different contrast characteristics. Thus, the number of data classes a designer should use for a specific map is a product of the data characteristics, the intended message of the map, the target audience and (importantly) the display medium. A discussion of how this diagnostic map was constructed and how it works follows.

THE MAP AS A DIAGNOSTIC TOOL

A well-known problem with choropleth maps is *simultaneous contrast*.[13] For example, a single enumeration unit (e.g. county) of medium lightness that is surrounded by dark enumeration units will appear lighter than it actually is. Thus, the map-reader will not be able to accurately match these outlying enumeration units with colours in the legend because they will appear lighter on the map than they do on the legend. A related problem involves the similarity in appearance of all light-coloured outliers surrounded by dark areas. For example, two different light yellows side-by-side would be easily differentiated. When physically separated and surrounded by dark colours, they will likely appear to be the *same* colour. As a general rule, the more complex the spatial patterns of the maps, the harder it will be to distinguish slightly different colours. To illustrate this, the maps in ColorBrewer present colours as both random distributions and well-ordered distributions (sequential banding). All other things being equal, a colour scheme will fail to be fully differentiable in the random portion of the map before it fails in the ordered portion.

Figure 6 is a portion of the 'diagnostic map' in ColorBrewer. The base map shows US counties, although the colours do not depict actual data. Colours are easy to differentiate when they appear in a nicely ordered sequence (such as a legend). The task of differentiating colours, however, becomes much harder when the patterns on the map are complex, such as in the lower left corner of the diagnostic map (Figure 1).

10. Muller, J.-C. (1976). 'Number of Classes and Choropleth Pattern Characteristics', The American Cartographer, 3, 169–76.

11. MacEachren, A. M. (1994). Some Truth with Maps: A Primer on Symbolization and Design, Association of American Geographers, Washington DC.

12. Slocum, 1999.

13. Brewer, 1997.

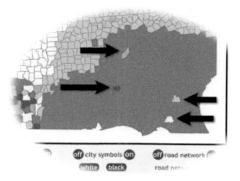

FIG. 6.   The image shows a portion of a five-class map where the green band has four outliers that can be easily seen. The outliers are the four other colours in the scheme. Users are encouraged to check that each one looks noticeably different than all of the other outliers in the band, and not just different from the band itself

There are two ways the 'diagnostic map' in ColorBrewer can be used to evaluate a colour scheme. First, can you clearly see every colour in the random section of the map (Figure 1)? For example, if you have chosen a ten-class colour sequence, make sure that you are able to clearly see ten *unique colours* in this random section. Second, within each large band of colour on the map, we placed one polygon filled with each of the other map colours (the outliers). For example, if you have selected a five-class map, there will be four outliers per band, demonstrating the appearance of all map colours with each as a surrounding colour (Figure 6). Can you see each outlier clearly in Figure 6? Do all pairs of outliers in the band look different? If not, perhaps a different scheme or fewer classes should be chosen.

COLOUR SPECIFICATIONS: OUTPUT FROM COLORBREWER

Once the user has identified a colour scheme they wish to use, ColorBrewer can display the numerical specifications of that scheme in five different colour specification formats (Figure 7): CMYK, RGB, hexadecimal, Lab, and AV3 (ArcView 3.x HSV). We provide specifications for each of these colour systems because different mapping and graphics packages require particular specifications or because users prefer to work with one specification. ColorBrewer colours were originally designed in CMYK in Adobe Illustrator 9.0. We looked up RGB conversions for each colour using Illustrator. We used Adobe Photoshop 5.5 to look up the hex and Lab conversions of these RGB colours from Illustrator. We chose these two software packages because a large share of the graphics market uses them, and we were familiar with and trusted the quality of their conversions between colour systems.

Red-green-blue (RGB) is fundamental for specifying onscreen colours (since these additive primaries are used to produce emitted colour). Cyan-magenta-yellow-black (CMYK) percentages are the standard ink specification for printing. Hexadecimal colour specifications are used to define colours on Web pages and in Macromedia Flash. Hexadecimal colour specifications are RGB specifications in base 16. Lab

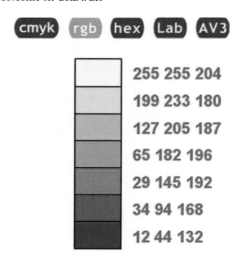

FIG. 7.   ColorBrewer allows the user to retrieve the exact numbers of any colour scheme in five different color specification systems. Since designers often work with multiple colour systems — and changing between systems can be confusing — we hope this functionality will speed their work

('L' for lightness, 'a' for red-green, and 'b' for yellow-blue axes) is a perceptually scaled colour system that is available in some graphics and mapping packages, so we have included it in the hope of moving toward perceptual specifications for perceptually ordered colours.

One of the frustrations we encountered as we worked with the five sets of specifications for each colour was that software packages convert colours differently (they use different algorithms to perform these conversions). For example, converting a dark red CMYK of 60C/100M/90Y/0K to RGB in Photoshop produced specifications of 103R/0G/13B. However, this same CMYK to RGB conversion in Illustrator produces 133R/48G/61B, which has a substantially different appearance. Even though the same company produces Photoshop and Illustrator, they apparently use different algorithms to make transformations between colour spaces. We found that Illustrator produced better conversions from CMYK to RGB because when displayed on-screen, these RGB conversions looked more like the original CMYK colours.

We provide an 'AV3' version of HSV for ArcView 3.x users (we calculated AV3 specifications from RGB using the algorithm from Hearn and Baker[14]) because ESRI uses an idiosyncratic version of HSV (hue, saturation and value are each rescaled to 0-255) in ArcView 3.x products. The next generation of ESRI mapping tools in ArcGIS 8.x use a more standard version of HSV. To specify custom colours in ArcGIS 8.x, use the RGB specifications from ColorBrewer. Do not use the ColorBrewer CMYK specs in ArcGIS 8.x for maps intended to be viewed onscreen. On the other hand, an ArcGIS map intended for print should look right when you use the ColorBrewer CMYK percentages for custom colour specification.

14. Hearn, D. and Baker, M. P. (1986). **Computer Graphics**, Prentice-Hall, Englewood Cliffs, NJ.

The colour specifications in ColorBrewer should never be treated as ironclad guarantees since colour reproduction (whether onscreen or in print) is an inexact science. As anyone knows who has done production print work before, we cannot guarantee that our CMYK percentages will produce exactly the same colours on the printed page that you see onscreen. It is best to think of these CMYK ink percentages as a good starting point but, because printers vary, some 'tinkering' with the exact ratio of inks will be necessary to create a satisfactory final printed product. Colour laser print quality is difficult to control and often produces marked shifts in colour appearance from screen to paper. From our own experiences, cyan pigments seem to be hard to control and light yellows, for example, can become light greens even on expensive colour laser printers. Worse, colour laser output can change over time (as toner amounts change and imaging drums degrade), and even the kind of paper used will influence the appearance of a colour map. Therefore, it should surprise no one that many trial prints may be required before the colours look right. Even if a designer has no intention of printing his or her maps, the situation is only slightly better with digital displays since computer monitors are calibrated differently, and cross-platform differences (e.g. from Mac to PC) can change the appearance of colours, although these changes are somewhat more predictable.

GUIDANCE FROM USABILITY ICONS

In an effort to help the user navigate the colour reproduction process, we have tested each of the permutations of every colour scheme in ColorBrewer across multiple display types, platforms and printed output. We systematically distilled these qualitative differences into 'use guidelines' based on media and display environments. These guidelines are included with ColorBrewer and represent a timesavings function of the tool for novice and expert cartographers alike. In other words, we offer to steer users away from potential problems because we have tried every possible combination of every colour scheme in multiple display environments.

We examined each of the colour scheme sets using two CRT screens (Mac and PC), two laptop LCD screens (old and new), two LCD projectors (old and new), and prints from a Tektronix colour laser printer and a black-and-white laser printer. We evaluated the schemes by deciding whether we could differentiate all of the colours on the multiple backgrounds offered in the ColorBrewer map display. If we had difficulty with the contrast in a scheme, the icon was marked with a red 'X.' If we saw a difference on one display but not another, or if the difference was weak but visible, the usability icon was marked with a red '?' The icons are shown in Figure 8.

The first icon in the usability set, the eye (Figure 8), indicates whether a scheme should be readable by people with red-green colour vision impairments. These approximate evaluations were made using both a theoretical understanding of colour-blind confusions throughout colour space[15] as well as an evaluation of the schemes by an individual with red-green colour impairment, although we would like to see more thorough testing of the schemes to confirm that they accommodate a wide range of colour vision impairments. Generally, the hue pairs for diverging schemes were

---

15. Olson, J. M. and Brewer, C. A. (1997). 'An Evaluation of Colour Selections to Accommodate Map Users With Colour Vision Impairments', **Annals of the Association of American Geographers**, 87, 103–34.

FIG. 8. Usability icons to guide map designer's choice of scheme. For each scheme choice, icons are each marked onscreen with an 'X' if we expect it to fail in the display situation or a '?' if we are unsure whether it will function well because of variability in colour differences in our testing. For example, many schemes that work well on a CRT (icon e) will not hold up when photocopied (b) or projected using an LCD digital projector (c). Icon (a) indicates suitability for colorblindness; icon (d) indicates suitability for laptop LCD displays; and icon (f) indicates suitability for colour printing. A description of the meaning of each icon appears when you click on it in ColorBrewer

selected to accommodate colour-blind readers, with the exception of red-yellow-green and full spectral schemes. Almost all of the qualitative schemes are difficult for colour-blind readers and all of the sequential schemes are useful to them, because they include visible lightness differences.

SOFTWARE DEVELOPMENT: FLASH 5

The list of functional requirements for ColorBrewer was long: it needed to be (1) Web-based, (2) download quickly, (3) run well on 'trailing-edge' computers, (4) work across different computer platforms (e.g. Mac, Windows, UNIX), (5) allow for a high degree of user-interactivity and (6) require little training time to use. The amount of control that users have in ColorBrewer is less than a full GIS package, but somewhat more than most multimedia Web content. The success of ColorBrewer depended upon providing visitors with enough flexibility and power to quickly find and 'test-drive' many colour schemes, without overwhelming them with a complicated interface.

Few Web technologies could fulfil all of our requirements. Although ColorBrewer could have been built in JAVA, this required a level of programming expertize beyond our abilities and would require that every component be built from the ground up. Instead, Macromedia Flash 5 was selected as it offered the right combination of graphic design capabilities, programming flexibility and rapid development.

Although originally conceived of as a tool for creating simple non-interactive animations for Web pages, Flash has matured into a powerful tool for creating dynamic and interactive Web material. Flash content has become the de-facto standard for dynamic multimedia online. Because they are vector-based, Flash Files are remarkably small and, hence, download quickly. The speed with which online applications download is directly proportional to their success with the public. Flash is attractive because it is compatible with all major Web-browsers (using the Shockwave plug-in) and runs well on lower-end computers. Moreover, current estimates are that over 97 per cent of Web users have the Flash plug-in already installed (Macromedia 2002). Those who need the Flash plug-in can find it for free on the Macromedia website (www.macromedia.com) and the introductory screen for ColorBrewer offers a link to the site. Flash's object-oriented programming language, called Actionscript, allowed us to add functionality such as zooming, panning, pull-down menus and 'pop-up' help windows. A portion of Actionscript code is shown in Figure 9.

Learning an interface consists of at least two critical steps: knowing *what* the buttons do, and knowing *the order* in which to use them. It never ceases to amaze us that software engineers often hide important interface controls or options deep within the interface. By labelling interface sections in ColorBrewer Step #1, Step #2, and so on, users should be able to quickly load colour schemes and understand the sequence of actions that are required to load those schemes. Simply put, the most important controls are highest in the visual hierarchy of the interface. To further assist users, some interface controls appear only once they are needed. For example, the Step 3 window that displays the mini colour legends is blank until the user has completed Steps 1 and 2 (see Figure 1). Thus, the user is 'directed' through the interface rather than left to wonder what button to click next.

ColorBrewer has a footprint of 800×600 pixels so that it will fit on smaller computer monitors typical of laptops and older machines. That is a very limited space in which to fit all of the interface elements, especially since we wanted to devote as much

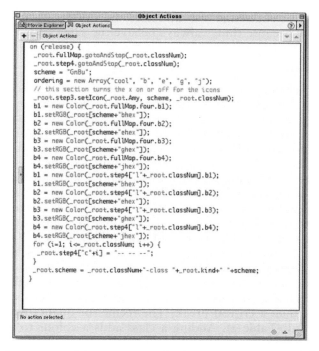

FIG. 9.   An example segment of Actionscript code. Actionscript operates behind the scenes and is used to load new colours into the diagnostic map. Flash changes the appearance of objects using hexadecimal codes, which are dynamically retrieved from a master list. This example sets states for a set of usability icons for a green-blue scheme

screen space as possible to the map. One solution to this problem was to build *zooming* and *panning* capabilities into the main map so that users can enlarge areas of interest or produce polygon sizes more similar to their intended application. Another solution was to use 'pop-up' windows for non-critical interface controls. For example, a number of 'learn more' buttons are embedded throughout the interface. These concise help files only appear on demand and provide explanations of how to use ColorBrewer, as well as some of the theory behind colour use and map representation. Importantly, these help buttons are context specific and the material they display is directly related to the portion of the interface from which they were initiated. This allows the user to retrieve the information they need with one button click, rather than force them to navigate through an online manual or interact with onscreen wizards. For the expert, or return user, these help features can be ignored and do not inhibit their work.

The Actionscript code that makes the system run has been optimized so that the controls will react almost instantly to user requests. All other things being equal, the faster a system reacts to user input, the more it will inspire confidence and encourage users to explore its functionality. Audio feedback is also built into the system to reinforce the message that clicking on buttons initiates actions. Consistent design elements, such as rollover states for buttons (that change colour or grow when they are touched by the cursor) further reinforce that the interface is 'hot' and at the control of the user. Consistent design also decreases the amount of time it takes the user to learn an interface because things always behave the same way.

MAKING THE SCHEMES

The sets of colour schemes in ColorBrewer were designed using both experience and trial and error. We needed a way to be efficient in scheme design and management of colour specifications in the Flash program. We accomplished this by selecting all of the colours in a scheme from a limited and shared set of colours. The 'recipe' for designing related sequential schemes with 3 to 9 classes from a common set of 13 colours is detailed in Table 1 below. For example, colour D is the fourth of the 13 colours designed, and it appears third in sequential schemes with 9 and 8 classes and second in the sequential schemes with 7 and 6 classes. The process of designing related schemes by pulling systematically from a limited set of colours for different numbers of classes was described by Judy Olson[16] in a talk at the 2001 annual conference of the Association of American Geographers in New York.

Notice that the recipe for the diverging and qualitative schemes is different (Tables 2 and 3). Diverging schemes from 3 to 11 classes are from a shared set of 15 colours. Notice that the middle colour H (usually a very light grey) alternates though the table, being used only for the odd numbered schemes that have middle critical class rather than a critical break. The recipe for qualitative schemes is much more straightforward (Table 2). Adding a class adds a colour to the bottom of the set, so schemes with up to 12 classes are pulled in order from 12 choices.

16. Olson, J. M. (2001). 'Strategies for Map Colour Selection in the Age of GIS', **Proceedings 97th Annual Meeting of the Association of American Geographers**, New York, NY. 27 Feb–3 Mar 2001, pp. 713.

Table 1. Colour sets for sequential schemes

Colour	Number of classes						
	9	8	7	6	5	4	3
A	A	A					
B			B	B	B	B	
C	C	C					C
D	D	D	D	D			
E					E	E	
F	F	F	F	F			F
G	G	G	G	G	G	G	
H	H	H	H				
I				I	I		I
J	J	J	J			J	
K	K			K	K		
L		L	L				
M	M						

Table 2. Colour sets for qualitative schemes

Colour	Number of classes									
	12	11	10	9	8	7	6	5	4	3
A	A	A	A	A	A	A	A	A	A	A
B	B	B	B	B	B	B	B	B	B	B
C	C	C	C	C	C	C	C	C	C	C
D	D	D	D	D	D	D	D	D	D	
E	E	E	E	E	E	E	E	E		
F	F	F	F	F	F	F	F			
G	G	G	G	G	G	G				
H	H	H	H	H	H					
I	I	I	I	I						
J	J	J	J							
K	K	K								
L	L									

CONCLUSION

ColorBrewer.org has been online since August 2001. Since that time, we have received numerous unsolicited emails from satisfied users around the globe. Judging from their reactions, ColorBrewer appears to be successful: they have told us that the system is easy to use, attractive and has been a tremendous help in their own design work. Not inconsequentially, ColorBrewer is available for free, 24 hours a day, to anyone with a Web connection.

ColorBrewer is an online tool that is designed to help mapmakers select effective colour schemes for thematic maps. ColorBrewer has been designed to fill existing gaps in the usability of GIS and graphics software by providing not only attractive colour schemes, but also recommendations on how to use those colour schemes most effectively. As such, it has found an audience with both map designers who rely on it to

Table 3. Colour sets for diverging schemes

Colour	\multicolumn{9}{c}{Number of classes}								
	11	10	9	8	7	6	5	4	3
A	A	A							
B	B	B	B	B	B	B			
C							C	C	
D	D	D	D	D					
E					E	E			E
F	F	F	F	F			F	F	
G	G	G	G	G	G	G			
H	H		H		H		H		H
I	I	I	I	I	I	I			
J	J	J	J	J			J	J	
K					K	K			K
L	L	L	L	L					
M							M	M	
N	N	N	N	N	N	N			
O	O	O	O						

speed their work and college-level educators who use it in their classroom to demonstrate important concepts in colour use and design. Some of the valuable features in ColorBrewer include: (1) multi-hued perceptually ordered colour schemes; (2) 'learn more' features which advise designers on some of the theory behind colour use and map making; (3) the ability to check how well different colours schemes will 'hold up' when additional map information is present, or the colours of the background and enumeration unit borders are changed; (4) the ability to check how well different colours schemes will perform when the mapped patterns are both ordered (e.g. banded) and complex (e.g. heterogeneous distributions); and (5) colour output specifications in five commonly used colour systems (CMYK, RGB, hexadecimal, Lab and ArcView 3.x HSV). Perhaps the most valuable feature in ColorBrewer is the use guidelines for every colour scheme. In total, 385 unique colour schemes have been evaluated across different computer platforms and monitors (i.e. Mac and PC, projected LCD, laptop LCD and CRT), for possible colour-blind confusions, as well as in printed formats (both as colour and black-and-white prints). Given the importance of designing maps for the medium in which they will be used, these guidelines will help cartographers avoid selecting colour schemes that look attractive in their *design medium* (usually a CRT monitor) but fail in their *display medium*.

ACKNOWLEDGEMENTS

This work was part of the Digital Government Quality Graphics project (www.geovista.psu.edu/grants/dg-qg/), which is funded by the National Science Foundation under Grant No. 9983451, 9983459, 9983461. We would like to thank members of the Penn State Department of Geography for their generous feedback and design advice. Special acknowledgement must be given to Penn State undergraduate student Amy Dean who tirelessly evaluated the usability of 385 colour schemes over a two-week period.

Reflections on 'ColorBrewer.org: An Online Tool for Selecting Colour Schemes for Maps'

ANTHONY C. ROBINSON
The Pennsylvania State University

In 2003, Mark Harrower and Cindy Brewer contributed what has now become the *de facto* colour scheme selection tool for contemporary cartographers and information designers of all sorts. Their ColorBrewer system leverages previous research on how people perceive colours (Brewer, 1994) in order to suggest maximally useful and usable colour schemes for thematic mapping. Their fusion of perceptually appropriate colour guidelines within a framework for easy practical application represented the first of many subsequent efforts by cartographers to develop focused tools that embody best practices in the science and art of map design.

The initial motivation behind the development of ColorBrewer was that federal thematic mapping in the USA was (and is) generally conducted by GIS analysts who not only do not have formal training in colour design for map-making, but are also using digital mapping tools that rarely have useful built-in colour schemes. Therefore, ColorBrewer filled a gap where advice for choosing thematic map colours was needed. Harrower and Brewer designed colour schemes for use in ColorBrewer by hand-selecting and evaluating colour variations across value and saturation for a given hue for sequential and diverging schemes, and by differences in hue and value for qualitative schemes (Brewer *et al.*, 2003). Their work was substantially influenced by the Munsell colour system (Munsell, 1905), which they referenced while creating each colour scheme.

ColorBrewer.org allows users to navigate a set of common cartographic design constraints in order to select, preview, and then apply a colour scheme for a choropleth map. Users can choose sequential, diverging or qualitative colour schemes and then select the number of categories they desire. These colour schemes are then previewed in real time on a realistic choropleth map. ColorBrewer also provides clear advice to mapmakers by suggesting which schemes would work well for LCD projectors, colour printing, colour-blind readers and for photocopy reproduction. When a user selects a large number of categories for a colour scheme, the tool reinforces best practices in cartographic design by telling the user that for most applications five to seven categories are sufficient.

Harrower and Brewer's ColorBrewer sparked a trend in academic cartography to develop tools that could translate best practices from research literature into their application in everyday map-making. Today, there is a wide range of similar *Brewer* tools to aid with other common cartographic design challenges. For example, you can now use TypeBrewer to choose fonts for map labeling (Sheesley, 2007), Map Symbol

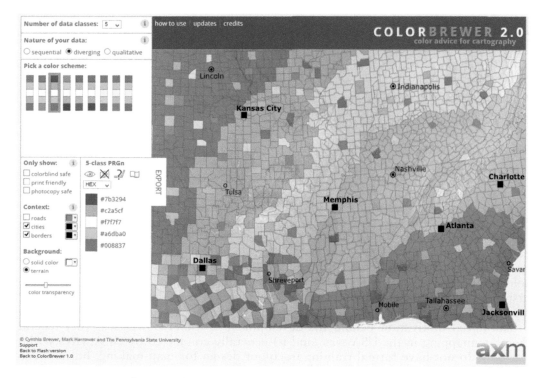

Fig. 1.   The revised ColorBrewer2 interface, implemented by Axis Maps

Brewer has been developed to design point symbols (Schnabel, 2005), the ScaleMaster was created to help users decide at which scales their data may be best utilized (Roth, 2011), MapShaper makes line generalization visually-accessible and interactive, and the Symbol Store provides a ColorBrewer-inspired interface for finding and previewing point symbols (Robinson *et al.*, 2013). A key common aspect of each of these *Brewer* tools is the ability for users to easily preview map design options before committing to a particular solution.

Since its debut in 2003, the original ColorBrewer.org tool has been revised into ColorBrewer2.org by the mapping firm Axis Maps. The new tool features additional map preview controls and colour scheme export options. The colour schemes themselves have been ported for use in R and D3 Javascript program development, in Esri style files for use in desktop GIS, and in an open source GeoTools Java library, among many other implementations.

ColorBrewer is particularly noteworthy in that it has had an impact on fields outside of the typical influence of most cartographic literature. The aforementioned adaptations of ColorBrewer schemes into the libraries of common programming languages is evidence of this, along with applications of ColorBrewer schemes in information visualization (Bendix *et al.*, 2005), public health science (Boulos, 2004) and the geosciences (Light and Bartlein, 2004).

The initial motivation for the development of ColorBrewer stemmed from the need to aid mapmakers who lacked any deep understanding of colour design and were using tools that provided little or no guidance when it came to suggesting default schemes. Today, the situation has changed, but not because current tools do a better job at providing useful default settings. Rather, the challenge today that was just beginning to emerge in the early 2000s is that the number and diversity of people who make maps has blossomed in an enormous way. Despite the fact that so many methods now exist to leverage ColorBrewer schemes, the rapid pace of mapping and information visualization tool development has also resulted in far more options than ever for novice designers to create thematic maps with poorly designed colour schemes.

ColorBrewer helps address these issues because it can be used to quickly and easily provide alternative colour solutions to these new and burgeoning audiences of map designers and tool developers. It has no doubt led to a great deal of better-designed maps, and it will surely continue to do so well into the foreseeable future. Ten years after its initial publication, it is clear that the true impact of this work has been felt not in its large number of academic citations, but in the improved map designs that have been made by the thousands.

REFERENCES

Bendix, F., Kosara, R. and Hauser, H. (2005). 'Parallel Sets: Visual Analysis of Categorical Data', in **IEEE Symposium on Information Visualization**, Minneapolis, MN, Oct 20–21, pp. 558–568.

Boulos, M. N. K. (2004). 'Web GIS in practice: an interactive geographical interface to English Primary Care Trust performance ratings for 2003 and 2004', **International Journal of Health Geographics**, 3, pp. 1–7.

Brewer, C. A. (1994). 'Color use guidelines for mapping and visualization', in **Visualization in Modern Cartography**, ed. by MacEachren, A. M. and Taylor, D. R. F., Pergamon, Oxford, pp. 123–148.

Brewer, C. A., Hatchard, G. W. and Harrower, M. A. (2003). 'ColorBrewer in print: a catalog of color schemes for maps', **Cartography and Geographic Information Science**, 30, pp. 5–32.

Light, A. and Bartlein, P. J. (2004). 'The end of the rainbow? Color schemes for improved data graphics', **EOS Transactions American Geophysical Union**, 85, pp. 385–391.

Munsell, A. H. (1905). **A Color Notation**, G.H. Ellis Company, Boston, MA.

Robinson, A. C., Pezanowski, S., Troedson, S., Bianchetti, R., Blanford, J., Stevens, J., Guidero, E., Roth, R. E. and MacEachren, A. M. (2013). 'Symbol store: sharing map symbols for emergency management', **Cartography and Geographic Information Science**, 40, pp. 415–426.

Roth, R. E., Brewer, C. and Stryker, M. (2011). 'A typology of operators for maintaining legible map designs at multiple scales', **Cartogrpahic Perspectives**, 68, pp. 29–64.

Schnabel, O. (2005). 'Map Symbol Brewer — A New Approach for A Cartographic Map Symbol Generator', in **International Cartographic Conference**, A Coruña, Spain, Jul 9–16, pp. 1–6.

Sheesley, B. (2007). TypeBrewer: design and evaluation of a help tool for selecting map typography. PhD thesis. University of Wisconsin-Madison. Madison, WI, USA.

Mapping the Results of Geographically Weighted Regression

JEREMY MENNIS

Originally published in *The Cartographic Journal* (2006) 43, pp. 171–179.

Geographically weighted regression (GWR) is a local spatial statistical technique for exploring spatial nonstationarity. Previous approaches to mapping the results of GWR have primarily employed an equal step classification and sequential no-hue colour scheme for choropleth mapping of parameter estimates. This cartographic approach may hinder the exploration of spatial nonstationarity by inadequately illustrating the spatial distribution of the sign, magnitude, and significance of the influence of each explanatory variable on the dependent variable. Approaches for improving mapping of the results of GWR are illustrated using a case study analysis of population density–median home value relationships in Philadelphia, Pennsylvania, USA. These approaches employ data classification schemes informed by the (nonspatial) data distribution, diverging colour schemes, and bivariate choropleth mapping.

INTRODUCTION

Local forms of spatial analysis have recently gained in prominence. For example, local adaptations have been developed for conventional summary statistics[1] as well as for the analysis of spatial dependency in both quantitative[2,3] and categorical data.[4] Because local spatial statistics often generate georeferenced data, maps and other graphics are typically used to present, and aid in the interpretation of, local spatial statistical results. And because these local statistics are generally exploratory, as opposed to confirmatory, in nature, they have much in common theoretically with recent research in cartography focusing on the use of maps and statistical graphics for data

1. Brunsdon, C., Fotheringham, A. S. and Charlton, M. E. (2002), 'Geographically weighted summary statistics: a framework for localized exploratory data analysis', **Computers, Environment and Urban Systems**, 501–524.

2. Anselin, L. (1995). 'Local indicators of spatial association — LISA', **Geographical Analysis**, 27, 93–115.

3. Ord, J. K. and Getis, A. (1995). 'Local spatial autocorrelation statistics: distributional issues and an application'. **Geographical Analysis**, 27, 286–306.

4. Boots, B. (2003). 'Developing local measures of spatial association for categorical data', **Journal of Geographical Systems**, 5, 139–160.

exploration (e.g. MacEachren and Ganter[5]; Andrienko *et al.*[6]; Carr *et al.*[7]). Few cartographers, however, have explicitly addressed the adaptation of conventional mapping techniques for local spatial statistics.

Geographically weighted regression (GWR) is a local spatial statistical technique used to analyze spatial nonstationarity, defined as when the measurement of relationships among variables differs from location to location.[8] Unlike conventional regression, which produces a single regression equation to summarize global relationships among the explanatory and dependent variables, GWR generates spatial data that express the spatial variation in the relationships among variables. Maps generated from these data play a key role in exploring and interpreting spatial nonstationarity.

A number of recent publications have demonstrated the analytical utility of GWR for investigating a variety of topical areas, including climatology,[9] urban poverty,[10] environmental justice,[11] and the ecological inference problem.[12] However, a standard approach for mapping the results of GWR has not yet been developed. This may be due to the relatively recent development of the technique itself, but is also likely a result of the complications in displaying the results of GWR. Note that each GWR analysis can produce a voluminous amount of spatial data, including multiple geo-referenced variables. Some of these variables can be considered ratio data while other variables can be interpreted as nominal. Numeric variables may be highly skewed and range over positive and negative values.

The purpose of this research is to review previous approaches to mapping the results of GWR and suggest methods to improve upon them. I focus on GWR as applied to the analysis of areal data, as opposed to data taken as samples of a continuous surface, as the vast majority of GWR research has been applied to socioeconomic data aggregated to census or other spatial units. As a case study, a number of mapping approaches are used to interpret the results of a GWR analysis of median home value in Philadelphia, Pennsylvania, USA using 2000 US Bureau of the Census tract level data.

5. MacEachren, A. M. and Ganter, J. H. (1990). 'A pattern identification approach to cartographic visualization', **Cartographica**, 27, 64–81.

6. Andrienko, N., Andrienko, G., Savinov, A., Voss, H., and Wettschereck, D. (2001). 'Exploratory analysis of spatial data using interactive maps and data mining', **Cartography and Geographic Information Science**, 28, 151–165.

7. Carr, D. B., White, D., and MacEachren, A. M. (2005). 'Conditioned choropleth maps and hypothesis generation', **Annals of the Association of American Geographers**, 95, 32–53.

8. Fotheringham, A. S., Brunsdon, C., and Charlton, M. E. (2002). **Geographically Weighted Regression: The Analysis of Spatially Varying Relationships**, Wiley, Chichester.

9. Brunsdon, C., McClatchey, J. and Unwin, D. (2001). 'Spatial variations in the average rainfall–altitude relationships in Great Britain: an approach using geographically weighted regression', **International Journal of Climatology**, 21, 455–466.

10. Longley, P. A. and Tobon, C. (2004). 'Spatial dependence and heterogeneity in patterns of hardship: an intra-urban analysis', **Annals of the Association of American Geographers**, 94, 503–519.

11. Mennis, J. and Jordan, L. (2005). 'The distribution of environmental equity: exploring spatial nonstationarity in multivariate models of air toxic releases', **Annals of the Association of American Geographers**, 95, 249–268.

12. Calvo, C. and Escolar, M. (2003). 'The local voter: a geographically weighted approach to ecological inference', **American Journal of Political Science**, 47, 189–204.

GEOGRAPHICALLY WEIGHTED REGRESSION

Because readers may not be familiar with the details of GWR, a brief explanation of it is offered here. The conventional regression equation can be expressed as

$$\hat{y}_i = \beta_0 + \sum_k \beta_k x_{ik} + \varepsilon_i \qquad (1)$$

where $\hat{y}_i$ is the estimated value of the dependent variable for observation i, β_0 is the intercept, β_k is the parameter estimate for variable k, x_{ik} is the value of the kth variable for i, and ε_i is the error term. Instead of calibrating a single regression equation, GWR generates a separate regression equation for each observation. Each equation is calibrated using a different weighting of the observations contained in the data set. Each GWR equation may be expressed as

$$\hat{y}_i = \beta_0(u_i, v_i) + \sum_k \beta_k(u_i, v_i) x_{ik} + \varepsilon_i \qquad (2)$$

where (u_i, v_i) captures the coordinate location of i.[13] The assumption is that observations nearby one another have a greater influence on one another's parameter estimates than observations farther apart. The weight assigned to each observation is based on a distance decay function centred on observation i. In the case of areal data, the distance between observations is calculated as the distance between polygon centroids.

The distance decay function, which may take a variety of forms, is modified by a bandwidth setting at which distance the weight rapidly approaches zero. The bandwidth may be manually chosen by the analyst or optimized using an algorithm that seeks to minimize a cross-validation score, given as

$$CV = \sum_{i=1}^{n} (y_i - \hat{y}_{i \neq i})^2 \qquad (3)$$

where n is the number of observations, and observation i is omitted from the calculation so that in areas of sparse observations the model is not calibrated solely on i. Alternatively, the bandwidth may be chosen by minimizing the Akaike Information Criteria (AIC) score, give as

$$AIC_c = 2n \log_e(\hat{\sigma}) + n \log_e(2\pi) + n \left\{ \frac{n + tr(S)}{n - 2 - tr(S)} \right\} \qquad (4)$$

where $tr(S)$ is the trace of the hat matrix. The AIC method has the advantage of taking into account the fact that the degrees of freedom may vary among models centred on different observations. In addition, the user may choose a fixed bandwidth that is used for every observation or a variable bandwidth that expands in areas of sparse observations and shrinks in areas of dense observations.[14]

13. Fotheringham, A. S., Brunsdon, C. and Charlton, M. E. (1998). 'Geographically weighted regression: a natural evolution of the expansion method for spatial data analysis'. **Environment and Planning A**, 30, 1905–1927.

14. Charlton, M., Fotheringham, S. and Brunsdon, C. (no date). Geographically Weighted Regression Version 2.x, User's Manual and Installation Guide.

Because the regression equation is calibrated independently for each observation, a separate parameter estimate, t-value, and goodness-of-fit is calculated for each observation. These values can thus be mapped, allowing the analyst to visually interpret the spatial distribution of the nature and strength of the relationships among explanatory and dependent variables. For more information on the theory and practical application of GWR the reader is referred to Fotheringham *et al.*[15]

CHALLENGES TO MAPPING THE RESULTS OF GWR

A survey of research incorporating GWR reveals that maps play a central role in interpreting GWR results. However, there are a number of issues that have led these maps to obscure the GWR results as much as illuminate them. One issue is that the spatial distribution of the parameter estimates must be presented in concert with the distribution of significance, as indicated by a t-value, in order to yield meaningful interpretation of the results. Some researchers have chosen to map only the parameter estimates and not associated t-values,[16,17,18] which can be very misleading as it may visually emphasize the areas of highest (or lowest, if the relationship is primarily negative) parameter estimation, regardless of the significance of the estimate. Thus, one may get the impression that the areas with the highest parameter estimates exhibit the strongest relationship between the explanatory and dependent variables, when those estimates may not, in fact, be significant. Clearly, maps of the spatial distribution of the parameter estimates must be accompanied by associated t-value data if spatial nonstationarity is to be interpreted effectively by the map reader.

A second issue concerns data classification. The equal step approach, where the data range is divided into classes of equal extent (Dent, 1999), appears to be the most common data classification technique for mapping the distribution of parameter estimates and t-values generated from GWR (e.g. Longley and Tobon, 2004). It should be noted, however, except in cases where exogenous classification criteria are used, the choice of data classification scheme for quantitative data is typically informed by the non-spatial data distribution.[19,20] The equal step classification is most appropriate for uniformly distributed data, which in the case of GWR-generated parameter estimates would occur when the frequencies of the estimates were approximately the same over the range of the estimates. While possible, this is certainly unlikely. Other classification schemes are likely to be more appropriate, such as the use of standard deviation classification for normally distributed data, or the use of optimal methods for maximizing within-class homogeneity (e.g. Coulson[21]; Cromley[22]).

15. Fotheringham *et al.*, 2002.

16. Fotheringham *et al.*, 1998.

17. Huang, Y. and Leung, Y. (2002). 'Analyzing regional industrialization in Jiangsu province using geographically weighted regression'. **Journal of Geographical Systems**, 4, 233–249.

18. Lee, S.-I. (2004). 'Spatial data analysis for the US regional income convergence, 1969–1999: a critical appraisal of b-convergence', **Journal of the Korean Geographical Society**, 39.

19. Evans, I. A. (1977). 'Selection of class intervals', Transactions of the Institute of British Geographers, New Series, 2, 98–124.

20. Dent, B. D. (1999). **Cartography: Thematic Map Design**, (5th Ed.), WCB/McGraw Hill, Boston.

21. Coulson, M. R. C. (1987). 'In the matter of class intervals for choropleth maps: with particular reference to the work of George Jenks', **Cartographica**, 24, 16–39.

22. Cromley, R. G. (1996). 'A comparison of optimal classification strategies for choropleth displays of spatially aggregated data', **International Journal of Geographical Information Science**, 10, 405–424.

In addition, the data classification for t-values should account for certain exogenous criteria that are of importance to the variable being mapped,[23] namely the threshold values that distinguish parameter estimates that are significant from those that are not. When a class interval extends across a significance threshold to encompass both significant and not significant t-values within one class, as it may be using an equal step classification scheme, it becomes impossible to visually distinguish significant parameter estimates from those that are not significant on the map.

A third issue is the choice of colour scheme. Many GWR researchers have employed a sequential no-hue colour scheme, which assigns a series of class intervals increasing shades of grey[24] for choropleth mapping of both parameter estimates and t-values.[25,26,27] Such a colour scheme gives the impression of a gradation of increasing influence (i.e. from a lighter to darker shade of grey) of the explanatory variable on the dependent variable.

In cases where the parameter estimates are all of the same sign, the sequential approach may be appropriate. However, this colour scheme is problematic in cases where the parameter estimate is positive in some locations and negative in others (which is not an unusual occurrence, e.g. Huang and Leung[28]; Lee[29]; Mennis and Jordan[30]), as it ignores the fact that the sign of the parameter estimate indicates an importance difference in the nature of the relationship of the explanatory with the dependent variable. In this case, a diverging colour scheme,[31,32] which indicates the magnitude of departure from a midpoint value (i.e. zero in the case of distinguishing positive from negative relationships), is most appropriate.

A fourth issue is the sheer number of individual maps required to report both the parameter estimates and tvalues for each explanatory variable. This is problematic in terms of cost of map production (e.g. physical space in a journal publication) and the cognitive effort in map comprehension required from the map reader.

Choropleth mapping has been extended to two variables simultaneously, as in a bivariate choropleth map.[33] Combining parameter estimates and t-values in a single choropleth map would reduce the volume of maps necessary for exploring the results of GWR.

CASE STUDY: GWR OF HOME VALUE IN PHILADELPHIA, PA

The case study concerns the GWR of median owner-occupied home value (US dollars) in Philadelphia, Pennsylvania, USA using population density (people km^{-2}) as the

23. Evans, 1977.

24. Brewer, C. (1994). 'Color use guidelines for mapping and visualization', in **Visualization in Modern Cartography**, MacEachren, A. and Taylor, D.R.F. (eds.), p. 123–147, Elsevier, New York.

25. Fotheringham *et al.*, 1998.

26. Longley and Tobon, 2004.

27. Lee, 2004.

28. Huang and Leung, 2002.

29. Lee, 2004.

30. Mennis and Jordan, 2005.

31. Brewer, 1994.

32. Brewer, C. A. (1996). 'Guidelines for selecting colors for diverging schemes on maps', **The Cartographic Journal**, 33, 79–86.

33. Olson, J. (1975). 'Spectrally encoded two-variable maps', **Annals of the Association of American Geographers**, 71, 259–276.

explanatory variable. These 2000 data were acquired from the US Bureau of the Census at the tract level. Note that the purpose of the case study is not to demonstrate anything novel about home values in Philadelphia per se, but rather to show and compare different strategies for mapping the results of GWR. The focus is on maps of parameter estimates and t-values as these are the most commonly reported maps in research using GWR. The use of only one explanatory variable in the case study keeps the volume of GWR results to a manageable level while generating interesting patterns of spatial nonstationarity that can be used to illustrate the benefits and pitfalls of various mapping strategies. Of the 381 tracts in Philadelphia, 24 were removed from the analysis because they represented very sparsely populated or unpopulated areas (i.e. parks, airports, and industrial land uses), leaving 357 tracts for use in the analysis. A map of Philadelphia neighbourhoods relevant to the case study is presented in Figure 1. Descriptive statistics and choropleth maps of the variables used in the analysis are presented in Table 1 and Figure 2, respectively.

The results of a conventional linear regression of home value are reported in Table 2. The model indicates that population density is negatively and significantly related to home value; as home values increase, population density decreases. Note, however, that the model is poorly specified, explaining only approximately 6% of the variation in home value. Reasons for this poor specification will be made clear in the GWR.

The data were entered into the GWR software using a variable bandwidth setting that minimizes the AIC. The variable bandwidth approach was chosen to account for the spatial variation in the size of the tracts, and hence the density of tract centroids. As noted above, the most common approach to presenting the results of GWR is to generate choropleth maps of the parameter estimates using a sequential no-hue colour scheme and an equal-step classification. Figure 3a presents such a map of the population density parameter estimate. One can immediately see that this map is problematic, as the imposition of this colour scheme and classification ignore relevant variations in the data that should be brought to the attention of the viewer. First, the sequential colour scheme suggests that the influence of population density on home value increases monotonically. In fact, in some tracts this relationship is negative and in others it is positive. Perhaps even more troubling is that the majority of the mapped area is occupied by a single class that includes both positive and negative parameter estimates (i.e. the class interval −7 to 12). Thus, it is impossible to tell within which areas the population density–home value relationship is positive versus negative. Finally, because no information on the distribution of t-values is provided, one cannot detect the areas in which the relationship between explanatory and dependent variables is significant. This last problem can be amended simply by creating a map of t-values (Figure 3a), presented here also using the conventional sequential nohue colour scheme and equal step classification, though similar problems regarding classification and choice of colour scheme apply.

Figure 4a presents a map that addresses the classification and colour scheme problems present in the choropleth map of parameter estimates presented in Figure 3a. In Figure 4a, the classification is based generally on a standard deviation classification scheme, as the data approach a normal distribution. In addition, manual adjustments to the statistically-derived data classification scheme are made to facilitate map

FIG. 1.   Important neighbourhoods of Philadelphia, Pennsylvania in the context of the case study, overlain with tract boundaries

Table 1.   Descriptive statistics

Variable	Minimum	Maximum	Mean	Standard deviation
Home value (US dollars)	9 999	843 800	75 860	70 362
Population density (people km^{-2})	120	21 168	6 618	3 853

interpretation.[34] The class breaks were shifted to distinguish positive from negative parameter estimates, and, because the range of negative parameter estimates is greater than the range of positive parameter estimates, the interval boundaries were set to allow the direct comparison of positive and negative parameter estimates of equivalent magnitude. Thus, of five classes, only one contains all the tracts with positive parameter estimates. A diverging colour scheme was also employed to differentiate negative from positive parameter estimates by hue, while expressing increasing magnitudes of the estimates using a combination of saturation and value. Unlike Figure 3a, Figure 4a clearly shows that the areas of positive relationship between population density and home value are largely limited to the greater Center City and University City neighbourhoods, as well as nearby Frankford. A negative population density–home value relationship of equal magnitude is evident in the remainder of

34. Monmonier, M.S. (1982). 'Flat laxity, optimization, and rounding in the selection of class intervals', **Cartographica**, 19, 16–26.

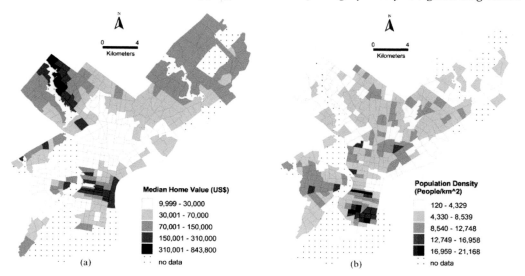

FIG. 2.   Choropleth maps of *a* median home value and *b* population density by census tract in Philadelphia, PA

Table 2.   Conventional regression of home value

Independent variable	Coefficient	t-value
Constant	−106 524.30***	−14.87
Population density	−4.63***	−4.96

*** Significance <0.005, N = 357, Adjusted R^2 = 0.062.

the city, with the exception of the Roxborough and Chestnut Hill neighbourhoods, within which stronger negative relationships occur.

Figure 4b presents a map that addresses the classification and colour scheme problems present in Figure 3b. Figure 4b has a classification scheme based on commonly used significance thresholds: 90, 95, 99, and 99.5%. A sequential colour scheme is used to represent different levels of significance. Unlike in Figure 3b, Figure 4b clearly indicates that in the majority of Philadelphia the relationship between population density and home value is, in fact, not significant at the 90% confidence level. It is significant primarily in University City, western Center City, Girard Estates, and a number of neighbourhoods in the northwestern part of the city. Clearly, this significance information is key to interpreting Figure 4a, as Figure 4a appears to suggest an equivalency between Center City and Frankford in the relationship of population density with home value. Figure 4b, however, clearly shows that in Frankford the relationship between the two variables is not significant at the 90% confidence level and, within those areas where the relationship between the variables is significant, the magnitude of the significance varies. Some parts of those areas show a significant relationship at the 99.5% confidence level (e.g. Chestnut Hill and Roxborough), while others only meet the 90% confidence level threshold (e.g. East Falls and West Oak Lane).

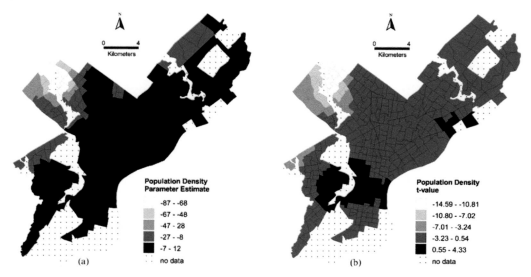

FIG. 3.   Choropleth maps of *a* parameter estimates and *b* t-values by census tract for the GWR of median home value using an equal step data classification and a sequential no-hue colour scheme for each map

FIG. 4.   Choropleth maps of *a* parameter estimates and *b* t-values by census tract for the GWR of median home value. In the parameter estimate map, a modified standard deviation data classification and a diverging colour scheme is used whereas in the t-value map, an exogenous data classification based on commonly accepted significance thresholds and a sequential no-hue colour scheme is used

The maps presented in Figure 4 are a marked improvement over those presented in Figure 3, as they allow for a much more accurate assessment of which areas have positive and negative relationships of the explanatory variable with the dependent variable, the magnitude of those relationships, and the significance of those relationships. However, given a regression with many explanatory variables, as opposed to just the one used in this case study, many maps are required to communicate this information, as each explanatory variable demands two separate maps — one for the parameter estimate and one for the t-value. Figure 5 offers a potential solution to this problem by encoding certain key characteristics of Figures 4a and 4b in a single area-class map. Here, tracts are classified according to their relationship between the explanatory and dependent variable, characterized as positively significant, negatively significant, and not significant (at the 90% confidence level). These classes are treated as nominal data and assigned varying lightness levels of grey in the map in a qualitative colour scheme that is intended to differentiate among classes without implying rank or quantity (Brewer, 1994). Note that the linework of the tract boundaries has been removed to reduce the visual complexity of the map. The advantage of this mapping approach is that one can easily see qualitative differences among areas in the sign of the relationship between the explanatory and dependent variable, as well as distinguish between areas exhibiting a significant versus not significant relationship. Another advantage is that a grey-scale, as opposed to colour, map may be used. Of course, the disadvantage of this mapping approach is that potentially interesting patterns may not be observed regarding the magnitude of the relationship between the explanatory and dependent variable as contained in the actual parameter estimate values, as well as in the magnitude of the significance.

Bringing colour back into the map allows for a compromise between Figures 4a and 5 as contained in a single map, presented in Figure 6a. Here, a map showing the parameter estimates in a manner similar to that of 3a is used, except that a significance threshold (at 90% confidence level) is used to mask out all those areas in which the relationship between the explanatory and dependent variables is not significant. Here, it is implied that distinguishing between positive and negative parameter estimates (and associated t-values) in these areas is unnecessary. These areas are given a neutral grey tone and their linework for the tract boundaries is removed, the assumption being that these areas are of less interest to an analyst than those areas that are significant.

Figure 6a can also be modified by using a bivariate colour scheme to simultaneously depict both the magnitude of the parameter estimate and the magnitude of the significance. In Figure 6b, a 4×4 class colour matrix is used to depict various combinations of parameter estimate and significance. A diverging colour scheme using two different hues is used to map the parameter estimate values, as in Figure 6a, because they range from positive to negative values. A sequential scheme using saturation is used to map significance, where increased saturation indicates higher significance, because the sign of the relationship is already captured by the hue in the vertical axis of the matrix. Thus, the map may be considered to use a diverging-sequential, bivariate colour scheme.

Because colours are only assigned to tracts with a significant relationship between the explanatory and dependent variables (at greater than or equal to 90% confidence), the matrix's class intervals are not continuous along the horizontal axis. All tracts that

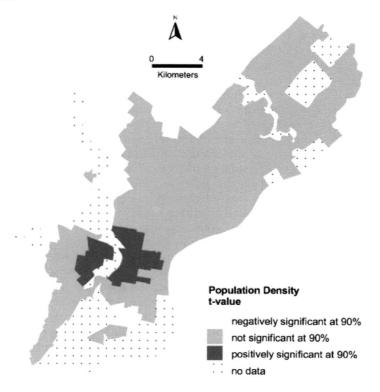

FIG. 5.   An area-class map of positively and negatively significant and not significant t-values,
for the GWR of median home value

do not exhibit a significant relationship between population density and home value (i.e. fall within the vertical class partition in the centre of the matrix) are assigned a neutral grey colour. Note also that the matrix is sparsely populated (i.e. there are a number of 'empty' cells) because the t-value and parameter estimate always share the same sign.

DISCUSSION AND CONCLUSION

Although the purpose of the case study concerns cartographic methodology and not the substantive topic of home values in Philadelphia, it is worth taking a moment to discuss the substantive results as a means to evaluate the various mapping approaches. First, the reason that the conventional regression was not specified properly is explained, at least in part, by the spatial nonstationarity indicated by the GWR. Clearly, a linear regression model that is global in nature will not be able to accurately characterize the relationship between explanatory and dependent variables when the relationship is positive in some portions of the study region and negative in others, as Figure 4a indicates. The negative relationship between population density and home value is perhaps one that could be expected; expensive homes are likely to occur in

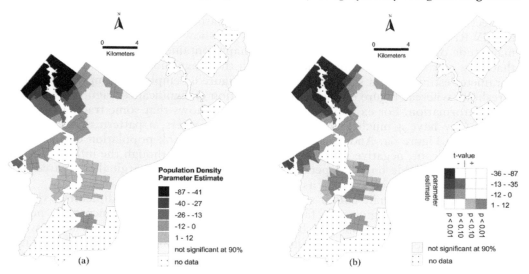

FIG. 6. Choropleth maps simultaneously displaying both the magnitude and significance of the parameter estimate by census tract: *a* a mask is applied to those tracts with a t-value with a significance less than 90%; *b* both the parameter estimate and associated significance are incorporated in a bivariate data classification and colour scheme

sparsely populated areas where single-family homes sit on large lots. This is indeed the case in certain Philadelphia neighbourhoods at the urban periphery, such as Roxborough, Chestnut Hill, and Overbrook, as Figures 4, 5, and 6 show.

The positive relationship between population density and home value exhibited in University City and western Center City is probably related to their historic roots as centres of wealth, high-end commercial activity, and higher education within the city core. Both neighbourhoods have maintained densely populated residential areas even as many nearby working-class neighbourhoods in North, South, and West Philadelphia have lost population in recent years. Population decline is associated with housing abandonment and marginal home appreciation (or even decline), thus creating the local positive relationship between population density and home value for University City and western Center City that can now be observed in Figures 4, 5, and 6.

This research demonstrates that the conventional approach of using an equal step classification and sequential no-hue colour scheme for choropleth mapping of GWR-generated parameter estimates is clearly inadequate. As Figure 3a shows, such a map is not only uninformative but can be downright misleading, even when paired with another map of t-values as an indicator of significance. Adjustments to the data classification and colour scheme to improve the cartographic representation of the sign, magnitude, and significance of parameter estimates, as in Figure 4, offer an improvement in interpreting the GWR results, but two maps are required for the representation of each explanatory variable.

The advantage of Figure 5 is that, because it is an area-class map with only three classes, it appears relatively uncluttered and is therefore easy to visually interpret. Yet it effectively communicates the basic pattern of spatial nonstationarity as captured by the GWR. On the downside, however, it does not show the spatial distribution of the magnitude of the parameter estimates. The maps contained in Figure 6 are unique in that they convey spatial information on both the magnitude and significance of the parameter estimates in a single map. Because Figure 6a employs a simple significance threshold, whereas Figure 6b maps the distribution of significance, Figure 6b contains more information. For example, Figure 6b clearly shows that some tracts in western Center City have a much higher significance than others, a pattern that cannot be observed in Figure 6a. And one can see that in Overbrook population density has a highly significant, negative relationship with home value, though the influence of the explanatory variable on the dependent variable is relatively marginal compared with its influence in other areas, such as Chestnut Hill.

However, the bivariate colour scheme used in Figure 6b can be difficult to visually interpret, particularly given the fact that additional colour assignments are needed for representing observations which are classified as not significant or which have no data. And while knowing the spatial distribution of significance values is certainly important, significance is typically treated as a threshold. For these reasons, I advocate the mapping approach taken in Figure 6a as a good rule-of-thumb for mapping the results of GWR. Or, an analyst may choose to use a map like that presented in Figure 5, if this reduced level of information communication is deemed sufficient.

It is worth noting that while the case study focuses on mapping the parameter estimate and t-value for GWR using a single explanatory variable, most GWR applications will have multiple explanatory variables. In such a situation, GWR may be used to interpret maps of parameter estimates and/or t-values to determine within which region(s) specific explanatory variables are particularly influential. Such an analysis demands a comparison of choropleth maps in a series, for which design criteria may differ from that used for a single map.[35] Mennis and Jordan[36] facilitate such a comparison by using area-class maps like that presented in Figure 5, thus supporting map comparison by standardizing maps according to a significance threshold applied uniformly to all explanatory variables. However, if choropleth mapping of parameter estimates is used to indicate the magnitude of influence of each explanatory variable, each parameter estimate must be standardized before being mapped (i.e. the standardized β). Likewise, standardization of the data classification and colour scheme across all maps in the series will facilitate map comparison, even if some maps contain data for only a subset of the classification range,[37] It is also worth noting that not all parameter estimates and attached significance values necessarily need to be mapped in order to generate an effective visualization of the overall quality and most relevant characteristics of a GWR model.

A software package devoted to automated mapping of GWR results would be a useful tool for assisting researchers in developing informative and useful maps for exploring spatial nonstationarity. Such a software package could ingest the output from GWR analysis and offer automated intelligent rules for cartographic display,

35. Brewer, C. A. and Pickle, L. (2002). 'Evaluation of methods for classifying epidemiological data on choropleth maps in a series', **Annals of the Association of American Geographers**, 92, 662–681.

36. Mennis and Jordan, 2005.

37. Brewer and Pickle, 2002.

based on the data classification, colour scheme, and bivariate mapping approaches described above. In addition, a software package whose purpose is to support the exploration of the results of GWR ought to include characteristics that have been developed for exploratory data analysis in other cartographic contexts, such as the use of small multiples for the visualization of many variables,[38] dynamically linked maps and other graphical displays,[39] and modes of interactivity.[40] For example, consider the significance threshold of 90% confidence used in Figure 6a to mask out tracts in which the relationship between population density and home value is considered not significant. A slider bar or other interactive device could facilitate the exploration of the effect of changing the threshold significance value on the interpretation of spatial nonstationarity. Interactive devices for dynamically altering class breaks for parameter estimates and/or significance values would be useful in exploring the maps presented Figures 4 and 6, as well as in transforming the t-values to nominal data in Figure 5.

It would be useful to provide choropleth maps of the explanatory and dependent variables, linked to the choropleth maps of the analogous parameter estimates and t-values so that panning, zooming, selection and other interactions in one map would be effective in all maps. In addition, dynamically linking statistical graphics, such as scatter plots and parallel coordinate plots (e.g. Gahegan *et al*.[41]), to the maps of parameter estimates and significance would facilitate the exploration of the multivariate 'signatures' associated with regions of homogeneity regarding the relationship between explanatory and dependent variables.

ACKNOWLEDGEMENTS

The choice of colour schemes used in this research were informed by ColorBrewer, an online mapping tool for choosing colour schemes for choropleth maps[42] and *Mapping Census 2000: The Geography of US Diversity*.[43]

38. Pickle, L. W., Mingle, M., Jones, G. K., and White, A. A. (1996). **Atlas of United States Mortality**, US National Center for Health Statistics, Hyattsville, Maryland, USA.
39. MacEachren, A. M., Wachowicz, M., Edsall, R., Haug, D., and Masters, R. (1999). 'Constructing knowledge from multivariate spatiotemporal data: integrating geographical visualization with knowledge discovery in database methods', **International Journal of Geographical Information Science**, 13, 311–334.
40. Crampton, J.W. (2002). 'Interactivity types in geographic visualization', **Cartography and Geographic Information Science**, 29, 85–98.
41. Gahegan, M., Takatsuka, M., Wheeler, M. and Hardisty, F. (2002). 'Introducing GeoVISTA Studio: an integrated suite of visualization and computational methods for exploration and knowledge construction in geography', **Computers, Environment and Urban Systems**, 26, 267–292.
42. Harrower, M. A. and Brewer, C. A. (2003). 'ColorBrewer.org: an online tool for selecting colour schemes for maps', **The Cartographic Journal**, 40, 27–37.
43. Brewer, C. A. and Suchan, T. A. (2001). **Mapping Census 2000: The Geography of US Diversity**. US Census Bureau Special Report, Series CENSR/01-1. US Government Printing Office. Washington DC.

Reflections on 'Mapping the Results of Geographically Weighted Regression'

LINDA BEALE

Imperial College London / Esri Inc

The importance of geography and spatial analysis is increasingly being embraced by a wide variety of disciplines and so the need for effective ways to visualize the results has never been more paramount. Data analysis can employ straightforward or complex techniques and increasingly, with advances in technology and data availability more advanced approaches are being sought. The key to any successful map lies in the ability for it to efficiently capture the message and convey it to the reader but when dealing with the results of complex spatial analysis this can be challenging.

Using a GIS, analysis includes a visual component but knowing about the science of your analysis and being able to operate the machinery to drive it are not always the same. Viewing data using location can certainly reveal new information, from highlighting possible errors in data collection by seeing where data are unexpectedly missing to visualizing continuous surfaces interpolated from sampled points. The analyst should take care, however, not be misled by the defaults in software which can inadvertently influence interpretation. For instance, displaying data using default classification schemes is a common error. Historically, there has been a divide between scientists whose realm is in data analysis, spatial or statistical, and those who make maps. For many, once the result of an analysis is calculated their job is done and if GIS was used, they additionally have a 'map'. Cartographers, by contrast, have often been guilty of overplaying the mapping dimension to the detriment of the quality of the analysis. The optimum, of course, is for analysis and cartography to sit hand-in-hand as two sides of the same coin and for those engaging in analysis to be cognisant of cartographic ways of visualizing their results, or, at the very least, building in time or availing themselves of a cartographer to help ensure that their work is well constructed in visual terms.

In this sense, Mennis, in his paper on mapping the results of geographically weighted regression (GWR), recognized the importance of the cartographic approach when mapping the results. The paper makes a strong statement of the value of joined-up thinking and combining strong analysis with strong cartography so that both aspects are well covered. This is no easy task for a subject matter that is not easily dealt with in cartographic terms because it has multi-dimensional output. In short, he asserts that multiple pieces of information should be read together to get the complete picture. Cartographically, this has more commonly been resolved by either using multiple maps each of which depicts a separate component of the analysis, or, unfortunately, by presenting only part of the story. Neither of these approaches is necessarily optimal since the use of multiple maps give rise to problems of comparison between them and the inevitable increase of misinterpretation that this can cause.

The idea that a measure of uncertainty should be mapped with results can be expanded to many analyses, particularly, as more advanced statistics are used. Modelling techniques can introduce error and this can vary spatially. This associated error is a key component of the results and interpretation. Although challenging, the map reader may be able to visually link information between two maps of known regions (e.g. counties). The challenge, however, becomes more complex if the areas are not familiar or are complex. Reading two interrelated pieces of information becomes very difficult. On the other hand, ignoring components of the analysis inevitably runs the risk of leaving out crucial pieces of information. The more challenging task — that of combining information into a meaningful cartographic product, has often been ignored. With reference to geographically weighted regression, Mennis states that parameter estimates should be represented in tangent with the distribution of significance to ensure that results are correctly interpreted. He suggests that visually, the highest parameter estimates may be misunderstood in the areas with the strongest relationship between explanatory and dependent variables, whereas these areas may not be significant. The simple solution is to combine the measure of uncertainty with the results to enable the map reader to understand not only the pattern of the data being mapped, but also where this might be more or less certain. Measuring uncertainty is not new but what Mennis successfully emphasized is its importance and explored how it can be achieved cartographically, allowing others to make better decisions in their own mapping.

A second issue that Mennis tackles is the choice of classification scheme. Equal classes are most suitable for uniformly distributed data; however, the frequencies of estimates are unlikely to be similar over the range of estimates. Classification is such a basic tenet of data analysis and mapping, yet so few give it the consideration it deserves. There are sufficient guidelines to ensure that reasonable choices can be made but these are dependent upon data and analysis and cannot be successfully automated for every case. It is, therefore, incumbent on those undertaking spatial analyses to pay attention to their data structure to ensure that they choose a classification scheme that fits.

The final comment Mennis makes is about colour and reveals what might be considered obvious to cartographers but which all too often is not for analysts. In mapping his results, Mennis observed that sequential colour schemes do not allow differences in the direction of the parameter estimate (from positive to negative), whereas a diverging colour scheme does. It is common practice for cartographers to illustrate a diverging dataset with a scheme that reflects that characteristic, e.g. by varying colour around a key value. Whether many analysts would know to do this is questionable but instead of ignoring the issue, Mennis discussed it and made recommendations. The key here is that he understood that a large proportion of those interested in mapping GWR are unlikely to want to read a text on cartography since they might see it as an unnecessary sledgehammer to crack a walnut. Knowing what nuggets of cartographic and cognitive theory can be usefully deployed to map the results of GWR was a useful guide in this paper. This approach might be more widely employed by cartographers who, instead of writing within their own field and being read by fellow cartographers, might publish in other fields to share some of their knowledge and expertize in a way that can be easily digested and understood by the non-cartographer. This will, of course, become increasingly important as more maps are made by more non-cartographers.

Despite the core message of the paper, the number of examples of mapping uncertainty as part of the mapping of results of spatial analysis remains few and far between. Before Mennis' paper, the *Atlas of United States Mortality* (Pickle *et al.*, 1996) includes mapping of sparse data, which would lead to uncertainty in results. Yet so few others followed this excellent example, particularly in important areas such as epidemiology, where uncertainty can have a large impact on reported rates and risks (Jarup, 2004). This, perhaps, owes more to the lack of software to automatically build-in these sorts of mapping requirements than to a lack of understanding. Indeed, another paper published in *The Cartographic Journal*, expands on the idea that the goal of GWR is to identify spatial and multivariate problems (Demšar *et al.*, 2008). In this paper, the authors explore the results in a geovisual environment to facilitate more in-depth analysis. This is helpful to the analyst, but does not facilitate conveying the results in one map, as proposed by Mennis. The suggestion here is that exploratory tools are not similarly matched by cartographic tools to ease the transition from exploration to presentation.

Mennis' paper threw down the challenge to appreciate the complexity of many statistical results and appreciate the need for uncertainty both in analysis and interpretation. This challenge has not been fully met by cartographic tools. Maps that effectively show uncertainty should be commonplace, but they are not. Statisticians need good exemplars yet these are hard to find. This paper boldly tackles an area that offers so much, even though lying between disciplines risks losing the value that each specialism brings. The key is for more researchers to acknowledge the expertize that another discipline can bring and to work alongside people who can help build the bridges required to both analyse and present effectively.

REFERENCES

Demšar. U., Stewart Fotheringham, A. S. and Charlton, M. (2008). 'Combining geovisual analytics with spatial statistics: the example of geographically weighted regression', **The Cartographic Journal**, 45(3), pp. 182–192.

Jarup, L. (2004). 'Health and environment information systems for exposure and disease mapping, and risk assessment', **Environmental Health Perspectives**, 112(9), pp. 995–997.

Pickle, L. W., Mungiole, M., Jones, G. K. and White, A. A. (1996). **Atlas of United States Mortality**, National Center for Health Statistics, Hyattsville, MD.

Cultures of Map Use

CHRIS PERKINS

Originally published in *The Cartographic Journal* (2008) 45, pp. 150–158.

Research into map use has so far largely focused on cognitive approaches and under-played the significance of wider contextual concerns associated with the cultures in which mapping operates. Meanwhile, cartography is being popularized and people are creating and employing their own maps instead of relying upon cartographers. Critical cartography has begun to offer new ways of understanding this cultural and social change, but research into map use has so far not engaged with this critical turn. It is argued that an approach informed by critical cartography is becoming more and more appropriate, stressing the need to rethink map use as a set of everyday activities practiced in real-world contexts and arguing map use is best interpreted using meth-odologies from the social sciences, employing a mixture of ethnographic and textual methods. Using case studies of community mapping, the mapping of golf courses, map collecting and mapping art, this paper shows how different insights into the nature of map use can flow from rethinking mapping. It is concluded that networks of practice of map use depend upon relations between many different artefacts, technologies, institutions, environments, abilities, affects, and individuals.

INTRODUCTION: MOVING ON FROM SCIENCE

The central argument of this paper is that research into map use has so far largely focused on cognitive approaches and underplayed the significance of wider contextual concerns associated with the cultures in which mapping operates. The aim is not to test a simple hypothesis, or separate off part of the complexity of mapping: instead, a broad, contextual and social approach to map use is deliberately adopted. The argument is not that a cultural and contextual approach is any better per se than a cognitive or semiotic approach to map use. Rather that a cultural approach can allow us to answer different questions about mapping and to explore different aspects of the ways in which our society deploys the map. Questions that are increasingly impor-tant, given the democratization of cartography, and that have been too little asked by academic researchers over the last two decades.

In the 1960s and 1970s, cartographic research focused upon communication of information. The emphasis was on how map design might be improved and the approach was underpinned by the belief that optimal maps might be produced to meet carefully specified user needs. Universal answers could be discovered through scientific investigation: users were presumed to exist outside of a social context. This kind of realist belief in progress and in scientific possibilities continued to be significant and was implicit in research sponsored by the International Cartographic Association Map Use Commission in the 1990s. Chaired by James Carter from 1991

221

Table 1.   James Carter's 2005 Many Dimensions of Map Use*

1	Users of maps
	• individual users as consumers
	• producers as users
2	Uses of maps
	• reading, analysis, interpretation
	• tasks in using maps
	• functions of map use
3	Environments in which maps are used
	• printed
	• projected
	• interactive
	• networked
	• operation
	• virtual
4	Nature of the map or maps being used
5	Communities of map users
6	Societal aspects of map use and abuse

*Source http://www.ilstu.edu/~jrcarter/mapuse/

until its rebranding as the Commission for Maps and the Internet in 1999, the Commission encouraged a particular approach to map use research.

It is instructive to examine Carter's[1,2] published attempts to construct an a priori list of the 'many dimensions of map use' underpinning the Commission's work (see Table 1 derived from Carter[3,4]). The individual user is categorized as either a consumer or as a producer who also uses, with motivations and abilities such as varying graphicacy. Map literacy requires understanding about availability and being able to know which map to use for what kinds of task. Carter then seeks to establish the use of a map, distinguishing uses from tasks and functions and isolating general reference use as against other more specialist roles. He accepts that the same map may be used in different ways and identifies levels of use (drawing on Muehrcke, Muehrcke and Kimerling[5]) and their distinction between reading, analysis and interpretation. Function apparently differs from generic use. He identifies cognitive, communicative, decision supporting and social functions (though how these are distinguished from some of his generic classes is unclear). Carter also draws on Olson's[6] notion of levels

1. Carter, J. (1999) 'Map use the many dimensions' from **http://lilt.ilstu.edu/jrcarter/icamuc/introduction.html** (accessed 15/10/07).

2. Carter, J. (2005) 'Map use: the many dimensions' from **http://www.ilstu.edu/,jrcarter/mapuse/** (accessed 15/10/07).

3. Carter, 1999.

4. Carter, 2005.

5. Muehrcke, Phillip C., Juliana O. Muehrcke, A. Jon Kimerling (2001). **Map use: reading, analysis, interpretation**, (4th ed.), J. P. Publications, Madison.

6. Olson, J. M., (1976) 'A Coordinated Approach to Map Communication Improvement', **The American Cartographer**, 3(2), 151–159.

of use, and Lobben's[7] emphasis upon reading tasks in the context of navigation. Carter also codifies the map use environment. Printed maps are distinguished from maps in projected environments such as news media or PowerPoint. The personal computer allows different kinds of interaction with maps displayed on a small screen. Networking allows real time update, easy distribution and sharing of Webserved maps. Specific operation environments such as those in navigational contexts are distinguished and the role of the virtual display considered. To Carter's list one might add whether the map is designed to be placed and read in the environment it depicts (e.g. a 'you are here' map).

Carter also argues that map use reflects genre. The classic distinction between topographic, general-purpose maps, and more specific thematic maps is extended through a consideration of how map user communities may determine mapped subjects, scales of display, designs and acceptable levels of accuracy and precision. This leads to the identification of 26 different broad categories.

This kind of enumerative approach to map use is seriously problematic. The enumeration changes over time — there are five dimensions in 1999, and six in 2005 but little explanation is given for the restructuring. The existence of the categories established in both listings is never properly justified. They dramatically oversimplify relations between people and mapping. Speculations are reified. Classes remain poorly related in any causal manner, and exist independent of time, place or context. I am reminded of Denis Wood's stinging critique of the Arthur Robinson's history of thematic mapping.[8,9] Wood rejects any such a priori classification arguing that the interests represented in a map must be understood in the cultural context in which the map is employed, rather than explained by an arbitrary classificatory grid.[10] In this view, cultures of map use mattered even in the era when cartographic specialists compiled fixed-formatted mapping: maps then, and now, are best understood as propositions, rather than representations.

In the period since the 1980s, technological change in mapping has increasingly called into question the fixed format and status of optimal designs and encouraged a profusion of mapping, which Morrison[11] termed a democratization of cartography. I would argue that this democratization further limits the scope of scientific approaches to map use, at the very same time as its tools have demonstratively altered the significance of mapping. Desktop mapping and GIS gave the general public tools to make their own maps. GIS allows users to change design specifications and content. Mapping is no longer tied to fixed specifications: users can interact and explore, rather than just employing the image as a final presentation.[12] To deal with this

7. Lobben, A. (2004) 'Tasks and Cognitive Processes Associated with Navigational Map Reading: A Review Perspective', **The Professional Geographer**, 56(2), 270–81.

8. Robinson, A. H. (1982) **Early thematic mapping in the history of cartography.** University of Chicago Press, Chicago.

9. Wood, D. (1983) 'Review: early thematic mapping in the history of cartography', **Cartographica 20**, 109–112.

10 Wood, D. (1992) **The power of maps**. Routledge, London.

11. Morrison, J. L. (1997) Topographic mapping in the twenty-first century. In: Rhind, D. **Framework for the world**. GeoInformation, Cambridge, pp. 14–28.

12. Rood, J., Ormeling, F., and Van Elzakker, C. (2001) 'An agenda for democratising cartographic visualisation', **Norsk Geografisk Tidsskrift**, 55(1), 38–41.

radical technological challenge scientific interest shifted towards representation instead of communication. MacEachren[13] demonstrated how science might still explain how maps worked, by fusing cognitive with more semiotic approaches.

In the decade since Morrison's work, the Web has encouraged a wide dissemination of this capability, and a remarkable sharing of mapping, which exacerbates the problems of approaches like Carter's. The medium becomes much more social and task-oriented, more ubiquitous, ephemeral and mobile. Users and producers are no longer separate. Pervasive technologies offer people possibilities of putting themselves on their own map, destabilizing the taken-for-granted representational neutrality of the image; new kinds of maps are being made; more people are making maps; more things are being mapped; and mapping is taking place in more contexts than ever before.

Scientific approaches to the use of maps as representations have subsequently investigated the best ways to design Web-served mapping, through a careful testing of a specified design variables in controlled conditions. The other papers in this theme issue all conform to this neutral view of scientific progress, in which the complexity of real world use might be slowly unpacked through rigorous experimentation. But it can be argued that this style of research is not well-equipped to deal with our changed times. Edsall[14] for example argues that we increasingly need to examine cultural differences in a dynamic user population, instead of simply designing systems. Understanding maps in terms of cartographic communication, semiosis, or scientific representation relies upon academic distance and underplays everyday practice. By distancing academic research from real world mapping practice, we risk missing the zeitgeist. People are making their own maps, and everyday map use is probably more common now than at any time in human history. Almost all of this map use is unresearched and beyond science.

Yet, despite the democratization of cartography, most research emphasizing cognitive approaches continues to stress 'correct' uses of mapping, drawing implicitly on many of the assumptions made by James Carter. Of course, in practice the map is employed for many different reasons: there is no single 'correct use', instead a multiple and often synchronous set of motivations are at play. Maps may reassure the lost, encourage debate, support arguments, keep the rain off, fire the imagination, help win or lose elections, sell products, win wars, catch criminals: an endless list of uses becomes possible, limited only by the imagination of its author: motivations may well be beyond science, even if most researchers investigating map use remain constrained by realist notions of scientific progress.

Science does not deal well with these complex social systems. It has problems with the unquantifiable and the unique. Experiments oversimplify. Functional explanations are overplayed, the irrational and feelings are marginalized. Wider social contexts outside of the experiment are not deemed relevant. Processes are hard to model.

There are, however, other ways of approaching mapping, which derive their insights from social–scientific and artistic understanding, and which might be particularly appropriate in the brave new world of collaborative cartography, mashups and

13. MacEachren, A. (1995) **How maps work**. Guilford, London.

14. Edsall, R. (2007) 'Cultural factors in digital cartographic design: implications for communication to diverse users', **Cartography and Geographic Information Science**, 34(2), 121–128.

map art. Different kinds of approaches to cartography emerged in the 1980s that sidestepped science: post-structural thought increasingly rejected the possibility of universal explanations and sought more local and contingent insights that welcomed local difference instead of rejecting it. New approaches to mapping were inspired by influential ideas from Brian Harley and Denis Wood and adopted a much broader and more critical approach.[15] Theory and practice became as important as mapping progress.[16] Approaches to mapping have been increasingly informed by approaches as diverse as Barthean semiotics[17]; Actor-Network Theory[18]; Foucauldian power-knowledge[19]; Derridean deconstruction[20]; post-colonial theory[21]; hermeneutics[22]; ethnomethodology[23]; affect[24]; emergence[25]; Deleuzian non-representational theory[26] and holistic performance.[27]

A unifying feature of these alternative approaches is a concern with culture, by which I mean something much broader than Edsall's[28] use of the term. Here, we treat culture as patterns of human activity and symbolic practices, including matters as diverse as social relationships and interactions; material and ideological consumption; systems of belief, norms and values and shared experiences; language; actions; artefacts; and regulatory frameworks and institutions.

In this myriad of different ways of thinking about mapping, attention shifts onto processes, institutions, social groups, power, interactions between different elements in networks, emotions at play in mapping, the nature of mapping tasks and a concern with practice, instead of focusing on one aspect of how an individual processes combinations of visual symbols on a screen, mobile device or paper sheet. In contrast to scientific approaches to map use, these diverse concerns reflect current real world and everyday uses of mapping in society and share a general focus on wider cultural concerns. They recognize that maps are capable of conveying authority, confirming

15. Crampton, J. W. and Krygier, J. (2006) 'An introduction to critical cartography', **ACME: An International E-Journal for Critical geographies,** 4(1), 11–33.

16. Perkins, C. (2003) 'Cartography: mapping theory', **Progress in Human Geography,** 27(3), 341–351.

17. Wood, 1992.

18. Perkins, C. (2006) 'Mapping golf: contexts, actors and networks', **The Cartographic Journal,** 43 (3), 208–223.

19. Joyce, P. (2003) **The rule of freedom.** Verso, London.

20. Harley, J. B. (1989) 'Deconstructiong the map', **Cartographica,** 26(2), 1–20.

21. Sparke, M. (1998) 'A Map that Roared and an Original Atlas: Canada, Cartography, and the Narration of Nation', **Annals of the Association of American Geographers** 88(3), 463–495.

22. Pickles, J. (2004) **A history of spaces: mapping cartographic reason and the over-coded world.** Routledge, London.

23. Brown, B. and Laurier, E. (2005) 'Maps and journeying: an ethnographic approach', **Cartographica,** 4(3), 17–33.

24. Kwan, M. P. (2007) 'Affecting Geospatial Technologies: Toward a Feminist Politics of Emotion', **The Professional Geographer** 59(1), 22–34.

25. Dodge, M. and Kitchin, R. (2007) 'Rethinking maps', **Progress in Human Geography,** 31(3), 331–344.

26. Crouch, D., and Matless, D (1996) 'Refiguring geography: the Parish Map Project of Common Ground', **Transactions Institute of British Geographers,** 21, 236–255.

27. Del Casino, V. and Hanna, S. P. (2006) 'Beyond the 'binaries': a methodological intervention for interrogating maps as representational practices.' **ACME: An International E-Journal for Critical Geographies,** 4(1), 34–56.

28. Edsall, 2007.

the subjective, or affirming the cultural taste or place associations of the user, but also recognize that like all iconic devices mapping is only ever able to tell a very partial story about the world, a story that depends upon the situated knowledge of the map reader and his or her culture. Maps are more than mere tools. Cognition works in historical contexts and is exercised by people in cultures. Even the more rigorously argued cognitive attempts to place map use into a coherently argued framework suffer from severe simplifications of the nature of the cultural process. Indeed, it has recently been acknowledged that 'many of the tasks, strategies and processes have yet to be identified, and possibly more importantly understood'.[29] So the key question becomes what new methods might be deployed to tease out tasks, strategies and processes in map use, but also how these practical methods might relate to some of the more abstract alternatives to scientific work on mapping.

METHODS FOR REAL WORLD MAPPING RESEARCH

There have been many different ways of 'doing' critical and contextual mapping research. No single prescriptive approach or method is 'best'. Instead, the aim for cultural research into real-world and everyday mapping is much more likely to be creating *different* insights, into often ambiguous, poly-vocal, mobile and changing mapping practice.[30] Cultural research into Web-served mapping is more likely to employ multiple methods, to tell a partial story of how the representations on a site could be employed in different contexts, instead of offering any universal generalizations about design or functionality. A shared characteristic amongst the many new ways of understanding mapping is that cultural and critical approaches are performed in different ways in different contexts. So case studies offer valid and different ways of interpreting these contexts, and may offer different kinds of explanations to more formal scientific experiment.

Qualitative research may well be more appropriate for this kind of work. There is some evidence of how researchers have used these approaches in cartography, e.g. Suchan and Brewer.[31] But the rich diversity of methods of understanding visual materials brought together in recent work by Rose[32] has, so far, been rather underplayed in mapping research. Rose documents the different modalities through which meaning might be constituted in the visual, and explores how different methods might be particularly appropriate for sites where the visual is deployed, be they the image itself, the institutions through which its work is done, the sites of its production, or the audiences deploying the mapping. Mapping is moving to new sites, no longer under the control of cartographers. It is time for mapping researchers to draw on Rose's impressive methodological catalogue, to escape the confines of the laboratory and ask broader questions.

29. Lobben, 2004: 270.

30. Perkins, C. (forthcoming) 'Performative mapping' In Thrift, N. and Kitchin, R. **International encyclopaedia of human geography**. Elsevier, London.

31. Suchan, T.A. and Brewer, C.A. (2000) 'Qualitative Methods for Research on Mapmaking and Map Use', **The Professional Geographer** 52(1), 145–154.

32. Rose, G. (2007) **Visual methodologies**, (2nd ed.), Sage, London.

Following the actors in different networks, and tracing the inscriptions they leave behind is one particularly appropriate recommendation that emerges from Actor-Network Theory,[33] and which is beginning to be applied in map use research (see for example Perkins[34] and Martin[35]). A related set of empirical suggestions also emerges from ethno-methodological research, where concern shifts towards observing the social practices of actors in real-world settings, and explores how practices lead to making sense of the world e.g. Brown and Laurier.[36] So a focus for cultural research into map use might shift towards participation and observation of real uses, as well as interviews, focus groups and read aloud protocols. A rich diversity of textual and visual methods needs to be deployed. Observation can and should mean more than just looking. It might encompass other senses; movement through the world; gesture; actions; or play. Research becomes carefully situated in specific times, places, institutions and spaces. Instead of deploying research methods as something somehow apart from the world, a neutral mechanism, we need a proper ethnography, that recognizes how messy is our understanding of mapping, and that acknowledges the interested and implicated role of methods in constructing knowledge.[37] Ethnographic approaches need to be *in* the field, not separate from the field. They need to bring together observations with rigorous recording of different kinds of speech act, ranging from informal conversation, through to formal interviews. An ethnographic approach to map use would see a *final* map as part of an ongoing cultural process. Initial informal discussions between a client and map designer may reveal fundamental influences on subsequent uses of a product that would be hidden from cognitive researchers. In this mapping process, many artefacts play different, changing and contested roles. Ethnographic approaches can allow us to investigate the social relations that strongly affect uses of mapping. For example, the design of Sat-Nav mapping interfaces reflects a commercial opportunity for system vendors and for those controlling digital map databases. It also depends strongly upon the development of locative technologies. On the other hand Sat Nav offers potential for altering driver behaviour: governments can deploy the same technologies to police road charging schemes, or encourage safer, or slower driving. The take up of systems of course also reflects individual driver preference and the social relations of driving. The Sat Nav becomes a consumer durable as well as a functional aid to navigation. By investigating *contexts* of use these processes can be clarified.

CONTEXTUAL CASE STUDIES

Four everyday contexts show the potential of a more cultural approach to map use. They illustrate how different methods may be appropriate in different contexts, focusing on Cultures of map use questions about mapping that are beyond science. In each of these contexts the cultural differences are as significant as any narrowly

33. Law, J. (1999) 'Traduction/trahison: notes on ANT', from **http://comp.lancs.ac.uk/sociology/papers/law-Traduction-Trahison.pdf** (accessed 15/10/07).

34. Perkins, 'Mapping golf: contexts, actors and networks', 2006.

35. Martin, E. (2000) 'Actor-networks and implementation: examples from conservation GIS in Ecuador', **International Journal of Geographical Information Science**, 14(8), 715–738.

36. Brown and Laurier, 2005.

37. Law, J. (2003) **After method: mess in social science research**. Routledge, London.

defined cognitive or information processing activity, in influencing how people deploy mapping.

Community mapping

The last decade has seen a significant growth in the amount of community mapping — local mapping, produced collaboratively, by local people and often incorporating alternative local knowledge.[38] Democratized mapping offers new possibilities for articulating social, economic, political or aesthetic claims. Data are increasingly available, free at point of use, often subsidized through advertising, and flexible. Software tools allow people to make their own maps. The Web encourages collaborative participation and cost-effective dissemination and can be used as an effective medium to organize opposition, at a time when the social context has shifted with the new orthodoxy of sustainable development encouraging local involvement. Social scientific research suggests, however, that community mapping is much less frequent or emancipatory than might be expected.[39] In the UK at least, there are still substantial barriers hindering participation: mapping is still perceived as a technical exercise, demanding expertize, carried out by others who know best. Ordnance Survey and other official mapping can be prohibitively expensive for community groups. The virtues of mapping have been lauded as a development tool (e.g. Chambers[40]), but its power for change has often been subverted into a safe participatory technique, rather than a radical alternative. Parker[41] offers one way of analysing the empowering potential of the medium, in a detailed ethnographic case study of one mapping agency, in which she explores the local processes underpinning making and using a Green Map in Seattle, and highlights the explicit transparency of this kind of mapping. On the other hand, a wider, more institutional focus in the UK suggests that practice depends on contexts beyond the immediate local concerns of participants, articulating contested notions of place, mediated through politics, practices, technology, and aesthetics.[42] Community empowerment is complex. Projects have different goals: cycling mapping is not the same as green mapping (see Figure 1), community artistic maps, or open-source collaborative cartography. Participants in the process will not buy into all these goals either. The same project may carry different meanings for different members, who are likely to engage in different ways with the mapping. Technologies can be deployed in complex ways by local groups, given umbrella support, for example in groundbreaking collaborative open-source projects such as OpenStreetMap, (www. openstreetmap.org, see Figure 2), but successful community mapping depends strongly upon shared communities of interest often extending well beyond a local scale. Perkins[43] concludes that tensions are always there to be negotiated in these community

38. Perkins, C. (2007) 'Community mapping', The Cartographic Journal, 44, (2), 127–137.
39. Perkins, 2007.
40. Chambers, R. (2006) 'Participatory mapping and geographic information systems: whose map? Who is empowered and who disempowered? Who gains and who loses?', Electronic Journal on Information Systems in Developing Countries, 25(2), 1–11.
41. Parker, B. (2006) 'Constructing community through maps? Power and praxis in community mapping', Professional Geographer, 58(4), 470–484.
42. Perkins, 2007.
43. Perkins, 2007.

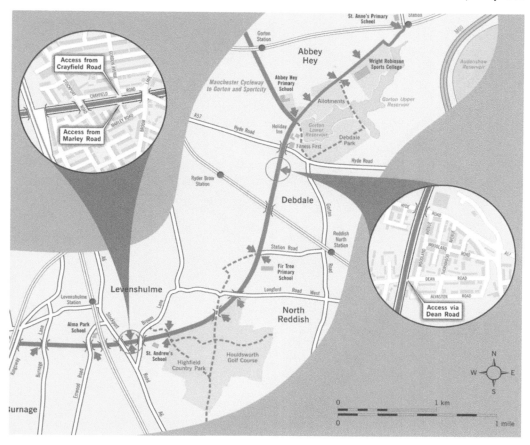

FIG. 1.   The Fallowfield Loop Cycle Map Manchester

initiatives. Scientific approaches may create the tools that allow community initiatives to be developed, but understanding the uses that are made of these maps depends upon more than science.

The mapping of golf

Recent work on the mapping of golf shows the importance of a contextual and network-oriented approach to mapping. There are now a very wide variety of contexts in which specialist maps are employed around the game of golf and political, cultural, and economic contexts strongly mediate how people read and engage with golf mapping. There is both continuity and change. Perkins[44] shows how the scale, inclusion criteria, and design of generic mapping fail to meet organizational, social and individual needs. Instead a network of specialist provision has emerged, in which

44. Perkins, 'Mapping golf: contexts, actors and networks', 2006.

FIG. 2. OpenStreetMap depiction of complex multi-layered junctions: The M1 Motorway
at Staples Corner North London
(Source: http://wiki.openstreetmap.org/index.php/Image:Road_junction.png)

numerous actors, contexts, products and resources interact (Figure 3). He argues that
we need to consider these products in the light of local contextual factors instead of
universal user design criteria, and deploys an Actor-Network-based approach to try
to explain how golf mapping is used. Perkins[45] shows that factors that come into play
when mapping is deployed include:

- mobility (whether the map is mobile or fixed);
- the viewing environment (projected, on screen, on paper, or fixed on a tee);
- where reading takes place (in clubhouse, at home, in the hand walking the course etc);
- the timing (before, during or after a round, or other activity linked with the game);

45. Perkins, 'Mapping golf: contexts, actors and networks', 2006.

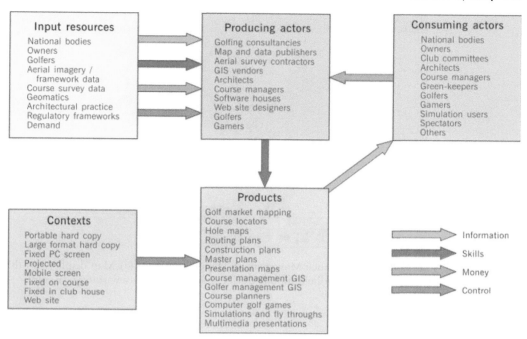

FIG. 3.   The actors in the golf mapping network

- the social context (read in a group, by an individual, or in a presentation);
- the interactivity of the medium;
- intertexuality and relations to other discourses or media; and
- rhetoric (arguments framing questions mapping might be able to answer[46]).

The mapping of golf has a series of practical uses: to inform, locate, navigate, orient, measure, compare, monitor change or establish relationships. A stable series of questions emerge in these contexts that are addressed by specialist products: to locate the course; to site the course and assess the market for golf; to plan the routing of the course layout; to map the holes; to manage the course assets; to measure distances on the course; or to market the facility etc. But these questions hide many different reasons for golf mapping: golf maps reassure; encourage debate; fire the imagination; sell a product; or help improve scores. They store information so that the unknown can be controlled. These reasons alter how mapping is deployed. The same maps may be employed for different roles, in different contexts. Architects may be players, golfers may also be owners, map makers may also be map users, course managers and golfers may use a GPS-based mapping system on a PDA for very different reasons (see Figure 4). Cognitive user studies would find it hard to unpack this complexity and would seriously simplify a richly complex context, best understood in a mix of ethnographic and textual approaches.

46. Denil, M. (2005) 'Practical map rhetoric', Paper presented at the 22nd International Cartographic Conference, La Coruna.

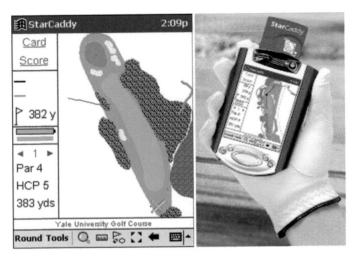

FIG. 4. The StarCaddy Golf Distance Management Mapping System: (a) Map display and (b) the context for the display: handheld mobile device (Source: http://www.starcaddy.com/)

Map collecting

If the mapping of golf reveals a complex network of functional deployment, then explorations of map collecting display the play of cultural capital. At the very time when the artefact is supposedly becoming supplanted by a fleeting digital image, it is increasingly an object of desire for collectors. Motivations are explored by Perkins[47] in a detailed comparative ethnographic study of recent map collecting behaviour in the UK. He distinguishes between antiquarian and everyday practices, showing how collecting identities are constituted around different spaces, social networks, artefacts and values and contrasts cultures of map collecting with those of other collecting fields.[48] Antiquarian map collectors are interested in different qualities to those that motivate people who collect more prosaic and mass-produced mapping. The values of antiquarian collecting are inherently conservative: authenticity, verified historical rarity, beauty, and the display of taste by a largely aging and wealthy group of collectors, who sometimes value their collections as investments, as well as collectable works of art: the aesthetic and economic merge in these collecting practices. Only the very rich are now able to amass significant numbers of antiquarian maps or atlases. Despite the impacts of globalization and modernization, dealers remain an important part of this map trade and social contact between collectors and others who love maps remains important. Collecting in this field requires specialist knowledge and is a well-regulated process. In contrast, the world of the everyday collector is altogether less aspirational. Here maps are cheaper and easier to acquire and the detail of amassing and completing a collection becomes more important. More maps are acquired. The pastime is much less regulated and the trade in maps is less of a business. Collectors

47. Perkins, C. (2006) 'Cultures of everyday map collecting', Sheetlines, 76, 35–45.
48. Elsner, J. and Cardinal, R. (1994) Cultures of collecting. Reaktion Books, London.

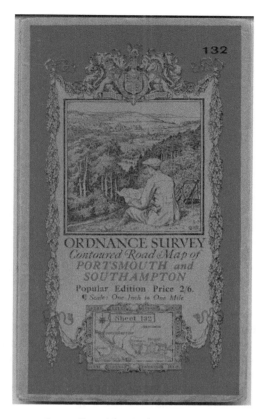

Fig. 5.   The map as everyday collectable: Ordnance Survey One-Inch Popular Edition
Sheet 132

value developing their own highly specialist knowledge, rather than relying upon expert opinion. Collecting activities and spaces are more local, and everyday collectors are disproportionately male, with a much wider spread of social backgrounds. The accumulated 'objects of desire' may be cheaper and more prosaic than for an antiquarian collector (see Figure 5), but the everyday collecting practices reveal just as much about the cultural capital involved in the quest for completion. Once again a cognitive or scientific study would be unlikely to advance this kind of understanding of map use. Cultures of collecting are best understood through the lens of social scientific and critical research.

Maps are artistic practice

The study of map collecting behaviour shows that mapping has *affective* qualities — people derive pleasure from engaging with mapping.[49] Maps move people in different

49. Kwan, 2007.

FIG. 6.   Mapping art: Joyce Kozloff 'Around the world on the 44th parallel: Sarajevo'
(Source: http://www.lib.mankato.msus.edu/News/Kozloff.html)

ways, and these emotive qualities are of increasing interest to modern artists. In the last
decade of the twentieth century, they have increasingly deployed mapping in their
work.[50] Maps are once again personal and subjective.[51] Surrealists, pop artists, situa-
tionists, land artists, conceptual artists, community artists, digital media artists and
live artists have all employed maps, encouraging a performative encounter. The forms
of the mapped image and media in which mapping is expressed vary. For example
Joyce Kozloff has often worked with ceramic tiles, using the map to bring together
the familiar with the alien (Figure 6). Common techniques include: fragmenting
known maps and rearranging them in novel ways; juxtaposing far with near; distorting
space into a relative or egocentric form; changing orientation; manipulating projection,
scale and generalization to infringe accepted mapping standards; drawing on standard
cartographic tropes such as the border, or naming to question social norms; abstracting
and over-coding a known form; employing recognizable country shapes in new
ways; shifting novel conceptual frames onto familiar icons such as the globe or tube
map; and mapping onto different media so as to ask questions about the world or our
identities.[52]

Many of these artistic encounters with mapping explicitly focus on the performative
potential of the medium, inspired by Situationist encounters with the modernist city
in the 1950s and the 1960s, when art moved from the galleries to the streets of Paris,
and artists encouraged people to take part in *dérives*, walking across urban space, to
subvert the controlled modernist dream of the city. Situationist maps were published

50. Schulz, D. (2001) **The conquest of space: on the prevalence of maps in contemporary art.** Henry Moore
Institute, Leeds.

51. Harman, K. (2003) **You are here: personal geographies and other maps of the imagination.** Princeton
University Press, Princeton.

52. Perkins, forthcoming.

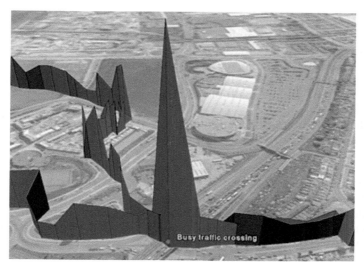

FIG. 7.   New visualization from Greenwich Emotion Map overlain on Google Earth
(Source: http://biomapping.net/new.htm)

as collages, fragmented, multi-mediated alternatives to the all-knowing map on which
they were based. This performative cartography has continued to inspire urban
opposition ever since. Live artists also enact mapping, for example by layering and
re-enacting one set of alien events against a familiar mapped place. Others devise
mapping puzzles to encourage urban exploration, casting the user as a player in an
evolving work. Locative art is emerging as particularly performative. Digital and
satellite mapping technologies supporting networked, portable, location-aware computing devices allow user-led mapping, social networking and artistic interventions,
to be enacted on the canvas of the real world.[53] People can track themselves, across a
landscape. In skywriting, for example, the old certainties of land art are subverted:
participants walk the shape of letters, or the outlines of animals, and, instead of the
landscape itself being shaped as art, unseen messages are overlain onto mapping.
Others deploy technologies to chart changing feelings in space: Christian Nold (http://
biomapping.net/, see Figure 7) for example, has devised bio-mapping technologies,
merging GPS and bio-sensors, to produce collaborative emotion mapping, and released this information in very diverse contexts, as exhibitions, in enabling workshops
for activists, as hard copy publications, and as Web-served Google Earth mashups.
Art shows how maps too evoke powerful feeling beyond science, but also that science
itself might be deployed in artistic and emotive ways.

CONCLUSIONS

Networks of practice of map use depend upon relations between many different
artefacts, technologies, institutions, environments, abilities, affects, and individuals.

53. Wood, D. and Krygier, J. (eds.) (2006) 'Art and mapping theme issue', **Cartographic Perspectives**, 53.

In this paper we have argued that we need different ways of approaching mapping and its use, that reflect this complexity and which are beyond the narrow hypothesis-testing of most contemporary mapping user studies. Paradoxically, it is the democratization of cartography, taking place as a result of scientific progress and technological change, which is fuelling these calls for plural ways of understanding map use. When local contexts of map use are explored the potential of alternative approaches beyond science becomes clear.

Case studies of community mapping reveal that its successes or failures depend upon much more than cognition of user interfaces. The local community context and institutional framework have significant impacts on how mapping is created, and deployed: local mapping and use has to be understood in the light of these wider influences. Other case studies show how the same maps may be employed in very different ways in different contexts: on and off the golf course representations of the golfing landscape mean different things according to the cultural context in which they are deployed: these may well be unpredictable. The player, course owner, manager, and architect have very different mapping demands, but owners may also be players and studying mapping as representation underplays the social complexity in which meaning is constructed.

A broader appreciation of cultural context also allows research to focus on wider uses of mapping. Map collecting might be regarded as being outside the remit of narrowly defined enumerations of map use informed by cognitive and scientific studies. Collecting is, however, arguably more popular than ever and mapping is deployed in collecting practices because it is *mapping*: pleasure in the artefact is no less important than more functional uses. The artistic impulse might also be excluded from narrowly defined scientific user studies. Once again, however, the mapping process is being deployed, the map is being used for a human purpose and modern art appears fascinated by the map. Critical, aesthetic and social uses of mapping in artistic endeavour reveal the subjective and personal emotive qualities and powers of using the map that are beyond science.

A scientific approach to mapping is certainly important, but it is only one of many ways of increasing our understanding of how and why maps are used. In addition to focusing on individual relations to a map, we should encourage investigation of the wider cultures of map use as a central concern for mainstream cartographic research and deploy the tools of social science and the humanities to help us in this endeavour.

Reflections on 'Cultures of Map Use'

ALEXANDER J. KENT

Canterbury Christ Church University

User-centred design is a constant theme in cartographic production and research in pursuit of a greater understanding of how maps are used, by whom, and under which environmental conditions. Modelling the communication of information via maps became a particular focus of cartographic research in the 1960s and 1970s, with the implicit belief that maps could be optimized by meeting the specific and defined needs of the user. The scientific reductionism inherent within these closed and narrowly defined systems isolated the 'user' from their social and cultural contexts, while formal cartographic training and education emphasized the user's importance, teaching that cartography is, in essence, about the communication of spatial information to the user in the most efficient way.

Changes in mapping technology since the heyday of the cartographic communication models have been dramatic. The capacity for users to make maps (e.g. desktop mapping) and explore and interact with them (e.g. GIS) has seen such a colossal shift that the fixed, optimal map is no longer assumed to constitute a final, finished product. This has meant that the processes of social interaction and engagement with maps and their deployment reflect an altogether different zeitgeist than that of the communication models. Appearing in a Special Issue of the Journal devoted to 'Use and Users', *Cultures of Map Use* offered a bold critique of the existing approaches to research into map use, providing a persuasive and well illustrated case for rethinking maps with regard to their wider cultural context. Perkins is especially critical of the largely cognitive methods employed by academic research into map use, and, in particular, the approach adopted by the ICA Map Use Commission in 1999, which continued to treat the user as either 'consumer' or 'producer' with varying abilities or graphicacy skills.

Perkins therefore urges us to be concerned with the process and practice of mapping 'instead of focusing on one aspect of how an individual processes combinations of visual symbols on a screen, mobile device or paper sheet' (Perkins, 2008, p. 152). Hence, science can provide only some of the tools required to examine the wider social and cultural context of map use, especially following a democratization of mapping. Yet rather than proposing a new method to supplant another, he demonstrates how a cultural approach can answer different questions about mapping and explore the different ways in which maps are deployed in society. In essence, Perkins argues that research into map use should embrace, rather than eschew, the complexities of the social and cultural dimensions and in so doing, seek inspiration from social-scientific and artistic ways of approaching mapping.

Far from providing a long-overdue rationale to ask wider questions surrounding map use and answer them more fully, various methods 'for real world mapping research' are introduced, from Actor-Network theory to Deleuzian non-representational theory. By way of four case studies: community mapping, the mapping of golf, map collecting and maps as artistic practice, a series of cultural approaches to map use are demonstrated, asking questions about mapping that are 'beyond science' (Perkins, 2008, p. 154). These case studies together provide an effective, concise and, perhaps most importantly, approachable illustration of how a deeper understanding of the context of map use can be gained. They reveal the complexity of community empowerment (e.g. the same community mapping project may have different meanings for different participants); deployment to actors (e.g. having multiple identities and interests); map collecting practices (e.g. different identities and aspirations); and artistic performance and response (e.g. affecting different people through different emotions).

According to Perkins (2013, pers. comm.), when the paper was first presented, at ICC Moscow in 2007, 'people were shocked that mapping could be about anything other than communication or representation'. Indeed, the paper was itself accompanied, in the same issue of the Journal, by a number of others adopting conventional approaches to investigate map use — a useful context that served to highlight their lacunae. As Perkins (2008, p. 152) states, 'like all iconic devices mapping is only ever able to tell a very partial story about the world, a story that depends upon the situated knowledge of the map reader and his or her culture'.

While the critical cartography of the 1980s and 1990s through, for example, Brian Harley and Denis Wood, sought to reveal the power-relations and interests served, arguably these approaches tended to paralyse (or dis-employ) the practising cartographer and strengthen academic geographers' ambivalence to maps. The timeliness and accessibility of Perkins' theoretical framework, coupled with his deft demonstration of how the wider, social and cultural context of map use can be better understood, have facilitated its application, from the use of sat-navs (Axon et al., 2012) to participatory mapping in China (Lin, 2013). Such an approach that has both theoretical and practical applications and encourages us to explore and celebrate the diversity of mapping and map use is noteworthy, but to argue that the cultural approach is intended to complement – not replace – existing scientific approaches to understanding map use ensures its success and will continue to do so.

REFERENCES

Axon, S., Speake, J. and Crawford, K. (2012). '"At the next junction, turn left": attitudes towards Sat Nav use', Area, 44, pp. 170–177.
Lin, W. (2013). 'When Web 2.0 meets Public Participation GIS (PPGIS): VGI and spaces of participatory mapping in China', in Crowdsourcing Geographic Knowledge: Volunteered Geographic Information (VGI) in Theory and Practice, ed. by Sui, D., Elwood, S. and Goodchild, M., pp. 83–103, Springer, Berlin.
Perkins, C. (2013). Personal communication.

Usability Evaluation of Web Mapping Sites

ANNU-MAARIA NIVALA, STEPHEN BREWSTER AND
L. TIINA SARJAKOSKI

Originally published in *The Cartographic Journal* (2008) 45, pp. 129–138.

To identify the potential usability problems of Web mapping sites, four different sites were evaluated: Google Maps, MSN Maps & Directions, MapQuest, and Multimap. The experiment comprised a series of expert evaluations and user tests. During the expert evaluations, eight usability engineers and eight cartographers examined the Web mapping sites by paying attention to their features and functionality. Additionally, eight user tests were carried out by ordinary users in a usability laboratory. In all, 403 usability problems were identified during the trial and were grouped according to their severity. A qualitative description is given of these usability problems, many of which were related to search operations that the users performed at the Web mapping sites. There were also several problems relating to the user interface, map visualization, and map tools. We suggest some design guidelines for Web mapping sites based on the problems we identified and close the paper with a discussion of the findings and some conclusions.

INTRODUCTION

Recent technological developments have provided new tools and techniques for designing interfaces and interacting with Websites. Web mapping sites, or simply Web maps, are interactive maps that are accessed through Web pages.[1] Consequently, many people use these sites for locating places and businesses, and for planning visits to unfamiliar places. Figures gathered from Web mapping sites' own Web pages give an indication of their popularity; one site states that it has over 40 million unique visitors each month,[2] while another maintains a unique user base of over 10 million, ranking consistently in the top 10 Websites by traffic in the UK.[3] Web maps are often freely available and not only provide the map, but different map tools and map-related services.

However, the use of Web maps is not always straightforward. One reason for this may be that Web maps are used by a large number and variety of people, and the sites may not always fulfil all of the users' needs. Another reason may be changes in information and communication technology, leading to new methods for visualizing geospatial data. Due to this, traditional map design and evaluation methods may no

---

1. Mitchell, T. (2005) **Web Mapping Illustrated**, O'Reilly Media INC., Sebastopol, CA.
2. MapQuest (2007), online at: **http://www.mapquest.com/** (accessed 1st April 2007).
3. Multimap.com (2007), online at: **http://www.multimap.com/** (accessed 1st April 2007).

longer always be valid. Koua and Kraak[4] crystallized this problem by stating that map use studies that have long been carried out in the field of cartography are not fully compatible with new interactive visualizations, which can have new representational spaces and user interfaces. So how can it be guaranteed that today's maps using different (new) technologies will fulfil user requirements?

Usability engineering — a term used to describe methods for analyzing and enhancing the usability of software — is an approach to help design products that take into account the new technical environments and user requirements.[5] Usability is defined in the ISO 9241 standard as 'the effectiveness, efficiency, and satisfaction with which specified users achieve specified goals in particular environments'.[6] The ISO 13407 standard gives instructions to achieve user needs by utilizing the User-Centred Design (UCD) approach throughout the entire life cycle of a system.[7] Making systems more usable may have noticeable benefits for users by guaranteeing easy-to-use systems, which are less stressful for the user and therefore more acceptable. A user-centred design can provide financial benefits for the system developer in reduced production costs, reduced support costs, reduced costs in use, and improved product quality.[8]

Several researchers have observed the lack of thorough usability engineering in cartographic visualization and geovisualization, for instance, MacEachren and Kraak,[9] Fuhrmann *et al.*,[10] van Elzakker[11] and Nivala *et al.*[12] The aim of the present study was to identify potential usability problems of Web mapping sites in order to provide guidance for the future design of such services.

Previous usability evaluations of on-screen maps

Previous research on the usability of Web mapping sites seems to be rare. However, several usability evaluations have been carried out in relation to other on-screen maps. Beverley[13] studied the benefit of a dynamic display of spatial data-reliability from the

4. Koua, E. L. and Kraak, M.-J. (2004) 'A Usability Framework for the Design and Evaluation of an Exploratory Geovisualisation Environment', **Proceedings of the 8th International Conference on Information Visualisation**, IV'04, IEEE Computer Society Press.

5. Nielsen, J. (1993) **Usability Engineering**, San Diego, Academic Press.

6. ISO 9241-1 (1997) 'Ergonomic Requirements for Office Work with Visual Display Terminals (VDTS) — Part 1: General Introduction', International Organisation for Standardisation, Geneva, Switzerland.

7. ISO 13407 (1999) 'Human-Centered Design for Interactive Systems', International Organisation for Standardisation, Geneva, Switzerland.

8. Earthy, J. (1996) 'Development of the Usability Maturity Model', **INUSE Deliverable D5.1.1(t)**, London: Lloyd's Register.

9. MacEachren, A. M. and Kraak, M.-J. (2001) 'Research Challenges in Geovisualisation', **Cartography and Geographic Information Science**, 28, 3–12.

10. Fuhrmann, S., Ahonen-Rainio, P., Edsall, R. M., Fabrikant, S. I., Koua, E. L., Tobon, C., Ware, C. and Wilson, S. (2005) 'Making Useful and Useable Geovisualisation: Design, and Evaluation Issues', in **Exploring Geovisualisation**, Dykes, J, MacEachren, A. M. and Kraak, M.-J. (eds.), Elsevier Ltd., pp. 553–566.

11. van Elzakker, C. P. J. M. (2005) 'From Map Use Research to Usability Research in Geo-information Processing', **Proceedings of the 22nd International Cartographic Conference**, A Coruña, Spain.

12. Nivala, A.-M., Sarjakoski, L. T. and Sarjakoski, T. (2007) 'Usability Methods' Familiarity among Map Application Developers', **International Journal of Human–Computer Studies**, 65 (9), 784–795.

13. Beverley, J. E. (1997) 'Dynamic Display of Spatial Data-reliability: Does it Benefit the Map User?', **Computers & Geosciences**, 23, 409–422.

user's point of view with a test using map data for decision-making that included both novices and experts. Harrower *et al.*[14] evaluated the design elements and communication quality of Internet maps for tourism and travel in a user survey. Studies have also been conducted on map animation and interactive tools (e.g. MacEachren et al.[15]), learnability, memorability, and user satisfaction with specific geovisualization tools,[16] and on the usability of zoomable maps with and without an overview map.[17]

Arleth[18] studied the problems of screen map design and listed a few of them, for example, the map area was too small and both the legend and instructions too dominating on the screen. Leitner and Buttenfield[19] investigated the effect of embedding attribute certainty information in map displays for spatial decision-support systems by having test users perform specific tasks with test maps. Harrower *et al.*[20] adopted a focus group method with structured user-testing to find out how novices understood and used the geovisualization tool that had been designed to support learning about global weather. Ahonen-Rainio and Kraak[21] described a study that included iterative design testing with map prototypes for visualizing geospatial metadata.

Agrawala and Stolte[22] studied how route maps are used, analyzing the generalization commonly found in hand-drawn route maps. Climate forecast maps were evaluated by Ishikawa *et al.*,[23] who concluded that in many cases, qualified and motivated test users failed to interpret maps in the way that the designer had intended. Richmond and Keller[24] carried out an online user survey to assess whether maps on tourism Websites met the expectations of users. Van Elzakker[25] carried out

14. Harrower, M., Keller, C. P. and Hocking, D. (1997) 'Cartography on the Internet: Thoughts and Preliminary User Survey', **Cartographic Perspectives**, 26, 27–37.

15. MacEachren, A. M., Boscoe, F. P., Haug, D. and Pickle, L. W. (1998) 'Geographic Visualisation: Designing Manipulable Maps for Exploring Temporally Varying Georeferenced Statistics', Infovis, **Proceedings of the 1998 IEEE Symposium on Information Visualisation**, pp. 87–94.

16. Andrienko, N., Andrienko, G., Voss, H., Bernardo, F., Hipolito, J. and Kretchmer, U. (2002) 'Testing the Usability of Interactive Maps in CommonGIS', **Cartography and Geographic Information Science**, 29, 325–342.

17. Hornbaek, K., Bederson, B. and Plaisant, C. (2002) 'Navigation Patterns and Usability of Zoomable User Interfaces with and without an Overview', **ACM Transactions on Computer-Human Interaction**, 9, 362–389.

18. Arleth, M. (1999) 'Problems in Screen Map Design', **Proceedings of the 19th International Cartographic Conference**, Ottawa, Canada, 1, pp. 849–857.

19. Leitner, M. and Buttenfield, B. P. (2000) 'Guidelines for the Display of Attribute Certainty', **Cartography and Geographic Information Science**, 27, 3–14.

20. Harrower, M., MacEachren, A. M. and Griffin, A. L. (2000) 'Developing a Geographic Visualisation Tool to Support Earth Science Learning', **Cartography and Geographic Information Science**, 27, 279–293.

21. Ahonen-Rainio, P. and Kraak, M.-J. (2005) 'Deciding on Fitness for Use: Evaluating the Utility of Sample Maps as an Element of Geospatial Metadata', **Cartography and Geographic Information Science**, 32, 101–112.

22. Agrawala, M. and Stolte, C. (2001) 'Rendering Effective Route Maps: Improving Usability through Generalisation', **Proceedings of the Conference on Computer Graphics and Interactive Techniques** (SIGGRAPH 2001), pp. 241–249.

23. Ishikawa, T., Barnston, A. G., Kastens, K. A., Louchouarn, P. and Ropelewski, C. F. (2005) 'Climate Forecast Maps as a Communication Decision-Support Tool: An Empirical Test with Prospective Policy Makers', **Cartography and Geographic Information Science**, 32, 3–16.

24. Richmond, E. R. and Keller, C. P. (2003) 'Internet Cartography and official Tourism Destination Web Sites', in **Maps and the Internet**, M. P. Peterson (ed.), Elsevier, NY, pp. 77–96.

25. van Elzakker, C. P. J. M. (2004) 'The Use of Maps in the Exploration of Geographic Data', **Netherlands Geographical Studies** 326, ITC Dissertation No. 116, Utrecht/Enschede.

user tests in order to investigate how maps were selected and utilized by users exploring geographic data. Similarly, Koua et al.[26] studied test subjects' ability to perform visual tasks in the data-exploration domain, and emphasized that use and usability assessment is an important part of understanding visual methods and tools for data exploration and knowledge construction. The UCD approach also played a central role in the development of the Atlas of Canada Website[27] and considered as the factor responsible for increased user satisfaction and growth in its overall use.

The usability evaluation of Web maps, similar to the study presented here, was carried out by Skarlatidou and Haklay,[28] who arranged workshops for assessing the usability of seven public Web mapping sites. In their method, users carried out six to seven tasks with the sites. Qualitative data was gathered through the 'thinking aloud protocol' and questionnaires and quantitative data by measuring the total time each user was performing each task, as well as the total number of clicks. Through measuring the users' performance, Skarlatidou and Haklay drew conclusions on which sites were the most and least usable and discussed the qualitative findings of their evaluation.

METHOD

The aim of this study was to identify potential usability problems with Web mapping sites and gather qualitative information to suggest guidelines for the design of future sites. Four different Web mapping sites were evaluated in this study: Google Maps (abbreviated in this paper as GM, available at http://maps.google.com/),[29] MSN Maps & Directions (MD, http://maps.msn.com/),[30] MapQuest (MQ, http://www.mapquest.com/) and Multimap (MM, http://www.multimap.com/). These well-known sites were chosen because they all consisted of an interactive 2D map application with zooming and panning options. Additionally, users were able to search for different locations and directions for routes.

Procedure

Several experiments were carried out in order to identify as many potential usability problems with the chosen Web maps as possible. First, a typical scenario for using these types of sites was drawn up: 'A tourist is planning to visit London and uses a Web mapping site for planning the trip beforehand'. Part of the evaluation was conducted as a series of user tests (with eight 'general' users), with the other part involving the evaluation of the maps by experts (eight cartographers plus eight

26. Koua, E. L., MacEachren A. and Kraak, M. J. (2006) 'Evaluating the usability of visualisation methods in an exploratory geovisualisation environment', **International Journal of Geographical Information Science**, 20, 425–448.

27. Kramers, R. E. (2007) 'The Atlas of Canada — User Centred Development', in **Multimedia Cartography**, Cartwright, W., Peterson, M. P. and Gartner, G. (eds.), Springer, Berlin, (2nd ed.), pp. 139–160.

28. Skarlatidou, A. and Haklay, M. (2006) 'Public Web Mapping: Preliminary Usability Evaluation', **GIS Research UK** 2005, Nottingham.

29. Google Maps (2007), online at: **http://maps.google.com/** (accessed 1st April 2007).

30. MSN Maps & Directions (2007), online at: **http://maps.msn.com/** (accessed 1st April 2007).

usability engineers). Altogether, 24 participants were involved and 32 different evaluations were carried out. Thus, each of the four Web maps was evaluated by eight separate participants (four test users and four experts). The experiments were run in a Windows environment using either desktop or laptop PCs. Evaluations were carried out from August–September 2006 and the results presented here are based on the content of the Web mapping sites at that time.

User tests

Before the test, users completed a background information questionnaire. Eight test users were involved in the evaluation (five males, three females), with ages ranging from 19 to 35. With the exception of one person, all users had previous experience of using several different types of maps (topographic maps, road maps, city maps, Web maps). All of the users regarded their map-reading skills to be fairly good or excellent.

The use scenario was described to the users at the beginning of the test. Following this, the test instructor gave the users one pre-defined task at a time, which they would try to complete by using the Web map (Table 1). The participants were given a Web map site that they had not used before. The users were then encouraged to 'think aloud' and describe the reasoning behind their actions. During the tests, the computer screens were recorded with a video camera to support subsequent data analysis.

Expert evaluations

Sixteen experts (eight cartographers and eight usability engineers) were involved in the evaluation (eight males, eight females), with ages ranging from 23 to 45. The term 'expert' here means a postgraduate student in cartography or usability engineering or a person who has already worked as a cartographer or usability specialist. The expert evaluators were given the use scenario and a list of typical user tasks (Table 1) and asked to go through the Web maps carefully, and, by using their own expertize, write down all the problems they encounter with when performing the same tasks as the

Table 1.   Usability evaluation tasks

	Task Description
1	You are planning to visit London during a weekend. Identify the most ideal location for a hotel by using the map site. Describe the reasons behind your choice.
2	Show the same place you chose during the previous task (the screen view returned back to the start page).
3	Find Roupell Street in London and point it out on the map to the test instructor.
4	Find the most northerly street in London with 'smith' included in its name.
5	What is the distance between Buckingham Palace and Piccadilly Circus?
6	Show the route that you would use if you were to walk from Sumner Road to Gresham Street.
7	Find London Bridge.

Table 2.   Usability problem classification according to its severity to the use situation

Rating	Description	Effect on map usability
1	A catastrophic usability problem	May even prevent the use of the application.
2	A major usability problem	Makes the use of the application significantly difficult.
3	A minor usability problem	Makes the use of the application somewhat difficult.
4	A cosmetic usability problem	Prevents the feeling of a finished design.

users in the user tests. The experts were asked to list all the usability problems found and send them to the conductors of the experiment as a text document.

Analysis

The video data from the user tests were analyzed by writing down everything that the users had problems with and/or commented as a problem in some way. The same was done with the expert evaluations and all the negative findings were picked up from the evaluation reports. In the following, the term 'usability problem' means an individual problem, which was identified either from the user test or from the expert evaluation.

Usability problems were grouped under four different categories (1–4) according to the severity of the problem (categories modified from Nielsen[31]) (Table 2). To make the rating more objective, a conductor of the experiment judged the severity of each problem together with one cartographer and one usability expert.

RESULTS

Altogether, 403 usability problems were found with different evaluation methods (Table 3). The number of problems here means the number of all the problems found with different methods. However, some were found with different methods, so the number of unique usability problems with each Web map is less than the total number

Table 3.   The number of usability problems found from different evaluation methods

	User tests	Cartographic experts	Usability experts	No. of problems	No. of unique problems
Google Maps (GM)	38	17	25	80	69
MSN Maps and Directions (MD)	57	21	18	96	83
MapQuest (MQ)	50	26	32	108	92
Multimap (MM)	71	32	16	119	99
Total	216	96	91	403	343

31. Nielsen, 1993.

of problems. In total, 343 unique problems were identified: 69 in Google Maps, 83 in MSN Maps & Directions, 92 in MapQuest and 99 in Multimap (Table 3).

Severity of the problems

Although the total number of usability problems gives an indication of the usability of the site, the severity of the problem also plays an important role. In total, 33 catastrophic problems were identified (severity category 1), in addition to 138 other major problems (category 2), 127 minor problems (category 3) and 44 cosmetic problems (category 4). From GM only one catastrophic problem was found, whereas MD and MM generated the same number of the most serious problems (13). GM also had the smallest amount of major problems (21) (Figure 1).

USABILITY PROBLEMS AND DESIGN GUIDELINE SUGGESTIONS

The usability problems were grouped under four different categories according to which part of the site they belonged to: 1) user interface; 2) map; 3) search operations; and 4) help and guidance provided to the users in an error situation. The following paragraphs give examples of the most typical problems, followed by a reference to the Web mapping site(s) in which the problem was encountered.

However, the discussion in the following does not take into account how many users experienced each of the usability problems, as it is not the focus of this paper. Preliminary suggestions for design guidelines are given at the end of each category. While some of the guidelines may be 'selfevident' among map designers, the fact that problems emerged during this evaluation suggests that some aspects were not as evident for the designers of these specific Web mapping sites. The words 'participant' and 'user' in the following mean either test users or expert evaluators who participated in the study.

The user interface

First impressions are important when entering Web mapping sites. Despite this, there were a lot of problems relating to 'start pages' and user interfaces (UIs).

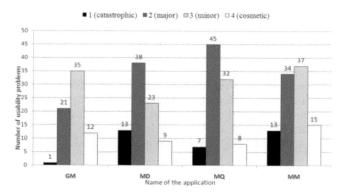

FIG. 1.   Distribution of the severity of usability problems in each Web mapping site

Layout

In many cases, the home pages of the Web maps appeared to be overloaded with different types of information (advertisements, links, images), and the users commented that these looked messy and prevented them from finding relevant information (MD, MQ, MM). Some of the links (e.g. school and insurance sites) were considered irrelevant for ordinary tourist users (MQ, MM). Users were also confused about inactive image links that did not seem to have any purpose (MD). Distractive animations were considered very annoying (MM). Some home pages were criticized for not indicating that they actually were about maps at all, i.e., there was no image or preview of a map (MD, MQ). It was also remarked that some of these sites seemed to be more interested in drawing user attention to different advertisements than actually helping them to find locations (MQ, MM).

The overall layout of the UI was also criticized. For instance, the search box was considered too small and its location wrong because it was not in the centre of the screen (MQ). Some users did not notice the search box during the first 5–10 minutes of trying to find something on the map (GM, MM). The grouping of the map tools, search boxes, and general UI tools was also criticized because function buttons were distributed all over the screen (MM) and advertisements placed disturbingly between some of the function buttons and the map window (MM).

Functionality

Where links in the UI opened in the same browser window as the map (MM), which was then easily lost, this was considered to be a problem. In the home page of one Web map there were three different maps that looked like links, but when the user clicked on the UI nothing happened, and there was no clear feedback as to why not (MM). In some cases there was no quick way back to the home page; the user had to click the back-button many times to get to the main menu (MD). With one Web mapping site, taking the bigger map view caused all the other functions to be pushed away from the screen, forcing the user to scroll the window to use them (MD).

From the analysis of feedback from this part of the experiment, several design guidelines are recommended. These are provided in the following section.

Design guideline 1: the user interface

Layout

- The home page should be clear and simple.
- Intuition is important; the user should be able to start using the map immediately when entering the page.
- There should be a modest number of adverts and animations and these should be located in such a way that they do not disturb the user.
- Information presented on the UIs should be placed logically; attention should be paid to the grouping of the various tools.
- The search box should be given a principal role in the layout.

Functionality

- Links in the UI should not be opened in the same browser window as the map.
- There should always be a short cut back to the home page.

The map

The map was naturally the main focus of these sites. However, in sites where the actual map took quite a small amount of the space on the Web page (MD, MM), this was criticized, as it made it difficult to get an overall picture of a location. There were also many problems related to map visualization and tools.

Map visualization

Maps were criticized for looking like they were designed to appear as paper maps instead of Web maps, because their visualization was messy, confusing, restless and awful to look at on a computer screen (MD, MM). Some maps were regarded as being quite sketch-like (MM), old fashioned (MD), or the map projection looked weird to the participants (MM).

The use of colours was also criticized. For instance, the background colour of built-up areas was considered unreadable and the text was not optimal in contrast (MD); similar colours made it difficult to distinguish between shopping areas and hospitals (MQ); and some colours were considered to be unintuitive (MQ: built-up area). Colours were also criticized in general (MM: scale 1:50 000) and in specific cases (MM: a motorway illustrated with a blue line — 'Looks like a river').

Some maps were overloaded with information and/or colours at certain scales (GM: overview map of London; MM), while others just looked unpleasant 'As if there were no cartographers involved!' (MQ). The categorization of streets for different scales was also thought inappropriate in some cases (MQ). On the other end of the scale were comments that the map was too general (MD, MQ) and, for instance, that the information on the overview map was not sufficient to support decision-making because street names, etc., would be needed (MM, MD: on the fifthclosest zoom, only the names of biggest sites and on the second-closest level, only a few street names were visible).

There were also problems with the text on the maps; the placement of the text was poor (GM, MD), the text was not legible (GM: hybrid map; MD), or the font size was too small for a Web map (MM). Some street and district names were also messy at the biggest scale, so the user could not read them (MD).

Some symbols caused problems because they 'stood out' in relation to the other symbols, especially if it was not clear what they were and why they were emphasized (MD, MQ, MM). Sometimes the users tried to click or point at the symbols to get information about them (MD). On a smallscale map, the names of towns, etc., looked like links from users' point of view, but did not work as such (GM, MM). Many symbols were also misinterpreted (red squares on MM; train tracks on MD).

Some maps also generated problems at different zoom levels. For instance, some symbols (MD) or text (MQ) appeared and disappeared randomly with different scales; the step between map scales was too large (MQ); and in many cases, the visualization between different scales was distinctively different (MD, MQ, MM) (Figure 2). This made it difficult to keep track of a specific location and to make a connection between different scales.

The information included in the maps was criticized as being insufficient, especially regarding public transportation (railway stations, airports, timetables, etc.) and

© Collins-Bartholomew
2006

© TeleAtlas NV/
Crown Copyright

© Crown Copyright,
Licence No 43513U

© Collins-Bartholomew
2005

FIG. 2.   Four different visualizations from the same location when zooming-in from zoom level
5 to level 2 in Multimap (maps from top to bottom; accessed April, 2007). Reproduced by
permission of Multimap, Tele Atlas NV, HarperCollins, and Ordnance Survey. Based on
Ordnance Survey mapping © Crown copyright. AM 53/08

different types of tourist attraction, points of interest, and landmarks (GM, MD, MQ,
MM). In terms of completeness, the data were considered to be inconsistent; some
airports and hotels were shown on the map from a specific location, while others were
not (GM, MM). This gave rise to questions such as 'Who decides what is included or
not in the map?'; 'Is it based on who is paying, e.g. their hotels to be listed for user
queries?' Some participants commented that because of this, they did not know
whether the data was valid. Data accuracy seemed to be insufficient also when one

participant commented that the 'hotel search' gave the same distance to several hotels, which in real life were not close to each other (MM). Sometimes it was impossible to find information about where the map data was from and when was it gathered (MQ).

Map Tools

There was either no legend for the maps, or the participants were not able to find it (GM, MQ, MM, MD). Some users had problems realizing that they could actually perform searches on the map (GM, MM). Estimating distances was also difficult, mainly because some of the users did not realize that there was a scale bar (MD, MM). One scale bar only showed miles, while some of the users only understood the metric system (MM). It was also criticized that the scale bar could only be used for a rough estimation of distance (MD, MM). Some users wanted a grid in the map for comparing different locations and estimating the distances between them (GM, MQ).

Mistakes in design were also observed. For example, the map-size buttons did not work if a route was shown on the screen, although they appeared to be active buttons (MD). At times, parts of the map were covered by zoom buttons and scale bars (GM). The scale bar was also considered to disappear on the map window because it was so tiny (MQ, MM) and/or poorly designed (MM). Some participants criticized the lack of an option to customize the map by checking 'boxes' to show or hide different data layers or symbols on the map (GM, MD), especially because some of the maps were overloaded with so many different objects (MM). An option to highlight various classes of object (e.g. tourist attractions, hotels, restaurants) was also called for (MQ).

In addition, a link to print the map was missing (MQ, MM), as was an option to save a search or system state and thus return to it easily (MQ). A route direction tool 'from here to' was also required (MM), as was an indication of north (MM). Users also wanted to add markers to the map in order to make re-finding a certain location easier (MQ). Some sites provided an option to change the map area, but either users did not realize this or did not understand how it worked (MD). It was also annoying for the participants that the setting for the map size was not retained for the next query (MD).

Panning was sometimes considered problematic and too slow when there was a discrete click to scroll the map (MQ, MM). If there was no feedback, users often thought that they had missed the button the first time and so they clicked it again. Participants were also confused about the different types of zoom setting and their relationship with each other (MM). One user did not realize that there was a zoom function at all (GM). Sometimes the zoom function was criticized as being old fashioned (with scale numbers) and confusing for ordinary users (MD, MM). Zooming was also considered problematic when there were neither steps nor animation when switching between different zoom scales (MD, MQ), because users lost the location that they were looking at earlier. With one site, zooming moved the search result out of the map window because the search result was not centred on the map when starting off (MM).

Some participants would have liked to point at the area into which they were interested in zooming (MD). It was found confusing that the map could be zoomed by clicking on it, as the cursor did not change when it was pointed at the map (MD). It was also considered annoying that clicking on the map did not just centre the view, because it always also zoomed in (MM). It was surprising to the users that clicking on the map re-focused and re-centred it, when they only wanted to point on it (GM). Accidental zooming also occurred when participants used the scroll wheel of the mouse when they wanted instead to scroll down the search results window (GM).

From these results, a series of design guidelines were developed that relate to the map. These are provided in the following section.

Design guideline 2: the map

Visualization

- The map should be visualized according to the properties of the computer screen.
- The map should be optimized for viewing on a computer screen.
- Maps should be simple and intuitive and pleasant to use. Colours should be in harmony.
- Each map scale should be considered separately: what information should be included and how it should be visualized at each of the scales.
- Information about data accuracy and validity should be provided.

Map Tools

- Map tools should be distinctive, but not obscure too much information on the map.
- A route-measuring tool would be beneficial (in addition to a scale bar).
- New tools would be beneficial for users: an option to add markers on the map; to click on different objects in order to get more information about them; to customize the map by checking 'boxes' to show or hide different data layers or symbols on the map (e.g. tourist attractions, hotels, restaurants); and incorporate an easy way to print and email the map.
- The scale bar (and other) units should be customisable.
- A continuous click-and-drag option would be best for panning.
- Scale increments should not be too great, allowing users to follow a specific location while zooming in and out.
- Scale numbers (ratios or representative fractions) should not be used. Instead, scale should be indicated by more commonly used terms (such as street level, city level, country level, etc.).

Search operations

A significant number of usability problems were found relating to queries and searches for different locations and objects on the maps.

Search Criteria/Logic

Of the four Web mapping sites used in this study, one site was different from the others in that it supported a 'free search', whereby the user could type their search

FIG. 3. Different types of search possibilities with Web mapping sites (accessed April, 2007): a) Google Maps (© 2007 Google); b) MapQuest (© 2008 MapQuest, Inc. MapQuest and the MapQuest logo are registered trademarks of MapQuest, Inc). Used with permission

criteria more liberally in one or two search boxes (GM) (Figure 3a). The other sites provided users with different search boxes, each requiring a certain kind of text, e.g. country, address, place name, etc. (Figure 3b).

Both search types had their positive and negative elements. The free search was liked because it is the way people normally find information when using search engines. However, it was also considered to be confusing: 'What can you really search? And how?'. For example, one user typed in the search box 'road1 to road2' and then pressed 'get directions', but got no results (GM). It was also commented upon that minimalist thinking is being taken somewhat too far; users may like to have access to at least some shortcut buttons (instead of always having to search).

The positive elements of the other search type (MD, MQ, MM) were that people are more used to having separate search boxes for 'location', 'directions' and 'businesses' with Web maps, and most of the time people also know what to type in each search box. On the other hand, the boxes were not very flexible and often required the data to be typed exactly in the correct way. For example, a specific operation such as 'Find a place' can be misleading when asked 'what does "a place" actually mean?' (MD).

It was observed that the users wanted to make not only one search at a time, but also several separate searches simultaneously (multi-searches), so that the different objects would appear on the same map at the same time (GM, MD, MM). Moreover, people did not know whether or not the search they carried out was only going to include the area currently shown on the map (GM, MM).

Another criticism was that the only way to search for addresses or directions was via entering text, whereas it would be helpful to have the map as an interface as well, i.e., to be able to click on the map for start/end points of a route (MD). Searching for addresses was also not always easy; for example, if an address was entered street name first, house number second — as is the norm in central Europe — no results were found. Hence, the user needs to know that in the UK, the house number is placed first for an address (MD). Users were also frustrated by not being able to search for anything else other than addresses (i.e. places, MQ).

Default Settings

The severe usability problems encountered most often related to the default settings of the Web mapping sites, which, in the worst case, prevented some of the participants from using the sites. For example, if the user typed 'London Bridge', the site would only give results from the USA (GM), because the participant did not notice the USA default, or did not know how to change it (MD, MQ) (Figure 3b). It was considered especially frustrating that the search box always went back to the default settings — even though users had already changed the country to something else (MD, MQ). However, some steps were in fact sought after as default settings, for example, when choosing 'UK' as a 'start' for a direction search, the country at the 'end' should automatically change to UK too (MD, MQ).

Search Results

Often, the participants did not know how the search results matched their search criteria. On one site, the user typed in 'The London Bridge' and the search returned 'The Bridge', which was not the required result. The user had to accept this because if only one result matches the search criteria, the result is shown on the map without any explanation of how it matched the search (GM). On another occasion, the user got a list of 'Londons' in the USA and did not realize that these were not in the desired country (MQ). The participants also tried to use two or more search criteria at the same time and often got a map with the result displayed (MM). However, this was not always the correct result, since it sometimes returned only one search criterion. The users did not always realize this, because it was not pointed out to them where the result came from and/or how it had been deduced.

The participants remarked that the search results were sometimes 'weird' and that there was no help available to explain where they had come from. The users had to be sure about what they were looking for because some searches gave a number of incorrect results, even though the search was very well defined (e.g. 'Big Ben' gave results everywhere else except London) (MM). Search results were even more confusing when they were based on similar-sounding place names; 'London tower' gave the result 'Lake Teterower' (a lake in Germany) and 'Longbridge' in Birmingham (MD). One user got 25 results for a simple search, because everything that included the searched name or even sounded the same was included ('Tussaud' resulted in 'Tosside' and 'Thickwood'; 'train' resulted in all the names starting with 'tr' and 'th') (MM). It was difficult for users to figure out why such results appeared in the results of specific searches.

Performing route queries became problematic when users did not know which of the search results was the start or end points on a map of their route (GM). It was also criticized that users could not easily change the start or finish of a route already shown on the screen (GM, MD). For 'directions search' the participants would have liked to have all the possible choices (results) for the end and start to be shown on the map to help choose between them (MM).

The users did not always like, or realize, the fact that one site made the route suggestion automatically based on the previous road search (MQ). It was also confusing that when users searched for a route they got the result as a text description, not as a map as they had expected (MQ) (to see the route, users had to scroll the view). Some participants also wanted multi-stop route searches, enabling them to search for routes from A to C via B (MM, MQ). More choice for customizing routes was also required: quickest, shortest, and with different transportation modes (GM, MD). It was also noted that it would be good to be able to search for businesses (e.g. restaurants) along a specific road or route (GM, MD).

Criticism of the visualization of search results, for instance, when a street result was visualized with a pin (usually used for a single location) instead of linear highlighting (GM, MD) was also present. Comparing different search results was considered difficult, because they were not shown on the map at the same time (GM, MD) or because they were shown in different scales (MD, MM). Sometimes, the users had to open another map window to compare a distance between two locations (MD).

The search results were occasionally shown on the map on top of each other, so that the users were not able to see them all (GM, MM). One site centred the map according to the result without any visual emphasis of its location (MD). The users did not realize that this was a result, especially because not even its name was visible (five zoom-in operations would have been needed to see the text) (MD, MM). The same problem occurred when searching for routes (MD) and roads (MQ) where the results were presented at a scale where they could not be seen.

The search results were also easily lost on a map (GM, MM). The users commented that there should be an option allowing a quick return to the search result instead of constantly having to click back to the search page (MQ, MM). With one site, the zooming did not work when clicking the search result, but only when clicking around it, and the user thought that zooming was impossible (MM). With one Web mapping site there was no route shown after performing a directions search, only points indicating where to turn, and these were considered difficult to read (MM, Figure 4). More dramatically, the route visualization changed between map scales (MM). Sometimes the search results were given on the same scale as the map preceding the search, therefore the whole route was not always shown on the map (MM). The lack of an option to print out the visualization of the route was criticized (MM), as it was considered a common task.

A number of guidelines related to the search operations were proposed as an outcome of these evaluation results and are provided in the following section.

Design guideline 3: the search operations

Functionality

- Different types of searches should be supported.
- Users should know with what type of criteria the search is carried out.
- A list of users' previous searches should be saved and provided to them.
- It should be made clear to users what the search results are based on and how they relate to the query.

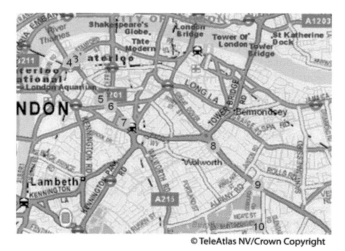

© TeleAtlas NV/Crown Copyright

FIG. 4.   Route visualization with turning points on top of each other in Multimap (accessed April, 2007). Reproduced by permission of Multimap and Tele Atlas NV. Based on Ordnance Survey mapping © Crown copyright. AM 53/08

Visualization

- The results should be centred on the map and distinctively visualized, taking into account the symbols that are already in use on the map.
- The result symbols should not cover the map too much and be on top of each other.
- The defaulting map scale should give enough information for the user to check whether or not the result is correct.
- It would be beneficial to show all the possible results on the map, so that the user can choose the correct option among them.
- Street and route search results should be visualized with a line.
- Route search results should be displayed on a tailored map scale so that the user sees the entire route.

Help and guidance in an error situation

Error situations are often inevitable with map sites because users may, for example, search something that does not exist in the database. It was observed that, in some situations there was no proper help available. Instructions on how to start using the Web map were missing from some sites, or the existing instructions were not considered useful (MQ, MM).

Some error messages did not look like a message and users did not notice them appearing on the screen (MM). If the error message was given clearly, it was not always informative (GM, MM). Some of the sites did not provide any help (or the users did not find it) for using the map (GM) or for looking for streets and directions (MQ). Some sites gave examples for help in using the searches, but they were also confusing: the help text 'in London' worked only for businesses (such as 'curry in London'), but not for street searches (GM). Sometimes the 'help' was not what the user expected; the user needed help for finding locations but only got a legend for tools (MM).

Design guideline 4: help and guidance

- The user should be provided with help in map use and in other functions in the site.
- Error messages should be clear, informative and distinctive.
- Users should be informed of current default settings and how they can be changed.

DISCUSSION AND CONCLUSIONS

By identifying pitfalls in existing Web maps, it is possible to offer recommendations on how to design Web mapping sites that are easier to use and attractive to different groups of users. A possible bias in this study, however, may be drawn from the fact that the Web maps included in the evaluation were well known and widely used. It might therefore be expected that such sites have fewer problems than more unfamiliar applications as a result of their popularity. The evaluation of the Web mapping sites nevertheless identified a considerable number of severe usability problems. If these were typical for Web maps that are in use every day and by large numbers of people, it would be interesting to investigate the usability of smaller, less familiar map applications. While this study did not seek to do so, the topic should nevertheless be investigated in future.

Even though many usability problems were identified, some of the problems may have been exacerbated by the tasks chosen for this study. Different sites can have different objectives, and the use-scenario may not have corresponded exactly to that for which these sites were originally designed. In fact, some of these Web maps may not have been designed for use by tourists. This uncertainty should be kept in mind when considering the results of this study.

As map sites are unquestionably visual in their nature, distractive advertisements and messy user interfaces were criticized. A map that frustrates the user from the very beginning may cause very negative feelings towards it. Some of the users actually stated that in a real-life situation they would have given up trying to complete the tasks with some of these sites and tried another Web map. This is important, as it may be that the product developer who can design the most usable application, will win the battle for market dominance.

Some of the Web maps have been in existence longer than others, which may have biased the results of this study. Some users may have been attracted by newer ideas and these might have received more positive comments because of that. On the other hand, some of the users valued traditional types of services because they are used to them. This was especially obvious when the different search criteria of the sites were discussed. Some people have been used to making Web searches with search engines, and they also wanted to carry out map searches in the same 'free' manner. Others needed more structured or guided searches. The challenge remains to design sites that different types of people can use without getting frustrated or without facing a lot of problems in using them.

Another challenge is that some of the participants had hardly used any types of maps at all and, for them, the use of these sites was especially difficult; some of the users did not even realize that the map scale could be changed or that searches could be carried out for different objects. This is understandable, since Web maps deal with complicated spatial data and may allow a high degree of interactivity between the user and the site. How can we help ordinary Internet users to realize the variety of map sites and their functionality and benefit from their use? The observed lack of guidance within these sites does not help in this situation.

The Web maps often offer links to different additional services (such as hotels and tourist attractions), which either have their own map interface or no map at all. If a user wanted information on how to use the underground rail network to get from one tourist attraction to a hotel, at least three different maps and services had to be opened at the same time: an underground route map, a map with hotels on it, a map with tourist attractions on it, and perhaps even a base map for combining all this information. If all of these have their own maps with different scales and visualizations, users will find it difficult to combine the information. The best solution would be to have all these embedded within the same map service, or to have harmonized maps between different services, but it is clear that this requires further study.

Reflections on 'Usability Evaluation of Web Mapping Sites'

CORNÉ P. J. M. VAN ELZAKKER
ITC, University of Twente

DAVID FORREST
University of Glasgow

The paper by Annu-Maaria Nivala *et al.* is itself an example of work that met the needs of the users (i.e. readers of *The Cartographic Journal*) at exactly the right moment in time. During the first decade of the twenty-first century, the results of the technological revolution became clear to everyone in cartography:

- maps are not static products any more; they are now dynamic and interactive;
- most maps are disseminated through the Web;
- increasingly, maps are displayed on mobile devices to provide location-based services;
- it has become relatively easy to produce Web mapping applications;
- it is no longer costly to iteratively produce several versions of a cartographic design solution;
- new methods and techniques of carrying out use, user and usability research are feasible.

At the same time, interest in users and usability increased enormously and cartography became more demand-driven, instead of supply-driven. It also became clear that the scope of use and user research in cartography had to be broadened; map use research could not be done in isolation, but now had to be undertaken in relation to the design and use of aspects such as the user interface, hardware and software, data and databases, and the context of map use (i.e. where and when map displays are being used). User-Centred Design approaches are on the rise and, rightfully, the focus is not only on usability research (testing design solutions), but also on use and user requirement analysis (before any prototype is made) (Figure 1).

Position and research papers related to the developments sketched above were avidly devoured, if mainly by researchers. But Nivala *et al.* (p. 129) claimed what was not only interesting to researchers, but was now clear to professional cartographers as well: 'traditional map design and evaluation methods may no longer always be valid'. At the time of publication of this paper, this was realized by professional cartographers but, of course, they had many 'how to' questions to ask. A wider readership was probably triggered by the clear description of the purpose of the study of

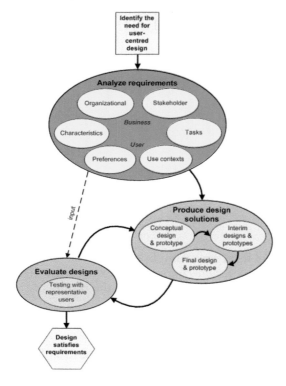

Fig. 1.   More attention for requirement analysis in user-centred design of mapping applications (van Elzakker *et al.*, 2008)

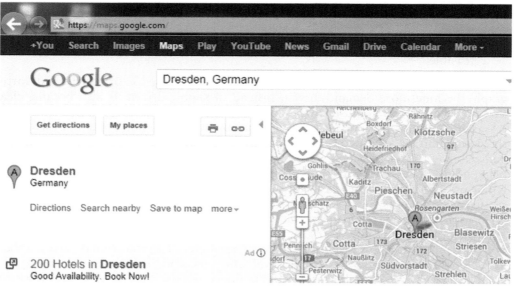

Fig. 2.   Required broadening of user research scope: not only the use of the map should be investigated, but also, for example, the user interface (© Google, 2013)

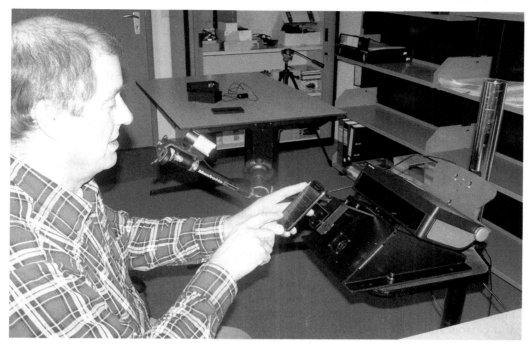

Fig. 3.   New user research methods are now feasible and are often applied in combination (here, the methods eye-tracking, video observation, thinking aloud and screen logging are 'mixed')

Nivala *et al.*: 'to identify potential usability problems of Web mapping sites in order to provide guidance for the future design of such services' (p. 129).

Indeed, Nivala *et al.* met this paper's purpose very well. Firstly, they provide very concrete and useful design guidelines. In doing so, in line with the required broadening of scope, they not only give guidelines for the design of the map image itself, but also for the design of the user interface and functions like searching and providing help to users of the Web mapping services (Figure 2). Secondly, the paper also sets a model for undertaking holistic, functional user research. At the time of publication not much research into the usability of Web mapping sites like Google Maps had been published. Nivala *et al.* mention one of the few exceptions: the research of Skarlatidou and Haklay (2006). There were still few published examples of applying (in holistic, functional user research) techniques other than surveys and interviews. Indeed, the study of Nivala *et al.* was one of those (along with the research of van Elzakker, 2004, for example) that marked the rise of qualitative research in the cartographic discipline at the beginning of the twenty-first century. Nivala *et al.* made use of both expert evaluations and actual user tests. In addition to a background questionnaire, their methodology was based on formulated use scenarios from which concrete tasks were derived that had to be executed by the participants. While completing these tasks, participants were video recorded and were asked to describe the reasoning behind their interactions with the Web mapping application by thinking aloud. The changes on the computer screen were recorded as well and later analysed simultaneously. As such, this is an early example of a so-called 'mixed methods' approach (see e.g. Bleisch *et al.*, 2010) that is much more common today than the application of a single-user research technique. The reason for this is that different techniques lead to different and complementary information about uses, users and usability (Figure 3).

Annu-Maaria Nivala wrote this paper as a result of a research visit to the Department of Computing Science at the University of Glasgow. There she learned a lot about what computer scientists already knew about human–computer interaction and about the methods and techniques that can be used to research that interaction. There is indeed a lot we can still learn from computer scientists, and *vice versa*. At a Workshop on Geographic Human–Computer Interaction held at CHI2013 in Paris (see http://geohci2013.grouplens.org/), it became clear, now that geographic locations become more and more important in mobile computing, computer scientists are eager to learn more from cartographers about the special characteristics of geographic information and linking a map representation of that information with reality and mental maps.

The research visit of Annu-Maaria Nivala to the University of Glasgow was part of her PhD research at the Department of Geoinformatics and Cartography of the Finnish Geodetic Institute (http://www.fgi.fi/fgi/research/department/geoinformatics-and-cartography). The usability research group of Tapani and Tiina Sarjakoski in this Department was and is one of the leading use, user and usability research groups in the world. The paper of Nivala *et al.* is a fine example of the type of applied research they were executing already for quite some time. Another example is the famous GIMODIG project (Sarjakoski and Sarjakoski, 2005). Perhaps one disadvantage of doing such concrete, applied research is that promising young scientists are captured by commercial companies: Annu-Maaria Nivala first became Senior Consultant in User Experience at Knowit Oy and is now Product Manager for Patient Data Capture Solutions at CRF Health in Helsinki, Finland.

REFERENCES

Bleisch, S., Dykes, J. and Nebiker, S. (2010). 'A Mixed Methods Research Approach for 3D Geovisualization Evaluation', Presentation at **Workshop on Methods and Techniques of Use, User and Usability Research,** London, Apr 13.
Sarjakoski T. and Sarjakoski, L. T. (2005). 'The GiMoDig public final report', GiMoDig-project, IST-2000-30090, Deliverable D1.2.31, Public EC report, http://gimodig.fgi.fi/deliverables.php.
Skarlatidou, A. and Haklay, M. (2006). 'Public Web Mapping: Preliminary Usability Evaluation', Presentation at **GISRUK 2005**, Glasgow, Apr 6.
van Elzakker, C. P. J. M. (2004). 'The use of maps in the exploration of geographic data', **Netherlands Geographical Studies,** 326.
van Elzakker, C. P. J. M., Delikostidis, I. and van Oosterom, P. J. M. (2008). 'Field-based usability evaluation methodology for mobile geo-applications', **The Cartographic Journal,** 45, pp. 139–149.

Unfolding the Earth: Myriahedral Projections

JARKE J. VAN WIJK

Originally published in *The Cartographic Journal* (2008) 45, pp. 32–42.

Myriahedral projections are a new class of methods for mapping the earth. The globe is projected on a myriahedron, a polyhedron with a very large number of faces. Next, this polyhedron is cut open and unfolded. The resulting maps have a large number of interrupts, but are (almost) conformal and conserve areas. A general approach is presented to decide where to cut the globe, followed by three different types of solution. These follow from the use of meshes based on the standard graticule, the use of recursively subdivided polyhedra and meshes derived from the geography of the earth. A number of examples are presented, including maps for tutorial purposes, optimal foldouts of Platonic solids, and a map of the coastline of the earth.

INTRODUCTION

Mapping the earth is an old and intensively studied problem. For about two thousand years, the challenge to show the round earth on a flat surface has attracted many cartographers, mathematicians, and inventors, and hundreds of solutions have been developed. There are several reasons for this high interest. First of all, the geography of the earth itself is interesting for all its inhabitants. Secondly, there are no perfect solutions possible such that the surface of the earth is depicted without distortion. Finally, factors such as the intended use of the map (e.g. navigation, visualization, or presentation), the available technology (pen and ruler or computer), and the area or aspect to be depicted lead to different requirements and hence to different optima.

A layman might wonder why map projection is a problem at all. A map of a small area, such as a district or city, is almost free of distortion. So, to obtain a map of the earth without distortion one just has to stitch together a large number of such small maps. In this article we explore what happens when this naive approach is pursued. We have coined the term *myriahedral* for the resulting class of projections. A *myriahedron* is a polyhedron with a myriad of faces. The Latin word *myriad* is derived from the Greek word *murioi*, which means ten thousands or innumerable. We project the surface of the earth on such a myriahedron, we label its edges as folds or cuts, and fold it out to obtain a flat map.

In the next section, some basic notions on map projection are presented, and related work is shortly described, followed by a section in which an overview of the approach employed here is given. Different solutions are obtained by using different myriahedra and choices for the edges to be cut, which are described in three separate sections. The use of graticule-based meshes, recursively subdivided polyhedra, and geographically aligned meshes lead to different maps, each with their own strengths. Finally, the results are discussed.

BACKGROUND

The globe is a useful model for the surface of the earth. Locations on a globe (and the earth) are given by latitude w and longitude l. The position of a point p(ϕ, λ) on a globe with unit radius is (cosλcosϕ, sinλcosϕ, sinϕ). Curves of constant ϕ, such as the equator, are parallels; curves of constant λ are meridians. A graticule is a set of parallels and meridians at equal spacing in degrees.

Compared with a map, a globe has some disadvantages, such as poor portability, and to obtain a more practical solution, the spherical globe has to be mapped to a flat surface. This puzzle has intrigued many researchers for two thousand years. John P. Snyder has provided a fascinating overview of the history of map projection. In the following, references for map projections are only given if not discussed in his book,[1] to keep the number of references within bounds. Introductions to map projection can be found in textbooks on cartography or geographic visualization.[2,3,4] Also on the web much information can be found, for instance in the extensive website developed by Furuti.[5]

The major problem of map projection is distortion. Consider a small circle on the globe. After projection on a map, this circle transforms into an ellipse, known as the Tissot indicatrix, with semi-axes with lengths a and b. If $a = b$ for all locations, then angles between lines on the globe are maintained after projection: The projection is *conformal*. The classic example is the Mercator projection. Locally, conformality preserves shapes, but for larger areas distortions occur. For example, in the Mercator projection shapes near the poles are strongly distorted.

If $ab = C$ for all locations on the map, then the projection has the *equal-area* property: Areas are preserved after projection. Examples are the sinusoidal, Lambert's cylindrical equal area and the Gall–Peters projection.

The problem is that for a double curved surface no projection is possible that is both conformal and equal-area. Along a curve on the surface, such as the equator, both conditions can be met; however, at increasing distance from such a curve the distortion accumulates. Therefore, depending on the purpose of the map, one of these properties or a compromise between them has to be chosen. Concerning distortion, uniform distances are another aspect to be optimized. Unfortunately, no map projections are possible such that distances between any two positions are depicted on a similar scale, but one can aim at small variations overall or at proper depiction along certain lines.

Besides these constraints from differential geometry, map projection also has to cope with a topological issue. A sphere is a surface without a boundary, whereas a finite flat area has to be bounded. Hence, a cartographer has to decide where to cut

1. Snyder, J. P. (1993). **Flattening the Earth: Two Thousand Years of Map Projections**, University of Chicago Press.

2. Robinson, A. H., Morrison, J. L., Muehrcke, P. C. and Kimerling, A. J. (1995). **Elements of Cartography**, Wiley.

3. Kraak, M.-J. and Ormeling, F. (2002). **Cartography: Visualization of Geospatial Data** (2nd edition), Prentice Hall, London.

4. Slocum, T. A., McMaster, R. B., Kessler, F. C. and Howard, H. H. (2003). **Thematic Cartography and Geographic Visualization**, Second Edition, Prentice Hall.

5. Furuti, C. A. (2006). 'Map Projections' **http://www.progonos.com/furuti/MapProj**.

the globe and to which curve this cut has to be mapped. Many choices are possible. One option, used for azimuthal projections, is to cut the surface of the globe at a single point, and to project this to a circle, leading to very strong distortions at the boundary. The most popular choice is to cut the globular surface along a meridian, and to project the two edges of this cut to an ellipse, a flattened ellipse or a rectangle, where in the last two cases the point-shaped poles are projected to curves.

The use of interrupts reduces distortion. For the production of globes, minimal distortion is vital for production purposes; hence gore maps are used, where the world is divided in for instance twelve gores. Goode's homolosine projection (1923) is an equal-area projection, composed from twelve regions to form six interrupted lobes, with interrupts through the oceans. The projection of the earth on unfolded polyhedra instead of rectangles or ellipses is an old idea, going back to Da Vinci and Dürer. All regular polyhedra have been proposed as suitable candidates. Some examples are Cahill's Butterfly Map (1909, octahedron) and the Dymaxion Map of Buckminster Fuller, who used a cuboctahedron (1946) and an icosahedron (1954). Steve Waterman has developed an appealing polyhedral map, based on sphere packing.

Figure 1 visualizes the trade-off to be made when dealing with distortion in map projection. An ideal projection should be equal-area, conformal, and have no interrupts; however, at most, two of these can be satisfied simultaneously. Such projections are shown here at the corners of a triangle, whereas edges denote solutions where one of the requirements is satisfied. Existing solutions can be positioned in this solution space. Examples are given for some cylindrical projections, with linear parallels and meridians. Most of the existing solutions, using no interrupts, are located at the bottom of the triangle. In this article, we explore the top of the triangle, which is

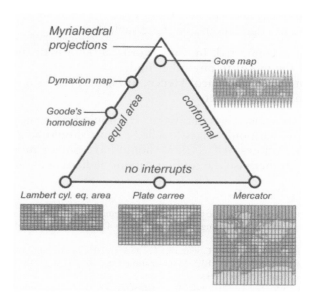

FIG. 1. Distortion in map projection

still *terra incognita*, using geographic terminology. Or, in other words, we discuss projections that are both (almost) equal area and conformal, but do have a very large number of interrupts.

Related issues have been studied intensively in the fields of computer graphics and geometric modelling, for applications such as texture mapping, finite-element surface meshing, and generation of clothing patterns. The problem of earth mapping is a particular case of the general surface parameterization problem. A survey is given by Floater and Hormann.[6] Finding strips on meshes has been studied in the context of mesh compression and mesh rendering, for instance by Karni *et al.*[7] Bounded-distortion flattening of curved surfaces via cuts was studied by Sorkine *et al.*[8] The work presented here has a different scope and ambition as this related work. The geometry to be handled is just a sphere. The aim is to obtain zero distortion, and we accept a large number of cuts. Finally, we aim at providing an integrated framework, offering fine control over the results, and explore the effect of different choices for the depiction of the surface of the earth.

METHOD

We project the globe on a polyhedral mesh, label edges as cuts or folds, and unfold the mesh. We assume that the faces of the mesh are small compared with the radius of the globe, such that area and angular distortion are almost negligible. We first discuss the labelling problem. A mesh can be considered as a (planar) graph $G = (V, E)$, consisting of a set of vertices V and undirected edges E that connect vertices. Consider the dual graph $H = (V', E')$, where each vertex denotes a face of the mesh, and each edge corresponds to an edge of the original graph, but now connecting two faces instead of two vertices (Figure 2). After labelling edges as folds and cuts, we obtain two subgraphs H_f and Hc, where all edges of each subgraph are labelled the same. The labelling of edges should be done such that

- N the foldout is connected. In other words, in H_f a path should exist from any node (face of the mesh) to any other node.
- N the foldout can be flattened. Hence, in Hf no cycles should occur, otherwise this condition cannot be met.

Taken together, these constraints imply that H_f should be a spanning tree of H. Also, the subgraph G_c of G with only edges labelled as cuts should be a spanning tree of G. This can be seen as follows. All vertices should have one or more cuts in the set of neighbouring edges (otherwise the foldout cannot be flattened), and cycles in the cuts

6. Floater, M. S. and Hormann, K. (2005). 'Surface parameterisation: a tutorial and survey', in **Advances in multiresolution for geometric modelling**, 157–186, ed. by Dodgson, N. A., Floater, M. S. and Sabin, M. A. (eds), Springer Verlag.

7. Karni, Z., Bogomjakov, A. and Gotsman, C. (2002). 'Efficient compression and rendering of multi-resolution meshes', **Proceedings of IEEE Visualization** 2002, 347–354, ed. By Moorhead, R., Gross, M. and Joy, K.I., IEEE Computer Society Press.

8. Sorkine, O., Cohen-Or, D., Goldenthal, R. and Lischinski, D. (2002). 'Bounded-distortion piecewise mesh parameterisation', **Proceedings of IEEE Visualization** 2002, 355–362, ed. by Moorhead, R., Gross, M. and Joy, K.I., IEEE Computer Society Press.

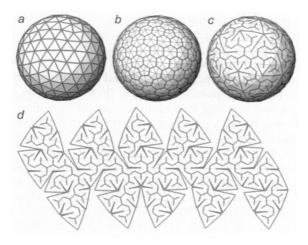

FIG. 2.   (a) Mesh G; (b) Dual mesh H; (c) Cuts and folds; (d) Foldout

would lead to a split of the foldout. The set of cuts unfolds to a single boundary, with a length of twice the sum of lengths of the cuts.

There is a third constraint to be satisfied: The labelling should be such that the foldout does not suffer from foldovers. The folded out mesh should not only be planar, it should also be single-valued. The use of an arbitrary spanning tree does lead to fold-overs in general. However, we found empirically that the schemes we use in the following almost never lead to fold-overs, and we do not explicitly test on this. The problem of fold-overs is complex, and we cannot give proofs on this. Nevertheless, it can be understood that fold-overs are rare by observing that the sphere is a very simple, uniform, convex surface; and also, the typical patterns that emerge are strips of triangles, connected to and radiating outward from a line or point, which strips rarely overlap.

The term spanning tree suggests a solution for labelling the edges: Minimal spanning trees of graphs are a wellknown concept in computer science. Assign a weight $w(e_i)$ to each of the edges e_i, such that a high value indicates a high strength and that we prefer this edge to be a fold. Next, calculate a maximal spanning tree H_f (or a minimal spanning tree G_c), *i.e.*, a spanning tree such that the sum of the weights its edges is maximal (or minimal). The algorithm to produce a myriahedral projection is now as follows:

1.   Generate a mesh;
2.   Assign weights to all edges;
3.   Calculate a maximal spanning tree H_f;
4.   Unfold the mesh;
5.   Render the unfolded mesh.

In the following sections, we discuss various choices for the first two steps, here we describe the last three steps, which are the same for all results shown.

For the calculation of the maximal spanning tree we followed the recommendations given by Moret and Shapiro.[9] We use Prim's algorithm[10] to find a maximal spanning tree. Starting from a single vertex, iteratively, the neighbouring edge with the highest weight and the corresponding vertex is added. This gives an optimal solution. The neighbouring edges of the growing tree are stored in a priority queue, for which we use pairing heaps.[11] The performance is $O(|E| + |V| \log |V|)$, where $|E|$ and $|V|$ denote the number of edges and vertices. In practice, optimal spanning trees are calculated within a second for graphs with ten thousands of edges and vertices.

Unfolding is straightforward. Assume that all faces of the mesh are triangles. Faces with more edges can be handled by inserting interior edges with very high weights, such that these faces are never split up. Unfolding is done by first picking a central face, followed by recursive processing of adjacent faces. Consider two neighbouring triangles PQR and RQS, and assume that the unfolded positions P', Q', and R' are known. Next, the angle a between RQS and the plane of PQR is determined, and S' is calculated such that the new angle is α', $|QS| = |Q'S'|$ and $|RS| = |R'S'|$. The use of $\alpha' = 0$ gives a flat mesh, use of (for instance) $\alpha' = \alpha(1 + \cos(\pi t/T))/2$ gives a pleasant animation (examples are shown in http://www.win.tue.nl/~ vanwijk/myriahedral).

The geography of the earth (or whatever image on a spherical surface has to be displayed) is mapped as a texture on the triangles. We use the maps of David Pape for this.[12] When the triangles are large compared with the radius of the globe, like in standard polyhedral projections, the triangles have to be subdivided further to control the projection in the interior. We use a simple gnomonic projection here.

Rendering maps for presentation purposes requires proper anti-aliasing, because regular patterns and very thin gaps have to be dealt with. For the images shown, 100-fold supersampling per pixel with a jittered grid was used, followed by filtering with a Mitchell filter.

All images were produced with a custom developed, integrated tool to define meshes and weights, and to calculate and render the results, running under MS Windows. Response times on standard PCs range from instantaneous to a few seconds, which enables fast exploration of parameter spaces. Rendering of high resolution, high quality maps can take somewhat longer, up to a few minutes.

GRATICULES

The simplest way to define a mesh is to use the graticule itself, and to cut along parallels or meridians. The results can be used as an introduction to map projection. A weight for edges, using the value of ϕ and λ of the midpoint of an edge, can be defined as

$$w(\phi,\lambda) = -(W_\phi|\phi - \phi_o| + W_\lambda \min_k|\lambda - \lambda_o + 2\pi k|),$$

9. Moret, B. M. E. and Shapiro, H. D. (1991). 'An empirical analysis of algorithms for constructing a minimum spanning tree', **Lecture Notes in Computer Science 555**, 192–203, Springer Verlag.

10. Prim, R. (1957). 'Shortest connection networks and some generalisations', **Bell System Technical Journal 36**, 1389–1401.

11. Fredman, M. L., Sedgewick, R., Sleator, D. D. and Tarjan, R. (1986). 'The pairing heap: A new form of self-adjusting heap', **Algorithmica 1(1)**, 111–129.

12. Pape, D. (2001). 'Earth images', **www.evl.uic.edu/pape/data/Earth**.

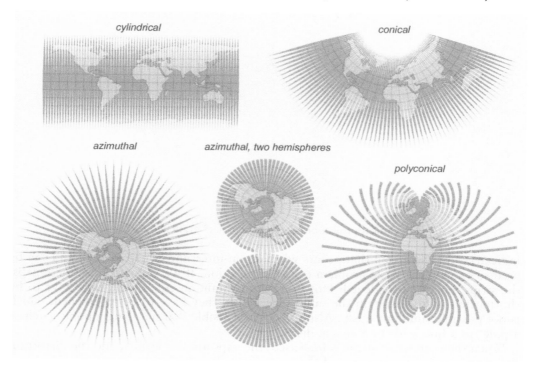

FIG. 3.   Graticular projections, derived from a 5u graticule. 2592 polygons: a) cylindrical;
b) conical; c) azimuthal; d) azimuthal, two hemispheres; e) polyconical

where W_ϕ and W_λ are overall scaling factors, and ϕ_o and λ_o denote where a maximal strength is desired. Different values for these lead to a number of familiar looking projections (Figure 3). The use of a high value for W_ϕ gives cuts along meridians. Dependent on the value of ϕ_o a cylindrical projection (0°, equator), an azimuthal projection (90°, North pole), or a conical projection (here 25°) is obtained when the meridian strips are unfolded. Use of a negative value for W_ϕ gives two hemispheres, each with an azimuthal projection. The meridian at which to be centred can be controlled by using a low value for W_λ and a suitable value for λ_o. The use of a high value for W_ϕ gives cuts along parallels. Unfolding these parallels gives a result resembling the polyconic projection of Hassler (1820).

The relation between a spatially varying weight w and the decision where to cut and fold can be understood by considering Prim's algorithm. Suppose, without loss of generality, that we start at a maximum of w and proceed to attach the edges with the highest weight. At some point, edges at the boundary will have approximately the same weight and, after a number of additions, a ring of faces is added, with cuts in between neighbouring faces in this ring. Hence, edges aligned with contours of w typically turn into folds, whereas edges aligned with gradients of w turn into cuts.

Each strip is almost free of angular or area distortion, however, a large number of interrupts occur with varying widths. These gaps visualize, just like the Tissot

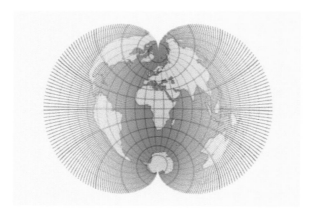

FIG. 4. Polyconical projection, derived from a 1° graticule, 64 800 polygons

indicatrix, the distortion that occurs when a non-interrupted map is used, and can be used to explain the basic problem of map projection. If we want to close these gaps, the strips must be broadened. However, to maintain an equal area, they have to be shortened, and to maintain the same aspect ratio they have to be lengthened, which is not possible simultaneously. Also, it is clearly visible that mapping a point (such as a pole) to a line leads to a strong distortion.

When the number of strips is increased, the gaps are less visible, and the distortion is shown via the transparency of the map (Figure 4).

RECURSIVE SUBDIVISION

For the graticular projections, thin strips of faces are attached to one single strip or face. This is a degenerated tree structure. In this section, we consider what results are obtained when a more balanced pattern is used. To this end, we start with Platonic solids for the projection of the globe, and recursively subdivide the polygons of these solids. This approach has been used before for encoding and handling geospatial data.[13]

At each level i, each edge is split and the new centres, halfway on the greater circle connecting the original endpoints, are connected. As a result, for instance each triangle is replaced at each level by four smaller triangles. Other subdivision schemes can also be used, for instance triangles can be subdivided into nine smaller ones.

The edge weights are set as follows. We associate with each edge three numbers w_o, w_1, and w_c, where the first two correspond with the endpoints and the latter with the centre position. For new edges, $w_o \leftarrow i$, $w_1 \leftarrow i$, and $w_c \leftarrow i+1$. If an edge e is split into two edges e' and e'', we use linear interpolation for the new values

$$
\begin{array}{lll}
w'_o \leftarrow w_o, & w'_1 \leftarrow w_c, & w'_c \leftarrow (w_o + w_c)/2; \\
w''_o \leftarrow w_c, & w''_1 \leftarrow w_1, & w''_c \leftarrow (w_c + w_1)/2
\end{array}
$$

13. Dutton, G. (1996). 'Encoding and handling geospatial data with hierarchical triangular meshes', in **Advances in GIS Research II** (Proc. SDH7, Delft, Holland), 505–518, ed. by Kraak, M.-J. and Molenaar, M., Taylor & Francis, London.

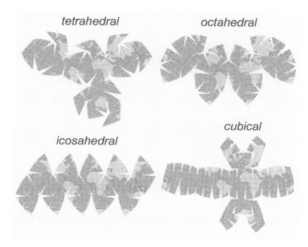

tetrahedral   octahedral

cubical

icosahedral

FIG. 5.   Recursive subdivision of Platonic solids, using five levels of subdivision,
4096220 480 polygons

As a result, the weights are highest close to the centre of original edges. Finally, we use w_c as the edge weight for the edges of the final mesh, plus a graticule weight w with small values for W_λ and W_ϕ to select the aspect.

The resulting unfolded maps are, at first sight, somewhat surprising (Figure 5). One would expect to see interesting fractal shapes, however, at the second level of subdivision the gaps are already almost invisible (Figure 6). Indeed, the structure of the cuts is self-similar, however, for higher levels of subdivision and smaller triangles, the surface of the sphere quickly approaches a plane, which has Hausdorff dimension 2. Only when areas would be removed, such as the centre triangles in the Sierpinski triangle, a fractal shape would be obtained.

As a step aside, fractal surfaces and foldouts do not match well either. Unfolding, for instance, a recursively subdivided surface with displaced midpoints leads to a large number of fold-overs (Figure 7).

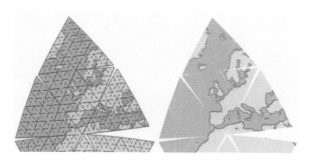

FIG. 6.   Close-up of icosahedral projection

269

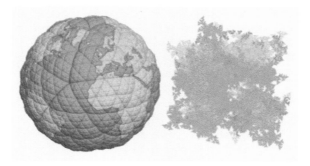

FIG. 7.   Folding out a fractal surface gives a mess

As another step aside, let us consider optimal mapping on Platonic solids. We consider a map optimal when the cuts do not cross continents. To find such mappings, we assign to each edge a weight proportional to the amount of land cut, computed by sampling the edges at a number of positions (here we used 25) and looking up if land or sea is covered in a texture map of the earth. Next, the map is unfolded using the standard method and the sum of weights of cut edges is determined. This procedure is repeated for a large number of orientations of the mesh, searching for a minimal value. We used a sequence of three rotations to vary the orientation of the mesh, and used steps of 1° per rotation. Results are shown in Figure 8.

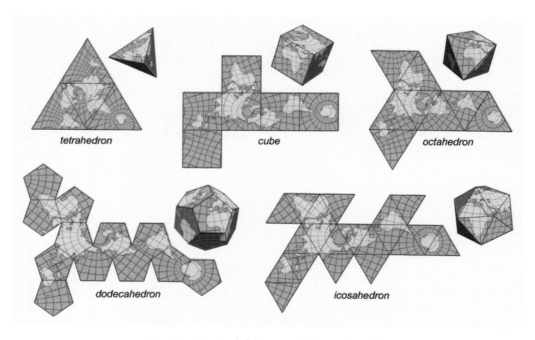

FIG. 8.   Optimal fold-outs of Platonic solids

For the tetrahedron a perfect, and for the other platonic solids an almost perfect, mapping is achieved. Except for the tetrahedron, the resulting layout of the continents is the same as the layout used by Fuller for his Dymaxion map. He used a slightly modified icosahedron for his best-known version, but the version shown here reveals that his modifications are not necessary per se.

GEOGRAPHY ALIGNED MESHES

Taking continents into account when deciding where to cut is an obvious idea. In this section, we explore this further. We generate meshes such that continents are cut orthogonal to their boundaries. First, we define for each point on the sphere a value $f(\phi, \lambda)$ that denotes the amount of land in its neighbourhood. High values are in the centres of continents, low values in the centres of oceans. This function is used to generate the mesh, and also to control the strength of edges. We use linear interpolation of a matrix of values F_{ij}, with $i = 0, \ldots, I\text{-}1$ and $j = 0, \ldots, J\text{-}1$ to calculate $f(\phi, \lambda)$. The corresponding values for λ and ϕ per element are $\lambda_i = 2\pi(i+0.5)/I$ and $\phi_j = \pi(j+0.5)/J - \pi/2$, respectively. The matrix F is derived from a raster image R of a map with the same dimensions as F via convolution with a filter m, i.e.,

$$F_{ij} = \sum_{k=K_j^-}^{K_j^+} \sum_{l=L^-}^{L^+} m_{jkl} R_{i+k,j+l},$$

where $i+k$ is calculated modulo I, $L^- = \max(-j,-L)$, and $L^+ = \min(J\text{-}1\text{-}j, L)$. We typically use $I = 256$, $J = 128$, and $L = 32$. A large weight mask m is used, because it is not only the edges that have to be blurred, but also areas far from coastlines must be assigned varying values. The convolution has to be done taking the curvature into account; therefore, the width and contents of the mask have to be adapted per scan line. For the width, we use $K_j^+ = -K_j^- = \lceil IL/2J\cos\phi_j \rceil$. We use a Gaussian filter, taking the distance r_{jkl} along a greater circle into account between a centre element $R_{0,j}$ and an element $R_{k,j+l}$, as well as the area a_{jl} of the latter. Specifically,

$$m_{jkl} = \sum_{k=K_j^-}^{K_j^+} \sum_{l=L^-}^{L^+} s_{jkl},$$

with

$s_{jkl} = a_{jkl} \exp (-r_{jkl}^2/2\sigma^2)/\sqrt{2\pi}\,\sigma,$

$a_{jl} = 2\pi^2 \cos \phi_{j+l}/NM$, and

$r_{jkl} = \arccos[\boldsymbol{p}(\phi_j,0)\cdot\boldsymbol{p}(\phi_{j+l},\lambda_k)].$

Figure 9 shows an example. As a result, for instance the value for the South Pole is similar to that of the centre of South America.

To obtain a foldout with cuts perpendicular to contours of f, the following steps are performed (Figure 10), inspired Unfolding the Earth: Myriahedral Projections by the anisotropic polygonal remeshing method of Alliez *et al.*[14]:

14. Alliez, P., Cohen-Steiner, D., Devillers, O., Levy, B. and Desbrun, M. (2003). 'Anisotropic polygonal remeshing', **ACM Transactions on Graphics** 22(3), 485–493. Proceedings SIGGRAPH 2003.

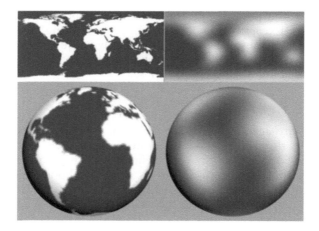

FIG. 9.   From R to F via convolution with a Gaussian

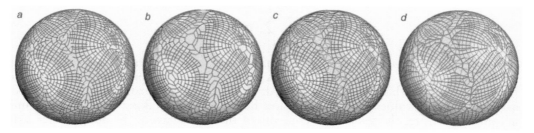

FIG. 10.   Use of contours and gradients to derive a mesh: a) Jobard and Lefer algorithm; b) finding polygons; c) triangulation; d) deciding on cuts

a.   Generate mesh lines along and perpendicular to contours of f with the algorithm of Jobard and Lefer[15];
b.   Calculate intersections of these sets of lines, and derive polygons;
c.   Tesselate polygons with more than four edges; and finally
d.   Use the standard approach to decide on folds and cuts.

These steps are discussed in more detail.

The algorithm of Jobard and Lefer[16] is an elegant and fast method to produce equally spaced streamlines for a given vector field. Starting from a single streamline, new streamlines are repeatedly started from seedpoints at a distance d from points of existing streamlines, and traced in both directions. If such a streamline is too close to an existing streamline or when a cycle is formed, the tracing is stopped. The time

15. Jobard, B. and Lefer, W. (1997). 'Creating evenly-spaced streamlines of arbitrary density', in **Visualization in Scientific Computing '97**, 43–56, ed. by Lefer, W. and Grave, M., Springer Verlag.
16. Jobard and Lefer, 1997.

critical step is to determine which points are close. The standard solution is to use a rectangular grid for fast look-up. Here streamlines are traced in (ϕ,λ) space, and the mapping to the sphere has to be taken into account. We therefore use horizontal strips of rectangles, where the number of rectangles per strip is proportional to $\cos\phi$.

To obtain mesh lines along contours, the vector field

$$c(\phi,\lambda) = (f_\lambda, - f_\phi) \left/ \sqrt{\cos^2 \phi f_\phi^2 + f_\lambda^2} \right.$$

is traced; lines perpendicular to contours follow from tracing the vector field

$$g(\phi,\lambda) = (f_\phi \cos \phi, f_\lambda/\cos \phi)/ \sqrt{\cos^2 \phi f_\phi^2 + f_\lambda^2} \text{ , where}$$

$$f_\lambda = \partial f(\phi,\lambda)/\partial\lambda \text{ and } f_\phi = \partial f(\phi,\lambda)/\partial\phi.$$

The factors $\cos\phi$ in the definition of c and g follow from the requirements that we want these fields to have a unit magnitude and to be orthogonal after projection on the sphere. Projection implies that components $\Delta\lambda$ of a vector $(\Delta\phi, \Delta\lambda)$ are scaled with a factor $\cos\phi$, whereas the $\Delta\phi$ components keep their length. For the tracing, we use a fourth order Runge–Kutta method with a fixed time step.

In the next step, crossings between these line sets are calculated and the lines are cleaned up. Streamlines without crossings are removed, neighbouring points of crossings are removed from the streamlines, and heads and tails are removed. Next, the resulting net is scanned and a set of polygons, covering the sphere, is constructed. This gives a regular, rectangular mesh for a large part of the sphere, but also and unfortunately, irregular polygons. This can be understood from the topology of vector fields, a wellknown topic in the visualization community.[17] Critical points are points where the magnitude of the vector field is zero. For the vector fields used here, these occur at maxima of f (centres of continents), minima of f (centres of oceans) and at saddle-points of f (for instance between South America and Africa). The domain of a flow field can be tessellated using streamlines between these critical points, the so-called separatrices, which gives a topological decomposition of the domain. For the vector fields used here, separatrices typically run through valleys of f. When f is used to decide which edges to label as cuts, the surface breaks along those valleys, which in turn appear as overall boundaries. Downhill gradient lines of f, following g, bend into such valleys with a sharp turn or stop because a line at the other side is too close, leading to irregular polygons.

We use a standard triangulation algorithm to tessellate polygons with more than four edges. First, the polygon is split into convex polygons, next, triangles are split off. Heuristics used are a preference for short inserted edges and avoidance of obtuse or very sharp angles. This is not perfect yet and leads to a somewhat fractured and irregular appearance of the map when unfolded. Improvement turns out not to be simple. In an image like Figure 10(c) it is easy to point at polygons where better choices could have been made, the hard part is to find methods that have no adverse effect at other locations. For instance, introducing extra points and edges often leads

17. Helman, J. L. and Hesselink, L. (1991). 'Visualizing vector field topology in fluid flows', **IEEE Computer Graphics and Applications** 11(3), 36–46.

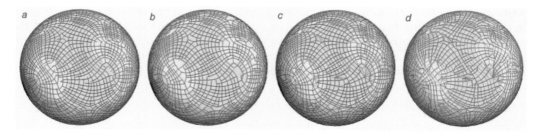

FIG. 11.   Same as Figure 10, using curvature tensor field

to more irregularities, and tracing lines between critical points in advance gives wide interrupts instead of multiple smaller ones.

We also tried the use of a tensor field based on curvature,[18] instead of a vector field (Figure 11). Here, at each point, the direction of minimum or maximum curvature is traced. This gives an orthogonal mesh, without singularities along lines and, indeed, the valley in the centre of the Atlantic Ocean is now filled in a more regular way. However, this does not necessarily lead to a more appealing tessellation, see for instance the small strip introduced in the centre of this valley. Tensor fields have two kinds of singular points: trisectors and wedges. Here, a trisector appears in the northern Atlantic Ocean, and a wedge in the Gulf of Guinea. This latter feature leads to irregularities in the resulting mesh.

Other solutions are to increase the density of the mesh, and, simply to accept the fractured boundaries. Visually, they show that the surface of the globe is torn apart, and they show that where this is done exactly is somewhat arbitrary.

Figure 12 shows results of this approach. Straightforward application leads again to the layout of the continents of Buckminster Fuller. A more familiar layout can be obtained by adding a graticular weight, and tuning W_λ and $W\phi$. The overall layout resembles a conical projection. The continents are shown with few interrupts and with correct shape and relative position. Instead of f, also $|f - f_c|$ can be used as a weight for the edges. As a result, the global boundary of the map is along contours $f = f_c$. This boundary is smooth, and divides the surface here into the main continents, the oceans, and Antarctica. The author does not know a similar map.

Also, $-f$ can be used as a weight for the edges. This results in a map where the oceans are central, surrounded by the coastline of the world. Ocean centred maps have been made before, such versions are available for Goode's homolosine map and Fuller's Dymaxion map. Closest is a map presented by Athelstan Spilhaus.[19] His map (and also Fuller's) is centred on Antarctica, showing the oceans as three lobes, and is, hence, somewhat less extreme than the version shown here. A map similar to Spilhaus's map can easily be generated with our method, simply by removing Antarctica from the map R.

18. Alliez *et al.*, 2003.

19. Spilhaus, A. (1983). 'World ocean maps: The proper places to interrupt', **Proceedings of the American Philosophical Society** 127(1), 50–60.

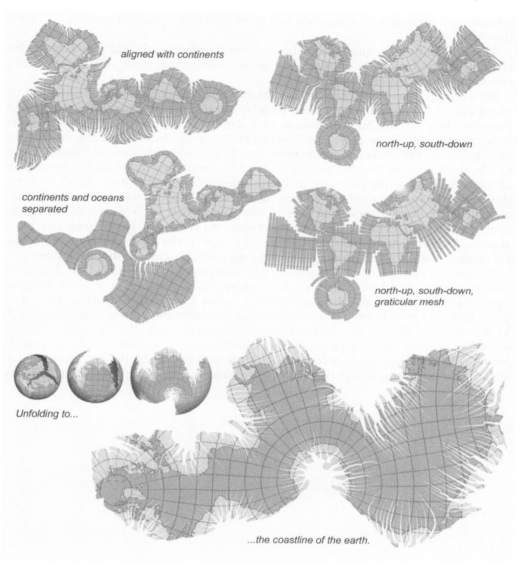

aligned with continents

north-up, south-down

continents and oceans
separated

north-up, south-down,
graticular mesh

Unfolding to...

...the coastline of the earth.

FIG. 12.   Myriahedral projections with geography aligned meshes, 5500 polygons

DISCUSSION

We have presented a new class of map projections, based on projecting the earth on myriahedra, polyhedra with many faces, and unfolding these. A general approach is presented to decide on cuts and folds, based on weighting the edges and calculating a maximal spanning tree. Three different choices for types of meshes and weighting schemes are presented, leading to a variety of different projections of the surface of the earth.

There remains one question to be answered: What is this all good for? Most resulting maps are highly unusual, and do not correspond with what on average is considered to be a useful map.

Furthermore, the complexity is high. Standard projection methods require, in the worst case, a few iterations per point to solve a transcendental equation; the methods presented here require implementation of a number of nontrivial algorithms. Hence, forward mapping is not easy, and also inverse mapping, from a location on the map to a point on the globe, is much more involved than with standard maps. Fortunately, there are also positive aspects that can be mentioned. From an academic point of view, a classic topic like map projection deserves an exhaustive exploration and this class of maps has not been addressed yet. What happens when many small maps are glued together is obvious and here an extensive answer is given. Hence, these maps could be used for textbook purposes. Furthermore, each class has its own interesting aspects. The graticular maps can be used to explain the basics of map projection. Polyhedral maps are entertaining, and here we have presented optimal versions.

We have investigated what happens when interrupts are removed. In Figure 13, two examples are shown, derived from maps shown in Figure 12. We matched corresponding vertices at a distance below a certain threshold, starting at the ends of gaps, followed by a finite element simulation to redistribute the points of the mesh. In the examples shown, we defined the stiffness matrix such that the equal-area property is satisfied. These steps are repeated until no corresponding vertices could be found. The maps are not conformal: Parallels and meridians do not cross at right angles. The hard boundaries of the maps without interrupts are somewhat arbitrary,

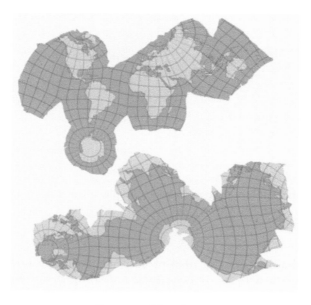

FIG. 13.   Closed gaps

but do attract attention, in contrast to the more fuzzy boundaries of their myriahedral counterparts. Finally, they reveal a quality of all myriahedral maps. The interrupts present in myriahedral maps show the inevitable distortion in a natural, and explicit way, whereas in standard maps it is left to the viewer to guess where and which distortion occurs.

Methodologically interesting is that here a computer science approach is used, whereas map projection is traditionally the domain of mathematicians, cartographers, and mathematical cartographers. Myriahedral projections are generated using algorithms, partially originating from flow visualization, and not by formulas. Implementation is not simple, but when the machinery is set up, a very large variety of maps can be generated just by changing parameters, such as W_λ, W_ϕ, F, f_o, σ, and the size of the faces used. This leaves much room for serendipity, and indeed, some of the maps shown here were discovered by accident.

Maps are not only used for navigation or visualization, but also for decorative, illustrative and even rhetoric purposes,[20] for instance on covers of magazines. The coastline map is an example, which can serve to emphasize the importance of oceans. Another example is shown in Figure 14. Here, we used a subdivided icosahedral projection, centred at 40°N, 100°W, and used an edge weight proportional to the distance from this point, plus a small random factor.

As a first step to a user study, the resulting projections were shown as static images and in animated version to about 20 people, ranging from laymen via computer scientists to cartographers. In general, the reception was very positive. Most people found the results compelling and intriguing. Computer science colleagues liked

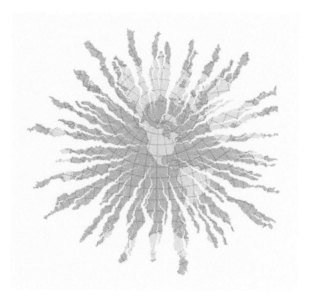

FIG. 14. Azimuthal projection, random weights added, 81 920 Polygons

20. Monmonier, M. (1991). **How to Lie with Maps**, University of Chicago Press, Chicago.

the general framework and the algorithms. Nevertheless, taking a utilitarian point of view, some cartographers argued that cuts are always more disturbing reading a map than having distortion, which is the reason that such projections have been discarded so far and are not useful in practice. Concerning usability for tutorial purposes, results were mixed again. Some cartographers found this a very strong feature; others argued that visualizing a distorted grid would be more effective. More elaborate usability tests are required to evaluate which approach is most effective here and, also, to see what the value is in general. Besides such a lab test, an interesting test is whether these results are interesting for, and find their way to, a large audience. So far the results were only shown under non-disclosure conditions, so we cannot report on that yet. It is encouraging, however, that many viewers asked when the results would become publicly available and if they could be notified on this.

There is room for more future work. We are considering alternative methods to produce geography aligned meshes. An interesting option is to use a physically based model to simulate crack formation.[21] Also, the methods presented here can be used for a variety of other purposes, for instance to show plate tectonics, voyages of discovery, or scientific data given for spherical angles. Such applications can be produced easily, just by varying the map used for mesh alignment.

ACKNOWLEDGEMENTS

The author thanks Michiel Wijers for coining the term 'myriahedral', and Jason Dykes and Menno-Jan Kraak for their encouragement and advice.

21. Iben, H. N. and O'Brien, J. F. (2006). 'Generating surface crack patterns', in **Proceedings of the 2006 ACM SIGGRAPH/Eurographics Symposium on Computer Animation**, SCA 2006, Vienna, Austria, 177–185, ed. by O'Sullivan, C. and Pighin, F., Eurographics association, Aire-la-Ville, Switzerland.

Reflections on 'Unfolding the Earth: Myriahedral Projections'

KENNETH FIELD

Editor, The Cartographic Journal / Esri Inc

Cartographic history is cursed by map projections. The necessity of translating the surface of the oddly shaped three dimensional Earth onto a two-dimensional plane for most mapping tasks simply creates problems. All projections have inherent distortions and so no matter the utility and benefit proposed, one can always point to some flaw or drawback that renders it ineffective for a particular mapping task. But that is to miss the point. The very fact that there is no perfect way to map the spheroidal surface onto a flat surface without some sort of compromise means there is fertile scope for reimagining, recalculating and creating new ways to warp the world.

When mapping at a global level, there's a natural tendency for map-makers to default to projections such as Mercator, Mollweide or Plate Carrée because they're familiar. That familiarity is reflected in the way in which map readers accept them as the 'correct' view of the world. They do not require as much interpretation; the barrier to understanding is lessened because of this even though, of course, many of the most familiar projections have gross distortions. Familiarity, of course, breeds contempt and projections suffer this too. Web Mercator is the prime culprit of the fashion and use of map projections. It's the default for web mapping for good reason given the computationally efficient way it stores and serves map tiles as tessellated squares. The fact that this causes such ire among cartographers is almost immaterial until we find an effective way to handle multiple, adaptive projections that are widely accepted and understood by the public. As a consequence, new or innovative work tends not to receive as much attention and rarely strikes a chord in the public conscience. Perhaps the most recent examples of alternative projections that have bucked this trend (no pun intended) are R. Buckminster Fuller's Dymaxion map projection from the mid-1960s and Arno Peters' controversial equal area projection unleashed as a 'new invention' in 1973. Neither replaced Mercator, but they gave it a good go.

First published as an article in the March 1st, 1943 edition of *LIFE* magazine (LIFE, 1943), Buckminster Fuller's compromise projection contains far less distortion than other flat maps. The map was printed as a pull-out section designed to allow readers to assemble the map. It divides the globe's surface into a continuous surface without bisecting major land masses and it is unique in that there is no right way up. It can be read from any orientation and rearranged in a number of alternative ways. Buckminster Fuller, himself used the map in his own Airocean map (Figure 1).

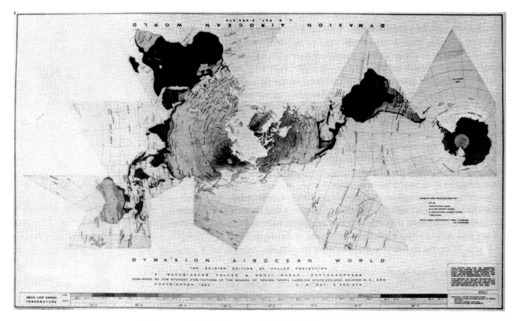

Fig. 1.   Buckminster Fuller's Airocean map

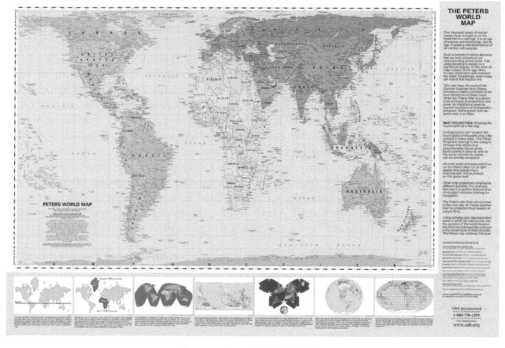

Fig. 2.   Peters' World Map

Though presented earlier, it wasn't until 1983 that Peters published his 'New Cartography' explaining the utility of the projection (Peters, 1983, Figure 2). The Peters projection was more a statement that caused considerable cartographic debate in the years following its publication and as such was a cartographic disruptor. The projection stimulated numerous discussions and acted to drive a wedge through the cartographic community. For a start, it was contentious as it apparently simply plagiarized an earlier projection and represented nothing more than a reincarnation of Gall's orthographic projection from the late nineteenth century. Arthur Robinson famously claimed it resembled 'wet, ragged long winter underwear hung out to dry in the Arctic Circle' (Robinson, 1985).

Cartography, though, is no different to any other scientific or artistic endeavour in that new work prompts, provokes and disrupts conventional wisdom. Without such work, we simply become followers of the default approach and the discipline stagnates. It seems, however, that newly proposed projections stir emotions in a way that other aspects of the discipline fail to achieve and new ones are considered with wariness. Perhaps they change the way we view the world in such a way that challenges us beyond our comfort zones? Perhaps we're just not willing to see things through different eyes? New projections, then, seem only to strike a chord if they are graphically arresting yet this often courts the most controversy. Whatever the reasons, anyone who pitches a new projection is likely going to start from a disadvantaged position.

A self-confessed non-cartographer, Jack van Wijk created a similar stir in 2008 when his work was published in *The Cartographic Journal* (and subsequently reported in *New Scientist*; Aron, 2009). He applied his computer science background and searched for algorithms that would deal with small steps to iterate a process for transforming the globe. He subsequently developed a new class of methods for mapping the Earth where the spherical surface was mapped onto a myriahedron. Myriahedral is his own term too — a brave move to coin a term for the work — as it is used to describe the multiple (or myriad) faces of polyhedra (the basic unit shape used). The method proposed veered away from most previous projections that attempted to minimize the number of cuts and interrupts in the surface of the globe by proposing multiple faces. In fact, the point was to create numerous interruptions, but the benefit is to preserve conformality and the relative size of areas. Van Wijk's projection defined an entirely new approach that owes more to the creation of gores (created during globe construction) than more traditional rectangular, conic or azimuthal projections. In fact, its lineage is more closely associated to Buckminster Fuller's approach in that it accepts interrupts in favour of other properties. The paper presents the detailed methodology for those interested in the algorithms yet the idea is a simple one. The algorithms simply divide the surface of the globe into many small polygons that can then be unfolded. In principle, this is no different to unfolding the six faces of a cube and laying them flat, except here, van Wijk deals with many faces. Buckminster Fuller did this, and so have many other cartographers, but rather than trying to limit the number of faces, van Wijk threw away the constraint and just upped the ante from a few faces to several thousand. That's a challenging concept for cartographers to accept since we tend to prefer our world maps to have minimal cuts and to take on a regular appearance.

The resulting projections are, however, beautiful in their own right, particularly when presented as unravelled rectangular, azimuthal and conic patterns. The maps contain multiple fronds that give the overall impression a feathered effect though the continents can still be seen clearly. But the way the Earth can be projected using this technique goes beyond the traditional. Using Platonic solids for the projection, the polyhedral mesh can be subdivided into a tetrahedral, octahedral, isohedral, dodeca-hedral and cubical pattern giving rise to a range of interesting world map shapes that support optimal foldout models similar to Buckminster Fuller's. By studying and proposing optimal interrupts, van Wijk also illustrates the creation of some elegant solutions to applying the projection to creating maps that are aligned with continents or a version that unfolds to create an almost continuous coastline for the earth. What he achieved was both mathematically innovative and represented a sound approach to the projections problem. His work was roundly accepted by the community. Yes, it was challenging, but the visualizations he created brought a certain aesthetic quality to the world map that had not been seen before. This undoubtedly helped to minimize the damage from detractors and showed that the shape of the world map can go beyond the rectangle and take on a shape that adds to its interest.

The flexibility of the methods van Wijk proposed are certainly one of the appealing facets of his work. As we return to thinking of the utility of relatively unseen projections, one cannot help but reflect on the fact that many projections remain obscure and consigned, unfortunately, to academic study rather than practical use. This, of course, is the curse of map projections; for those who devise such intriguing and wonderful new ways for us to make the world flat to simply face the brick wall of the familiar. Mercator rules, still... but map projections remain a fascinating avenue of cartographic research that every now and then yields some surprisingly innovative work; new visualizations that engage and a fresh perspective on geography, and van Wijk's approach is among the best.

REFERENCES

Aron, J. (2009). 'Clever folds in a globe give new perspectives on Earth', New Scientist, 10 December 2009, http://www.newscientist.com/article/dn18264-clever-folds-in-a-globe-give-new-perspectives-on-earth.html.
LIFE. (1943). 'Life presents R. Buckminster Fuller's Dymaxion world', LIFE, 41–55.
Peters, A. (1983). Die Neue Kartographie/The New Cartography (in German and English), Carinthia University, Klagenfurt/ Friendship Press, New York.
Robinson, A. H. (1985). 'Arno Peters and his new cartography', American Geographer, 12, 103–111.

Stylistic Diversity in European State 1:50 000 Topographic Maps

ALEXANDER J. KENT AND PETER VUJAKOVIC

Originally published in *The Cartographic Journal* (2009) 46, pp. 179–213.

To what extent do European state topographic maps exhibit unique styles of cartography? This paper describes an investigation to classify and analyse stylistic diversity in the official 1:50 000 topographical mapping of 20 European countries. The method involves the construction of a typology of cartographic style, based upon the classification of distinct graphical legend symbols into mutually exclusive thematic categories. In order to identify stylistic similarities between national symbologies, hierarchical cluster analysis was performed to compare the relative proportions of symbols within each category. This was complemented by a qualitative analysis of various aspects of cartographic design: colour, 'white' space, visual hierarchy, and lettering. The results indicate a high degree of stylistic diversity throughout Europe, with the symbologies of Great Britain and Ireland demonstrating the strongest example of a supranational style. The typology of cartographic symbologies is shown to be an effective method for determining stylistic association among maps of differing geographical (and potentially historical) origins and it is suggested that the cartographic language paradigm should be revisited as a means for understanding why national differences persist in state cartography. A version of this paper was presented at the Twenty-third International Cartographic Conference in Moscow.

INTRODUCTION

Topographic maps are among the most familiar — and most trusted — of all cartographic products, maintaining a unique status in the history of cartography and possessing especial relevance to the growth and development of nations. Maps aiming to provide detailed and accurate observations of topography were among the first maps to be made, and according to Piket,[1] they form the ancestral line in the cartographic family tree from the origins of mapmaking to the present. The spread of the idea of drawing to a fixed scale revolutionized topographical mapping in the sixteenth century[2] and maps became increasingly valuable tools for exploring, understanding,

---

1. Piket, J. J. C. (1972). 'Five European topographic maps: a contribution to the classification of topographic maps and their relation to other map types', **Geografisch Tijdschrift**, 6, pp. 266–276. Page 267.
2. Harvey, P. D. A. (1980). **The History of Topographical Maps: Symbols, Pictures and Surveys**, London, Thames & Hudson. Page 14.

and controlling our environment. The relevance of topographic information for military and cadastral purposes provided the motivation for the first initiatives in state mapmaking and gradually, the methods of survey and cartographic production developed and improved to allow topographic maps to be regarded as the 'supreme achievement of the modern age of cartography'.[3] More recently, David Rhind, formerly Director-General of Ordnance Survey, proclaimed that 'the core of all mapping is that of the topography of the Earth'.[4]

Topography is defined by *The Oxford English Dictionary* primarily as 'the science or practice of describing a particular place, city, town, manor, parish, or tract of land; the accurate and detailed delineation and description of any locality'.[5] It is intriguing that the words 'science', 'accurate', and 'detailed' are mentioned in this definition, as in pursuing a tradition of accuracy from scientific survey and institutionalized production, topographic maps have come to be regarded as authoritative, objective, and truthful representations. Indeed, as Tyner[6] claims, many people react to maps as though they were not representations but objectively the world itself, and, as Dorling and Fairbairn[7] add, routinely accept them as 'truth'.

This view may lead to the assumption that official topographic map symbols are internationally standardized, perhaps within the European Union in particular, where homogenizing principles pertaining to currency, law, and agricultural produce, for example, might suggest the existence of such a regulation. However, the degree to which topographic maps vary in style remains a contentious issue. While Board[8] exclaims that 'there is no question' that official topographic maps demonstrate different styles, Taylor's[9] foreword to Eric Böhme's three-volume *Inventory of World Topographic Mapping* insists that 'the general style adopted for the representation of topographic features is virtually universal'. If different national mapping organizations (NMOs) use cartographic conventions (e.g. surrounding the use of colour, such as blue for water, brown for contours, green for vegetation, and black for 'cultural' features) on topographic maps, should this not give rise to a 'supranational' cartographic style, where variations in the appearance of such maps simply correspond to variations on the surface of the Earth? Indeed, according to Dorling and Fairbairn,[10] 'Standardization of content is such that topographic maps from widely differing landscapes, produced by different national mapping agencies, employ notably similar

3. Jervis, W. W. (1936). **The World in Maps: A Study in Map Evolution**, London, George Philip & Son. Page 171.

4. Rhind, D. (2000). 'Current shortcomings of global mapping and the creation of a new geographical framework for the world', **The Geographical Journal**, 166, pp. 295–305. Page 296.

5. Simpson, J. A. and Weiner, E. S. C. (Eds.) (1989). **The Oxford English Dictionary**, 2nd ed., Oxford, Oxford University Press. XVIII, p. 257.

6. Tyner, J. A. (1987). 'Interactions of culture and cartography', **The History Teacher**, 20, pp. 455–464. Page 456.

7. Dorling, D. and Fairbairn, D. (1997). **Mapping: Ways of Representing the World**, Harlow, Longman. Page 80.

8. Board, C. (1981). 'Cartographic communication', **Cartographica Monograph**, 27, pp. 42–78. Page 63.

9. Taylor, D. R. F. (1989). 'Foreword', in **Inventory of World Topographic Mapping**, Vol. 1, Western Europe, North America and Australasia, ed. by Böhme, R., Barking, Elsevier Applied Science Publishers. Page vi.

10. Dorling and Fairbairn, 1997, page 92.

representation'. It would be easy to propose that the basis for any fundamental difference simply lies in the contrasting characteristics of physical landscapes.

However, maps of the same geographical area covered by different NMOs do vary in appearance, as may be demonstrated by the 1:50 000 topographic map series of Austria and Slovenia (Figure 1). Although the two maps share a similar colour scheme, their subjects have been symbolized differently. For example, road and vegetation classifications contrast prominently, labels are different in both content and in appearance (the Slovenian example also includes Slovene settlement names as exonyms), the Austrian map generally seems to include more topographic detail, and it is also clear that different grid systems have been used.

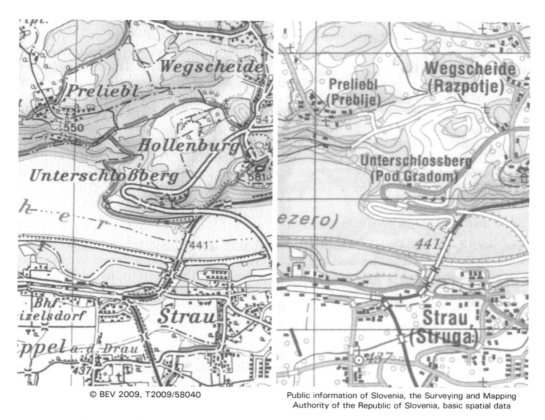

© BEV 2009, T2009/58040

Public information of Slovenia, the Surveying and Mapping Authority of the Republic of Slovenia, basic spatial data

FIG. 1.   Left: extract from sheet 202: Klagenfurt from the Austrian 1:50 000 topographic paper map series,[11] and right: an extract of the same area shown on sheet 12: Jesenice from the Slovenian 1:50 000 topographic paper map series,[12] (both extracts enlarged to 240%)

11. Bundesamt Für Eich- und Vermessungswesen. (1998). '202: Klagenfurt' from **Öesterreichische Karte 1:50 000**, Vienna, Bundesamt für Eich- und Vermessungswesen.

12. Geodetska Uprava Republike Slovenije. (2003). '12: Jesenice' from **Državna Topografska Karta 1:50 000**, Ljubljana, Geodetska Uprava Republike Slovenije.

The cartographic style of a national series of topographic maps is derived from how the landscape is symbolized, in terms of both content and appearance. Through the processes of abstraction and generalization, its character is suggested through a particular selection of features and their expression through the creation of symbols, according to a customary use of the graphic variables available. In addition, the coordinated relative emphasis of map symbols (reflecting the interests of the national mapping organization and with them the perceived needs of society) conveys an apparent hierarchy of features and suggests the character of the national landscape on a more holistic level. Topographic maps portray a socially constructed landscape; a 'good view' of the land which is based on certain ideas and conventions.[13]

This paper explores the variation in how state topographic maps produced by different countries symbolize their subject — the national landscape. It investigates the way in which this landscape is classified through an analysis of state 1:50 000 topographical mapping in 20 European countries (Austria, Belgium, Czech Republic, Denmark, Finland, France, Germany, Great Britain, Iceland, Ireland, Italy, Latvia, The Netherlands, Norway, Poland, Portugal, Slovenia, Spain, Sweden, and Switzerland). As all topographic maps involved are those produced by national mapping organizations, these are the smallest entities with which particular styles might be associated. However, while it may seem plausible to define these as 'national styles', such a definition would require a thorough investigation into the heritage and evolution of cartographic symbology and this will not be attempted here. What this paper aims to provide, however, is an exploration of state cartographic style with a view to developing an understanding as to why any diversity persists.

DEFINING CARTOGRAPHIC STYLE

The use of the word 'style' usually depends upon the recognition of certain, defined similarities within a group and the recognition of differences to these outside that group. In cartography, style seems to be used to describe a certain manner, or form of expression, which leans towards the general — as opposed to the specific — results of symbolization. As in the realm of publishing, maps often exhibit a 'house style' that is demonstrated through the particular use of colour, typography, and symbols, for example.

The idea that individual countries might produce topographic maps in a particular style has a great deal of significance because it suggests that the landscape may be symbolized in a way that is definable and recognizable as belonging to a country. John Keates (1925–1999) was especially keen to recognize the existence of style and asserted that 'many modern topographic map series fall into a stylistic group, in the sense that they are based on an accepted and effective method of design and representation', and identifying the 'Swiss manner' as 'probably the best example'.[14] Conversely, he also suggested the existence of a supranational style:

---

13. Kent, A. J. (2008). 'Cartographic blandscapes and the new noise: finding the good view in a topographical mashup', **The Bulletin of the Society of Cartographers**, 42, pp. 29–37. Page 34.

14. Keates, J. S. (1996). **Understanding Maps**, 2nd ed., Harlow: Longman. Page 252.

Virtually all medium-scale topographic map series are based on a 'classical' style which evolved with the introduction of lithographic printing. This made possible the use of different hues to represent and contrast the major feature categories. In its simple form this is based on black planimetry, blue water, brown contours and green vegetation.[15]

Basing the notion of style on the use of colour makes a link between the appearance of a map and cartographic production capabilities and limits. If current methods of cartographic production, at least, allow for radically different styles, it is nevertheless plausible to suggest that these could provide the technical basis necessary for the evolution of topographic maps. It also defines a particular manner, a convention, which has been derived from certain schema (rules for structuring the unity of experience) over time, the use of blue for water and green for vegetation, for example. Moreover, with the acknowledgment of a basic style comes the possibility of recognizing departures from it, leading to a notion of distinctiveness. For example, Knowles and Stowe[16] refer to the style of Swedish 1:100 000 maps as 'legible and aesthetically satisfying' and that of the Belgian 1:50 000 series maps as 'very bold'. More recently, Nicholson[17] went so far as to detect Art Nouveau, Art Deco, and 'the wilder shores of Italian Futurism' in early twentieth century motoring maps, but these observations were based on cover designs, which, by their nature, do not fall under the same design constraints as the map itself.

Topographic maps also retain a particular choice of features which appears to be the most useful, i.e. having the highest number of potential functions (or the highest significance), to the greatest number of users. They are consequently rich in complexity and exhibit the application of skill (particularly in the ordering of information), involving many different individuals and meeting several levels of approval, so the evolution of style in topographic maps is relatively slow. Changes require the collective judgment of many involved in the design process within the native NMO and it would therefore seem that in major departures from the established style, the input of individuals is hard to trace. But there are some rare exceptions, as in the case of Eduard Imhof and his contribution to Swiss topographical mapping. The development of his own style of mountain relief cartography in the 1920s[18] and subsequent role in the creation of what Keates (1996, p. 252) refers to as the 'Swiss manner' or what Collier *et al.*[19] call the 'Swiss style', suggests one instance. In the cartography of topographic maps, style therefore has more to do with the general quality of expression than the introduction of innovative ways of visualizing the landscape that may be introduced by an individual.

FORMER APPROACHES

Given their unique role in the development of cartography, there is a surprising lack of empirical research into the design of topographic maps, particularly the similarity

15. Keates, 1996, page 256.

16. Knowles, R. and Stowe, P. W. E. (1982). **Western Europe in Maps: Topographical Map Studies,** Harlow: Longman. Pages 26, 56.

17. Nicholson, T. (2004). 'Cycling and motoring maps in Western Europe 1885–1960', **The Cartographic Journal**, 41, pp. 181–215. Page 194.

18. Dorling and Fairbairn, 1997, page 112.

19. Collier, P., Forrest, D. and Pearson, A. (2003). 'The representation of topographic information on maps: the depiction of relief', **The Cartographic Journal**, 40, pp. 17–26. Page 20.

and difference between those currently produced by national mapping organizations. *Foreign Maps*, the early contribution by Olson and Whitmarsh,[20] whose publication was brought forward by the imperatives of war, describes the symbologies of topographic maps produced outside the USA, complete with comparative tables. Studies in landscape interpretation such as those by Sylvester,[21] Wood,[22] and Knowles and Stowe[23] suggest an awareness of the differences in symbology utilized by different NMOs, but their focus is on an appreciation of diversity among various physical and human landscapes. For authors such as these, however, topographic maps appear as a function of the Earth's surface and no more. Of the few studies that offer interpretations of cartographic style, most involve thematic maps, such as public transport maps (e.g. Morrison[24]; Morrison[25]) and cycling and motoring maps (e.g. Nicholson[26]).

Piket,[27] however, was able to differentiate between types of topographic map from a classification of the legend contents by dividing the number of features by the type of feature, raising questions surrounding the design of the maps themselves. Using topographic maps of the same scale (1:25 000) from five different European countries (Belgium, The Netherlands, West Germany and Denmark, Italy, and Switzerland), five types of phenomena were selected — built-up areas, roads, ground cover, orography (relief), and hydrography — to form five 'range classes'. The number of features represented in the map legend for each was then counted to indicate the variations in selection of feature type, summarizing the overall character of each map. These results led to an identification of a 'type' (which could easily be read as 'style') of topographic map, based on the treatment of particular features, e.g. the 'Italian type', with the accent on relief and a remarkably narrow range for built-up areas and ground cover.[28]

Rather than providing an illustrated stylistic comparison between the topographic maps of NMOs at an identical scale, the methodology adopted for the ensuing series of studies by Forrest et al.,[29] Collier et al.,[30] and Collier et al.[31] encompassed a range of scales from 1:10 000 to 1:1 000 000 and incorporated both state and privately owned map producers. As scale and type of mapmaker (i.e. state and commercial) greatly influence factors such as generalization and audience, this method is unsuitable as the basis for a pan-European comparison of topographic maps, however, because it allows too much flexibility to draw any firm conclusions. Furthermore, despite

20. Olson, E. C. and Whitmarsh, A. (1944). **Foreign Maps**, New York, Harper & Brothers.

21. Sylvester, D. (1952). **Map and Landscape**, London, George Philip & Son.

22. Wood, M. (1968). **Foreign Maps and Landscapes**, London, George G. Harrap & Co.

23. Knowles and Stowe, 1982.

24. Morrison, A. (1994). 'Why are French transport maps so distinctive compared with those of Germany and Spain?', **The Cartographic Journal**, 31, pp. 113–122.

25. Morrison, A. (1996). 'Public transport maps in Western European cities', **The Cartographic Journal**, 33, pp. 93–110.

26. Nicholson, 2004.

27. Piket, 1972.

28. Piket, 1972, page 276.

29. Forrest, D., Pearson, A. and Collier, P. (1997). 'The representation of topographic information on maps — the coastal environment', **The Cartographic Journal**, 34, pp. 77–85.

30. Collier, P., Pearson, A. and Forrest, D. (1998a). 'The representation of topographic information on maps: vegetation and rural land use', **The Cartographic Journal**, 35, pp. 191–197.

31. Collier et al., 2003.

offering an insight into the different approaches that different mapmakers adopt to map certain phenomena, the methodology was based on the premise that these were objective phenomena receiving subjective treatment; they posed certain problems of cartographic representation which were overcome in different ways. Different societies map their landscape according to the needs and values of that society and these affect the choices over what to show and how to show it. In other words, not only would the representation of features be different but also the survey of features and the basis for their inclusion or omission. An ideal comparison would therefore need to derive from a situation where different NMOs were involved in mapping the same land at the same time, so that variations in symbology and style would result from the choices made in symbolizing the landscape.

THE METHOD

Choice of criteria

An evaluation of approaches suggests that the approach outlined in Forrest *et al.*[32] could perhaps be regarded as a 'vertical' technique, in which the symbolization of a particular set of features was examined under the variables of scale and author. It is crucial to ensure as much consistency as possible between samples, because factors such as scale, subject, and relief would greatly influence the degree of symbolization and therefore the choices involved. Moreover, it is important to bear in mind that it is not the local landscape under investigation but the symbology, allowing a comparison of the range of symbols presented to the user for interpreting the landscape.

The methodology to be developed for this investigation draws instead on the more 'horizontal' system adopted by Piket[33] in his comparison of topographic maps at a single scale.[34] It involves the construction of a typology of European 1:50 000 topographic symbology that concentrates on comparing the 'vocabulary' used for expressing each national landscape — as shown in the legend (or key) — rather than the local landscapes as shown by individual sheets. However, instead of analysing and comparing the graphical variables of every individual symbol, the approach taken here focuses on elements which together have the greatest effect in ordering the stylistic character of a national series of topographic maps and allows a feasible pan-European investigation to be conducted. The typology incorporates five criteria: classification, colour, 'white' space, visual hierarchy, and lettering, yielding both quantitative and qualitative data and aiming to provide a means for comparing stylistic similarity and difference in state cartographic symbologies, through which a deeper investigation can be made.

Classification

The backbone of the typology is its system for classifying the map symbologies, where each distinct graphical symbol within the map legend is sorted into one of a number

32. Forrest, D., Pearson, A. and Collier, P. (1996). 'The representation of topographic information on maps — a new series', **The Cartographic Journal**, 33, pp. 57–58.

33. Piket, 1972.

34. See Kent (2009) for more details on the strategy leading up to this investigation.

of mutually exclusive categories. In addition to a simple symbol count per category, this allows the measurement of the number of legend symbols used to represent a feature type as a proportion of the total number of symbols used to express the national landscape. The qualifying factor for a symbol to be included relates to the nature of its design: it must comprise of a separate, complete, and unique means of graphic expression within the symbology. As such, each discrete symbol should therefore have the capacity to act as a whole and unified design that is intended to denote a particular feature without necessarily involving other symbols. The count avoids all labels (both words and abbreviations) that are not incorporated as part of a cartographic symbol. A combination of symbols, such as a 'cluster' showing a hypothetical arrangement of features to simulate their interplay on the map or denote a particular habitat, for example, is therefore exempt; every discrete symbol included in the cluster is counted separately. The count also avoids all labels (both words and abbreviations) that are not incorporated as part of a cartographic symbol. Road labels, for example, are not counted if they consist solely of a text label (e.g. 'E35') but an identical label placed within a rectangular symbol would be accepted as constituting a cartographic symbol and counted. Each individual and discrete cartographic symbol within a map legend is sorted according to the 'theme' to which it belongs, at three levels of categorization (Figure 2).

In compiling the typology, an *a posteriori* approach was taken to develop an original range of classes, which gives more attention to the diversity of features represented on the maps of Europe and allows a more consistent classification of symbols into mutually exclusive categories. To ensure that the final results of the classification

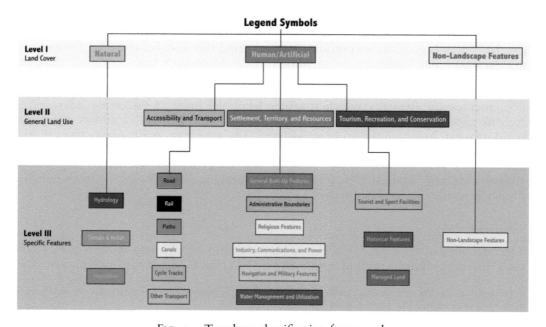

FIG. 2.   Typology classification framework

are consistent, the number of classes steadily evolved to reduce ambiguity and the symbols re-sorted where necessary.

Three levels of symbol classification were developed to allow various levels of analyses. At the broadest level, the classification distinguishes between human/artificial, natural, and non-landscape features (e.g. grid ticks appearing alongside other features in the legend), to allow a very general assessment of the state symbologies. This ensures that 'physical' elements of the landscape (e.g. streams and contours) are separated from 'human' elements (e.g. canals and triangulation pillars). Of course, there will always be difficulties with such a distinction. Vegetation, for example, provides much ambiguity regarding instances of plantation and cultivation and in this classification appears under the 'Natural' class. The first level, therefore, essentially provides a basic dichotomy of land cover as opposed to a broader categorization of land use, while further levels of symbol classification acknowledge land use as opposed to land cover, where human activity carries significance. Level II provides a distinction between types of natural symbol but also divides the human symbols into three classes.

The most detailed classes in level III therefore allow even more specific analyses and rely on a closer reading of the feature being symbolized to allow the particular similarities and differences between national map series to emerge (Table 1). This represents an attempt to classify every discrete symbol in the sample set into classes which are precisely defined, in order to generate data for making detailed and valid comparisons between countries. The approach thus aims to reach a solution whereupon the greatest number of symbols was classified into as many mutually exclusive groups representing the same level of detail as possible. Classification has its limitations, however, and the typology aims to be objective in the sense that a classification of the same set of symbologies by the same sequence of methods will produce the same analysis. As with any classification system, there is always room for future development.

Colour

In order to provide a more holistic comparison of cartographic style, the classification of symbols is complemented by a comparison of elements which have a considerable impact on the map's overall appearance: colour, 'white' space, visual hierarchy, and lettering. According to Robinson *et al.*,[35] 'Even a small amount of color [sic] can make an enormous difference in a map's appearance'. Colour plays an important role in the symbolization of features in topographic maps; as a graphic variable, it can be used to increase the levels of classification such as land use and road type and determine the more holistic impressions of the topographic map, influencing the mood of the reader and the reader's emotional response to the landscape. The importance of colour in the design of topographic maps is affirmed by Imhof,[36] who also suggests that the

35. Robinson, A. H., Morrison, J. L., Muehrcke, P. C., Kimerling, A. J. and Guptill, S. C. (1995). **Elements of Cartography**, 6th ed., New York, John Wiley & Sons. Page 381.

36. Imhof, E. (1982). **Cartographic Relief Presentation**, trans. By Steward, H. J., Berlin, Walter de Gruyter. Page 74.

Table 1. Examples of legend symbols classified at the most detailed level of classification

Level III class	Examples of features symbolized
Road	Motorways, roads, tracks, bus stations, parking, junctions, tree-lined roads, and road tunnels/bridges
Rail	Railways, railway stations, cargo railways, and railway tunnels/bridges
Paths	Footpaths, bridleways, passes, ski-tracks, and footbridges
Canals	Canals, locks, canal beacons, trafficable dykes and aqueducts, and canal water level gauges
Cycle tracks	Cycle tracks, cycle routes, and cycle bridges
Other transport	Trams, ferries, ports, docks, airports, and helipads
General built-up features	Residential buildings, schools, hospitals, post-offices, police stations, town halls, farms, towers, fences, walls, and sheepfolds
Administrative boundaries	International, national, district, province, canton, and county boundaries
Religious features	Cathedrals, monasteries, churches, chapels, and shrines
Industry, communications, and power	Quarries, peat-cuttings and huts, factories, fish farms, oil/gas stores, radio masts, windmills, watermills, pylons, and power stations
Water management and utilization	Reservoirs, fountains, dams, dykes, levees, irrigation canals, weirs, water towers, groynes, sluices, and sewage treatment facilities
Navigation and military features	Triangulation pillars, cairns, isolated objects as reference points, beacons, lighthouses, shipwrecks, and military camps
Tourist and sport facilities	Hotels, campsites, golf courses, ski-lifts, cable-cars, sports centres, and football pitches
Historical features	Castles, ruins, ancient earthworks, burial mounds, and monuments
Managed land	National parks, nature reserves, cemeteries, gardens, and parkland
Hydrology	Bodies of water, submerged rocks, rivers, streams, springs, currents, and bathymetric depths
Terrain and relief	Contours, spot heights, escarpments, natural escarpments, rocks, scree, sand, cliffs, caves, glaciers, and snowfields
Vegetation	Woods, forests, grassland, open land, shrubland, heathland, meadows, hedges, orchards, vineyards, and arable land
Non-landscape features	Grid and graticule intersections

choice of colours should be somewhat mimetic:

> One differentiates between and explains the most diverse features in a topographic map by colors [sic]. As far as the choice of color permits, one should retain the colored appearance of the landscape, in a generalized form, where it is part of the map. Examples are lowland green, ocean blue, white or light grey ice, the yellow or brown color of fields or desert, the green of grassland and the darker blue green of forest.

It would be interesting to investigate the degree to which the use of colour in topographic maps is mimetic of the features represented by its symbolization of landscape, but this would be inherently problematic. While a method might involve investigating the correlation between map symbol colours and reflectance values or digital numbers in natural colour imagery, the approach would inevitably run into theoretical as well as practical difficulties. The physical environment is comprised of constantly changing phenomena and this is not represented in the idealized landscape of the topographic map, where all trees have their leaves, the sea is blue, and the interpretation of features depends on their recognition and influences their appearance (e.g. road classes). Additionally, given that the typology is concerned with printed maps, colours can be inconsistent between print runs and so exact colour matching between a sample sheet and the style it represents may be inconsistent, however small the discrepancies. Besides, the fact that light conditions can affect map use would mean that the results would perhaps have a limited application, even if a constant light source were used for the technique.

The method used here allows the various types of symbol to be grouped — albeit more subjectively — according to a printed set of pre-determined colours. As opposed to measuring colours according to a continuum, these are taken from a discrete and logical range of printed colours (Figure 3). The colours of particular symbols on the map are matched by eye according to their closeness to a certain colour square and this colour name is recorded. This method is to some extent subjective, as are the colour names used in the descriptions of symbols, but it perhaps better serves the overall purpose of the typology than interpreting digital samples or classifying colours used from the rudiments of hue, value, and chroma, which, in anything other than a detailed study of colour, would have little wider meaning.

A series of representative groups of symbols are examined, consisting of point, line, and area features, which together constitute the majority of the map surface:

- land use: general built-up features, roads, railways, and arable and/or pastoral land
- land cover: forests, rivers, rocks/scree, and vegetation
- other: tourist point symbols, borders, and any other features.

As colour is such a powerful variable, it is important for this aspect of the typology to be flexible and accommodate symbols that may not fall easily into the other, more broadly defined categories.

'White' space

'White' space is the base colour of a map, which may not necessarily be white, or explained in the legend, but which, by its nature, forms the ultimate 'ground' and can influence style. 'White' space retains meaning through its power of suggestion and assumption and may use other colours to suggest the omnipresence of a certain characteristic of the landscape. A question Keates[37] curiously asks is 'from the user's point of view, does blank space on the map indicate "no information" or "no description"?'. How likely this feature is to exist on a topographic map is interesting in itself to investigate; is everything assigned to a category according to a legend criterion, e.g. to land use or terrain, or is the user left to assume the existence of a feature, e.g. grassland or bare rock? 'White' space is present on the map rather than absent — it

37. Keates, J. S. (1972). 'Symbols and meaning in topographic maps', **International Yearbook of Cartography**, 12, pp. 168–181. Page 172.

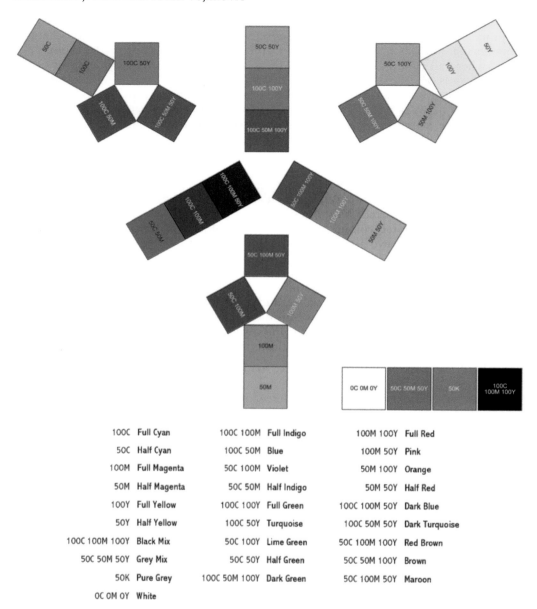

100C	Full Cyan	100C 100M	Full Indigo	100M 100Y	Full Red
50C	Half Cyan	100C 50M	Blue	100M 50Y	Pink
100M	Full Magenta	50C 100M	Violet	50M 100Y	Orange
50M	Half Magenta	50C 50M	Half Indigo	50M 50Y	Half Red
100Y	Full Yellow	100C 100Y	Full Green	100C 100M 50Y	Dark Blue
50Y	Half Yellow	100C 50Y	Turquoise	100C 50M 50Y	Dark Turquoise
100C 100M 100Y	Black Mix	50C 100Y	Lime Green	50C 100M 100Y	Red Brown
50C 50M 50Y	Grey Mix	50C 50Y	Half Green	50C 50M 100Y	Brown
50K	Pure Grey	100C 50M 100Y	Dark Green	50C 100M 50Y	Maroon
0C 0M 0Y	White				

FIG. 3. Colour chart and descriptions based on CMYK (cyan, magenta, yellow, black) values

symbolizes something — so it is important to stress that while it may appear blank, it is not an ontological vacuum. Pure 'white' space offers no definitions and lets the reader assume the basic nature of the landscape, but its 'colour' (or pattern) may suggest the existence of a particular landscape and can help to construct this in the reader's mind. As Keates[38] asserts,

> *the unqualified general category indicates a degree of uniformity or simplicity which may be far from the truth. Unless this is stated, the user has no means of distinguishing between lack of information, omission of information, and the terms of symbolization.*

The ontology of 'white' space, i.e. what can be known from its presence or exactly what it is meant to symbolize, is by its nature, perhaps the most indefinable or enigmatic of any symbol. In each country, a prevalent type of vegetation may be so universal that this is to be included in its 'white' space and one map's 'white' space may well appear as a more tangible symbol in the legend of another. In orienteering maps, for example, white space conventionally depicts the most easily traversable areas of forest. The use of orthorectified satellite imagery as a background, on the other hand, perhaps offers an alternative to 'white' space in which everything on the map has some implied meaning. Whether 'white' space is symbolized on the map and if so, how, is recorded in the typology.

Visual hierarchy

Although Robinson *et al.*[39] assert that a strong visual layering is to be avoided on general reference maps, given that paper topographic maps are limited in their capacity to treat all features equally, the selective inclusion and omission of features and their subsequent emphasis can suggest levels of importance through the use of variables such as size and colour. Larger, darker symbols appear to be closer and more significant than smaller, lighter ones. However, as visual hierarchy is affected by the presence and/or absence of features within the landscape, this quality cannot be based on a judgment of the map sheet alone. For example, if major roads are symbolized as light green lines and if railways are symbolized with thicker black lines, a particular map sheet covering an area that happens to have several major roads but not railways would be likely to produce a biased result. A topographic map of a country that heavily subsidizes public transport could therefore be poorly interpreted. In compiling the typology, the visual hierarchy is judged using the legend symbols for the same classes of features as under colour above, noting any particular characteristics. The top three feature types judged to be the most prominent on the legend and reinforced on the map (and thus the highest in the visual hierarchy) are recorded in order in the typology. A comparison of these features between maps should therefore give an idea of the relative importance of these features between the different countries.

Lettering

The term 'lettering' incorporates the various aspects of labelling on maps, including typographic design and the placement of text. The use of varying sizes and styles of

38. Keates, 1972, page 176.
39. Robinson *et al.*, 1995, page 327.

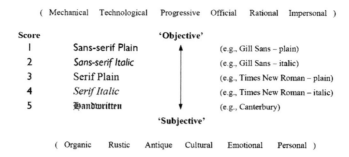

FIG. 4. A continuum for classifying type styles on topographic

label can also suggest a visual hierarchy, and, in conjunction with symbol design, is often used to indicate the relative importance of settlements. In this investigation, typographic design is classified according to certain criteria (Figure 4), which allow for a flexible and holistic appreciation of lettering, and regards the differences as a continuum based on general appearance.

This range is applied to describe five different feature labels in the typology: settlements, relief features, rivers, and the legend explanation text itself. By allocating a score to each of these styles ranging from 1 to 5, it is also possible to generate a total value, which provides a quantitative comparison of the overall typographic style of the national map series in relation to the continuum above. This suggests the degree of deviation from a uniform appearance and the 'objective' connotations that may be implied through their use.

Scale

To achieve as much consistency as possible, only current, standard, official, civilian edition 1:50 000 paper topographic maps are included in the typology, all of which will have been designed according to pre-determined specifications and for general release. The scale of 1:50 000 is selected for numerous reasons. First, this scale provides some equilibrium between representation and abstraction, being small enough to allow an appreciation of its rendition of the landscape but sufficiently large to render distortions caused by the map projection to have a minimal effect on the representation of features. Maps of this scale are typically derived from larger-scale products, giving rise to potentially different approaches to the generalization of features. Second, different scales also suit different purposes and here, the choice of 1:50 000 also represents a balance. Scales larger than 1:10 000, for example (which are often referred to as plans), usually relate to cadastral or other land administration issues such as utilities management and with smaller scales (such as 1:100 000), the purpose of the map leans towards navigation. Third, the balance of purposes that the 1:50 000 scale is intended to serve means that the resulting maps are perhaps the most 'general purpose' of general purpose maps and consequently, topographic maps at this scale are designed for a wider market, serving more users than other scales. All European countries have been mapped at 1:50 000, with a greater number of native

NMOs producing dedicated mapping at 1:50 000 than at any other scale.[40] Moreover, as this scale is employed in topographic map series worldwide, another reason for this choice is that it offers scope for the comparison of map series from further afield.

Format

Another important decision concerns which format — paper, digital, or both — should be included in the typology. For centuries, paper has formed the traditional medium for the production of topographic maps, with Lawrence[41] recently reiterating their value: 'The first choice for millions of walkers, cyclists, and motorists, Ordnance Survey's paper map series are a reflection of our commitment to national consistency'. Printed official topographic maps are not always for sale outside the country of production, although there is a drive to make digital versions of official 1:50 000 scale mapping more accessible through the websites of NMOs (where a 'geoportal' provides vector topographic data in the form of an Internet GIS, for example). But while most NMOs have published a countrywide paper 1:50 000 topographic map series, some do not have digital mapping available at this scale and many that do provide limited datasets (i.e. with fewer themes than their paper counterparts).

There are, however, more fundamental reasons for using paper maps for this investigation. As this medium limits the number of themes that can be shown simultaneously, the selective inclusion and omission of features is more crucial maps to the design process. Moreover, in order to manage the simultaneous presentation of various types of information at a fixed scale, paper maps must therefore embody a visual hierarchy in order to provide a successful interpretation. This inflexibility means that the attachment of certain values to features is both unavoidable and unhidden in the symbology of the map. Paper maps therefore present a particular challenge to the cartographer because, once published, they are 'finished'. They are limited in the quantity of detail they can show and the consequences of selective choice — and hierarchies between different types of feature — remain a product of the values of the institution involved in the map's creation.

By contrast, the added functionality of digital products, such as the availability of more layers of data, flexibility of scale and presentation — where layers of information (themes) can be turned on or off — provides more freedom. While the procedures of selection are still involved in the creation of digital maps, the user is often given more control over the presentation of the geographic information: data are stored in a series of independent layers, allowing the user to select a preferred combination of different types of information. However, the visual ordering of components is unlikely to be performed in digital topographic cartography because of the isolation of layers from one another, leading to the absence of a holistic sense of design. As each data layer is a map within itself, considerations of visual hierarchy are therefore not exercised, as Kraak and Ormeling[42] suggest:

40. Parry, R. B. and Perkins, C. R. (2000). **World Mapping Today**, 2nd ed., London, Bowker-Saur.

41. Lawrence, V. (2004). 'The role of national mapping organizations', **The Cartographic Journal**, 41, pp. 117–122. Page 119.

42. Kraak, M. and Ormeling, F. (1996). **Cartography: Visualization of Spatial Data**, Harlow: Longman. Page 44.

In a digital environment the differentiation between topographic maps and thematic maps is less relevant, [. . .] a topographic map would be a combination of separate road and railway layers, a settlement layer, hydrography, a contour-lines layer, a geographical names layer and a land cover layer. Each of these layers would be a thematic layer in itself, and a combination of layers, in which each data category had the same visual weight, would be a topographic map. If one category were to be graphically emphasized or highlighted, and the others thereby relegated to the status of ground, then it would again change into a thematic map.

The post-Fordist JIT (Just In Time) supply of digitally-derived geographic information that some NMOs offer, such as Ordnance Survey's MasterMap® service, is therefore not subject to the same design considerations or suffering the same design constraints as a paper map (while the strength of GIS lies in their ability to present and interrogate a combination of datasets, the combination of layers in a single map with each struggling for visual supremacy in the ensemble view, is what causes many cartographers to deride the output of GIS). Where the goal is the provision of topographic data rather than a topographic map, the challenge lies (as it always has done) in making an effective map from these data — turning spatial data into spatial information is the quintessential purpose of cartography. As maps that are designed to reach the user as finite and finished products, the paper medium therefore helps to ensure that what is on the map is there as a result of choice. These limitations therefore have several advantages for this investigation and it is for this reason that the focus will be on paper maps rather than their digital counterparts.

Geographical scope

Europe offers unique opportunities for investigating stylistic diversity in topographic mapping, offering sufficient diversity in climate, vegetation, economy, and culture among a high number of NMOs. The selection of sample sheets was initiated by identifying areas of approximately similar topography and urban cover using a small-scale topographic map of Europe and subsequently chosen from country indexes provided in Parry and Perkins[43] and the relevant NMO websites. Requests for maps meeting the criteria above were sent to 38 European NMOs, using contact information supplied in Parry and Perkins,[44] EuroGeographics,[45] and NMO websites, with the aim of including one sample map from as many countries as possible from within and outside the European Union. A few NMOs were not contacted as it was discovered through various sources (i.e. Böhme[46,47,48]; Parry and Perkins[49]; EuroGeographics[50]) that current official 1:50 000 maps available for civilian use were unlikely to be

43. Parry and Perkins, 2000.
44. Parry and Perkins, 2000.
45. EuroGeographics. (2003). Members (NMO Statistics), **http://www.eurogeographics.org/eng/01_ members.asp** (accessed 2003–2006).
46. Böhme, R. (1989). **Inventory of World Topographic Mapping**, Vol. 1, Western Europe, North America and Australasia, Barking, Elsevier Applied Science Publishers.
47. Böhme, R. (1991). **Inventory of World Topographic Mapping**, Vol. 2, South America, Central America and Africa, Barking, Elsevier Applied Science Publishers.
48. Böhme, R. (1993). **Inventory of World Topographic Mapping**, Vol. 3, Eastern Europe, Asia, Oceania and Antarctica, Oxford, Elsevier Applied Science Publishers.
49. Parry and Perkins, 2000.
50. EuroGeographics, 2003.

acquired. This included the Republika Geodetska Uprava of the former Yugoslav Republic of Macedonia, where only military editions of 1:50 000 maps exist. Northern Ireland, which has its own mapping agency, Ordnance Survey of Northern Ireland, was not included because it does not share the same level of responsibility as other NMOs. Although in Germany the 16 Länder are each responsible for producing their own topographic maps, these still come under the centralized jurisdiction of the Bundesamt für Kartographie und Geodasie and so a sample from Nordrhein-Westfalen was still valid for the purposes of this research.

QUANTITATIVE ANALYSIS

Of the 38 NMOs contacted, only 14 returned the 1:50 000 map sheet as specified (one sheet did not include a legend), with a further two NMOs sending acceptable alternatives (Figure 5). This result perhaps serves to highlight the unavailability of certain topographic map series outside their countries of origin. The low level of responses from the former Soviet satellite states and former republics of the Soviet Union is particularly distinctive and not dissimilar to that experienced in the surveys of map production and publication of these countries by Collier *et al.*[51,52] Further efforts were made to acquire specific map sheets (either by contacting the NMOs or by visiting their headquarters) to ensure that the sample was as large as possible and fulfilled the selection criteria.

While not exhaustive in its representation of NMOs in Europe, the sample comprises maps from 20 different NMOs and representing countries that exhibit considerable variation in population size, land area, climate, economic and industrial development, political heritage, and culture. Apart from Greece and Luxembourg, where topographic maps at 1:50 000 are not published by the native NMO, all countries forming the European Union before the major enlargement on 1st May 2004 (the EU 15) are represented in the sample, along with accession countries joining the EU on this date, i.e. from the Czech Republic, Latvia, Poland, and Slovenia, and those outside the EU, i.e. from Iceland, Norway, and Switzerland.

Classification of national symbologies

Compiling the typology involved the classification of some 2388 cartographic symbols from 20 symbologies, with legend explanations appearing in 17 different languages. As all map sheets in the sample are printed using at least four colours, a substantial range of expression is available (within the constraints suggested by scale). This grants NMOs the potential to demonstrate commonality or uniqueness through the choices they make regarding the symbolization of their national landscape, allowing their stylistic associations to emerge through the five elements of the typology: classification, colour, 'white' space, visual hierarchy, and lettering. The main findings of the analysis of each element will be summarized together with an interpretation of the results, which will provide an understanding of the extent to which stylistic diversity is exhibited in European state 1:50 000 topographic maps.

51. Collier, P., Fontana, D., Pearson, A. and Ryder, A. (1996). 'The state of mapping in the former satellite countries of Eastern Europe', **The Cartographic Journal**, 33, pp. 131–139.

52. Collier, P., Pearson, A., Fontana, D. and Ryder, A. (1998b). 'The state of mapping in the European republics of the former Soviet Union', **The Cartographic Journal**, 35, pp. 165–168.

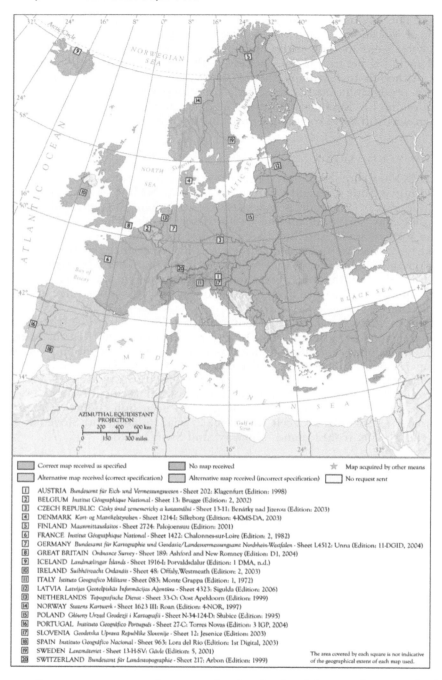

FIG. 5. Responses to requests for maps, showing details of the map sheets involved in constructing the typology

Table 2. Total number of legend symbols in European state 1:50 000 topographic maps ranked by country

Country (rank)	Symbology total	Country (rank)	Symbology total
Slovenia (1)	218	Belgium (11)	117
Italy (2)	161	France (12)	109
The Netherlands (3)	159	Germany (13)	108
Switzerland (4)	155	Spain (14)	100
Great Britain (5)	149	Finland (15)	91
Poland (6)	132	Norway (16)	86
Latvia (7)	131	Denmark (17)	80
Sweden (8)	125	Czech Republic (18)	76
Austria (9)	124	Iceland (19)	74
Portugal (10)	120	Ireland (20)	73

Stylistic analysis of the total number of symbols

The first step in the analysis[53] is to compare the total numbers of different discrete symbols used in the symbolization of national landscapes, as determined from the map legends (Table 2). If a higher number is used, the resulting maps will be more comprehensive in their symbolization of landscape (though more symbols may be employed in describing particular features than others). With a standard deviation of 36.78 from a mean of 119.40 for the sample, the degree of variation in the total number of symbols that NMOs use to express the landscape is not small.[54] Slovenian topographic maps use 57 more symbols than the country using the next highest total, Italy, which itself uses more than double the count of the Irish legend symbology.

As a national framework of spatial information, state topographic maps are produced for a variety of purposes. But it is clear that NMOs with a legacy of military impetus (i.e., Belgium, France, Great Britain, Italy, The Netherlands, Slovenia, Sweden, and Switzerland) generally adopt a greater number of discrete symbols for their civilian topographic maps than others. Moreover, members of the EU or NATO tend to use a more extensive 1:50 000 symbology. As Figure 6 shows, the geographical distribution of total symbol counts indicates a degree of regional bias, with many neighbouring countries using similar amounts of symbols, i.e. Slovenia and Italy; France, Belgium, and Germany; Sweden, Poland, and Latvia; and Denmark and Norway. Additionally, countries located towards the geographical fringes of the sample tend to utilize fewer symbols, especially Iceland, Ireland, and the Czech Republic (and to a lesser extent, Denmark, Finland, Norway, and Spain). The distribution pattern also reveals two small pockets where the most extensive cartographic vocabulary is employed: one comprising The Netherlands and Great Britain, and the other Switzerland, Italy, and Slovenia. As these countries are served by mapping organizations founded in the eighteenth and nineteenth centuries and are among the oldest in Europe, it seems plausible to suggest that their symbologies have grown to incorporate more inter-textual elements resulting from a longer history of cultural and cartographic exchange.

53. Computer software packages (Excel 2000 and SPSS 13) were used to perform statistical analysis.

54. Leaving aside the symbol count from the Slovenian symbology, the standard deviation of the sample is still 29.31 and the mean is 114–21.

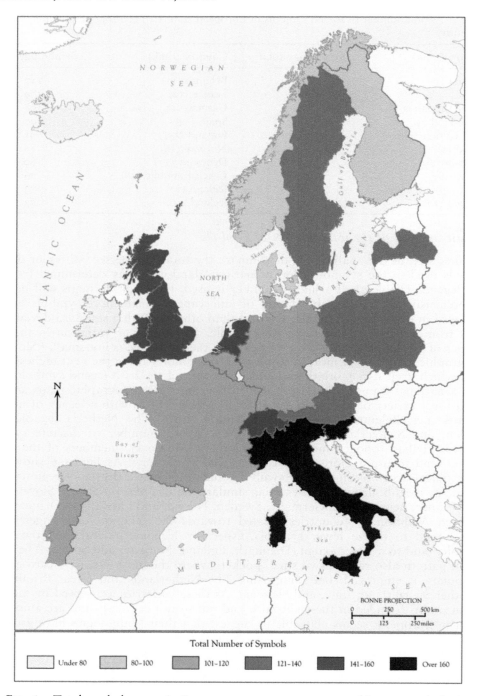

Fɪɢ. 6.   Total symbol counts in European state 1:50 000 topographic map symbologies

Stylistic analysis of the classification of symbols — levels I and II

As explained above, the main component of the typology is the classification of discrete cartographic symbols within the topographic map legend symbologies into three hierarchical levels. Symbols were classified individually and annotated in sequence at the most detailed level (level III), where the exact demarcations of each class evolved *a posteriori*, and these data were subsequently aggregated to supply data for the broader levels (levels I and II).

At the broadest level of classification (level I), most symbols are allocated to the human/artificial features class (a mean of 74.97%), with 24.93% being the mean devoted to the natural features class. Within the more detailed classification (level II), the settlement, territory, and resources category tends to use the most symbols, and with a mean percentage of 41.15%, this class employs over three times as many symbols as tourism, recreation, and conservation, which with 12.1%, typically uses fewest symbols. These data are presented in Tables 3 and 4.

Although it is interesting to compare symbologies according to the number of symbols counted within each class, what offers more insight in exploring stylistic

Table 3. Extract from the typology indicating the level I classification expressed as the number of symbols and as a percentage of the total symbol set

Country	Human/artificial features total		Natural features total		Non-landscape features total	
	Symbol count	Percentage of total symbol set	Symbol count	Percentage of total symbol set	Symbol count	Percentage of total symbol set
Austria	97	78.23	27	21.77	0	0.00
Belgium	89	76.07	28	23.93	0	0.00
Czech Republic	59	77.63	17	22.37	0	0.00
Denmark	58	72.50	22	27.50	0	0.00
Finland	57	62.64	34	37.36	0	0.00
France	86	78.90	23	21.10	0	0.00
Germany	78	72.22	30	27.78	0	0.00
Great Britain	127	85.23	21	14.09	1	0.67
Iceland	44	59.46	30	40.54	0	0.00
Ireland	60	82.19	12	16.44	1	1.37
Italy	133	82.61	28	17.39	0	0.00
Latvia	87	66.41	44	33.59	0	0.00
The Netherlands	128	80.50	31	19.50	0	0.00
Norway	64	74.42	22	25.58	0	0.00
Poland	89	67.42	43	32.58	0	0.00
Portugal	97	80.83	23	19.17	0	0.00
Slovenia	153	70.18	65	29.82	0	0.00
Spain	82	82.00	18	18.00	0	0.00
Sweden	93	74.40	32	25.60	0	0.00
Switzerland	117	75.48	38	24.52	0	0.00
Mean	89.90	74.97	29.40	24.93	0.10	0.10
Standard deviation	28.695	6.800	11.465	6.930	0.300	0.325

Table 4. Extract from the typology indicating the level II classification expressed as the number of symbols and as a percentage of the total symbol set

Country	Human/artificial I: accessibility and transport		Human/artificial II: settlement, territory, and resources		Human/artificial III: tourism, recreation, and conservation	
	Symbol count	Percentage of total symbol set	Symbol count	Percentage of total symbol set	Symbol count	Percentage of total symbol set
Austria	29	23.39	55	44.35	13	10.48
Belgium	49	41.88	34	29.06	6	5.13
Czech Republic	27	35.53	23	30.26	9	11.84
Denmark	22	27.50	27	33.75	9	11.25
Finland	15	16.48	37	40.66	5	5.49
France	35	32.11	36	33.03	15	13.76
Germany	25	23.15	41	37.96	12	11.11
Great Britain	72	48.32	25	16.78	30	20.13
Iceland	12	16.22	29	39.19	3	4.05
Ireland	27	36.99	15	20.55	18	24.66
Italy	67	41.61	56	34.78	10	6.21
Latvia	30	22.90	47	35.88	10	7.63
The Netherlands	60	37.74	62	38.99	6	3.77
Norway	23	26.74	30	34.88	11	12.79
Poland	36	27.27	43	32.58	10	7.58
Portugal	44	36.67	48	40.00	5	4.17
Slovenia	44	20.18	89	40.83	20	9.17
Spain	27	27.00	45	45.00	10	10.00
Sweden	33	26.40	37	29.60	23	18.40
Switzerland	56	36.13	44	28.39	17	10.97
Mean	36.65	30.21	41.15	34.33	12.10	10.43
Standard deviation	16.375	8.640	15.976	7.055	6.580	5.442

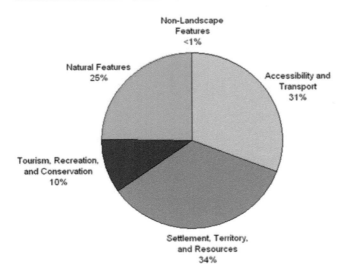

FIG. 7. Pie chart of mean symbol counts for levels I and II classes for the whole sample

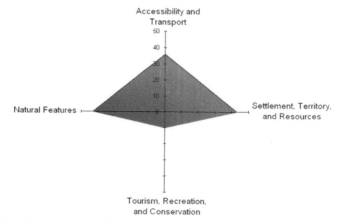

Fɪɢ. 8.　Star plot of Poland's symbology, indicating the number of symbols allocated to four classes from the levels I and II classification (non-landscape features omitted)

similarities and differences is the comparison of how a feature has been treated as a proportion of the entire symbol set. It is possible to graphically present these relative proportions very simply using a pie chart (Figure 7), where the level I human/artificial features class is split into its three level II subclasses and placed alongside the natural features and non-landscape features classes.

While pie charts can be used to visually compare the relative proportions of all the country symbologies, a more effective alternative is the star plot (Figure 8), where a constant value for all axes would indicate a perfectly continuous topographic range (all types of feature being symbolized using the same number of symbols).

Using a star plot for each country — where proportions are expressed as shapes — it is easier to identify similarities among the whole sample, especially when the symbol count of non-landscape features (exclusive to Great Britain and Ireland) is omitted.

The star plots in Figure 9 are arranged into groups that might be distinguished visually, with their spatial arrangement being an interpretation of their similarity. These have each been plotted on the same axes as that of Figure 8 above and use an identical set of scales (0–90 on each axis in this case) to allow comparisons between the total symbol counts as well as proportions, allowing a more refined distinction. It becomes possible to identify groups of countries based on their characteristic balance of these five variables. Iceland, for example, with its peculiar symmetry of natural features and settlement, territory and resources, is visibly similar to Finland and Latvia, to a lesser extent, while Great Britain and Ireland also exhibit a similar balance. Furthermore, the symbologies of Denmark, Norway, and the Czech Republic are closer together because of their similar size of symbology and form a loose association with the group containing Portugal, Italy, France, and The Netherlands, being separated with a dashed line. The symbology of Latvia similarly occupies a position between two other groups, whereas the extraordinary balance exhibited by the symbology of Sweden renders it completely separate. If, according to Piket,[55] the legend

55. Piket, 1972, page 271.

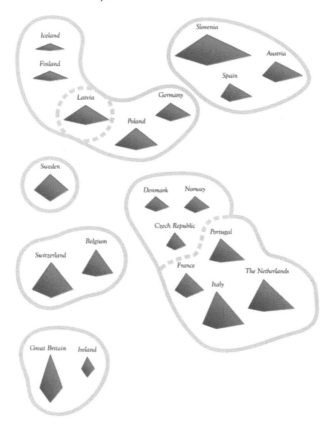

FIG. 9.   Star plots of each country in the sample, constructed from the number of symbols allocated to four classes from the broader levels of classification (non-landscape features omitted) and arranged according to a visual interpretation of similarity

symbology acts as a filter that withdraws certain topographic data from the map image, the various shapes of the star plots make it easier to see how some features may filter through into some symbologies more than others.

Although pie charts and star plots provide a visually immediate means of comparing symbologies, the characteristics that would define any supranational style nevertheless remain subtle. In order to analyse and classify the various symbologies according to the relative proportions of these feature types, it is therefore necessary to introduce a rigorous method of multivariate analysis. Hence, the method of cluster analysis — a technique often applied to explore the similarities between species — is introduced in order to classify the symbologies by dividing them into groups according to similarities in their percentage values for each class. Everitt[56] describes the

56. Everitt, B. S. (1993). **Cluster Analysis**, 3rd ed., London, Arnold. Page 6.

technique as an automatic and numerical measurement of relative distances between points that attempts to mimic in higher dimensions the eye–brain system for identifying clusters in two-dimensional data. It is therefore a useful complement to the purely visual method of identifying patterns as illustrated in Figure 9 above. The technique applied here is agglomerative (or hierarchical) cluster analysis, which explores the data to determine the quantity of groups.

Put simply, the method starts with n clusters, where n is the number of objects (the symbologies in this case), and proceeds by merging the two which are most similar so that $n-1$ clusters remain. The closest objects continue to fuse at each stage until only one cluster is left, fused at a level that encompasses all the other clusters, and thus objects, which gives rise to its hierarchical structure. As each fusion remains in place throughout the process, the stage at which each cluster is merged is especially relevant as it marks the degree of efficiency in making the fusion. Those merging later in the process require more 'effort' to fuse together, and, for this particular case, would therefore indicate symbologies that were stylistically less similar than those fused at earlier stages.

All clustering procedures use distance measures, or dissimilarities, or some kind of surrogate[57] to combine clusters, but as differences between methods arise because of the different ways in defining distance (or similarity) between objects, it is possible that different hierarchical clustering procedures may give different solutions. In using such a technique, therefore, it is important to try different methods and note which results exhibit the most consistency. Hansen and Tukey[58] advise against asking too much of cluster analysis and so it should be used in collaboration with other methods to gain a clearer understanding of the data. This is especially relevant with this application; if style is described accurately by Munro[59] as 'a recurrent trait-complex; a distinctive cluster or configuration of interrelated traits', cluster analysis might appear to offer the perfect solution as a tool for identifying supranational styles in the sample. The results of this analysis should therefore be evaluated in comparison with subsequent analysis of the more detailed classes (i.e. level III) and observations from the qualitative aspects of the typology.

In performing cluster analysis using the percentage data in Tables 3 and 4, country symbologies are therefore grouped according to variations in the proportion of symbols allocated to each symbol class. A variety of clustering techniques (i.e. between groups-linkage, within-groups linkage, and Ward's method) was applied in order to explore the possible outcomes. These techniques were favoured over other methods such as single linkage (nearest neighbour clustering) and complete linkage (furthest neighbour clustering) as they are based on information about all inter-cluster pairs, not just where the distance between groups is defined as that of the closest or most distant pair of individuals respectively.

The coefficients resulting from hierarchical cluster analysis are usually shown in a proximity matrix and represented visually in a dendrogram. In an agglomerative

57. Hansen, K. M. and Tukey, J. W. (1992). 'Tuning a major part of the clustering algorithm', **International Statistical Review**, 60, pp. 21–44. Page 22.

58. Hansen and Tukey, 1992, page 42.

59. Munro, T. (1946). 'Style in the arts: a method of stylistic analysis', **The Journal of Aesthetics and Art Criticism**, 5, pp. 128–158. Page 129.

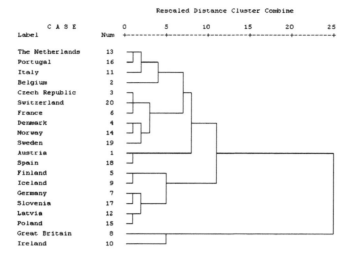

FIG. 10. Dendrogram from cluster analysis of countries based on percentage values for levels I and II classes using the betweengroups linkage method

dendrogram, the horizontal scale indicates the distance between the groups clustered together, so that the fusions made first (and most efficiently) appear nearest to this axis. Figure 10 therefore shows how countries with proportions of symbols that are deemed to be most similar are joined first and these links appear nearer to the left, with a distance closer to zero. Subsequent fusions between clusters of countries therefore occur further along the horizontal scale, to the right of the dendrogram.

The interpretation of dendrograms to determine plausible groups of individuals is based upon examining the structure and placing a division before a fairly large horizontal range where the number of groups does not change.[60] In this investigation, the dendrograms which result from the three different clustering techniques (i.e. between-groups linkage, within-groups linkage, and Ward's method of linkage), produce similar results. Several groups are distinguished and membership within these is, on the whole, consistent between the three clustering methods[61] below a 'rescaled distance' of 5:

- Group 1: Belgium, Italy, The Netherlands, and Portugal;
- Group 2: Czech Republic, France, and Switzerland;
- Group 3: Denmark, Norway, and Sweden;
- Group 4: Austria and Spain;
- Group 5: Finland and Iceland;
- Group 6: Germany, Latvia, Poland, and Slovenia;
- Group 7: Great Britain and Ireland.

As these groups contain few categories, it is easy to summarize the traits of each group by referring to the balance of particular types of feature. This, however, results in

60. Rogerson, P. A. (2001). **Statistical Methods for Geography**, London, Sage. Page 206.
61. Exceptions: the within-groups linkage fuses Groups 3 and 6 together very early, and Ward's method of linkage fuses Belgium to the cluster containing France, Switzerland, and Czech Republic.

more general descriptions of difference. Hence, Group 1 (e.g. Italy) has dominant but roughly equal proportions of accessibility and transport and settlement, territory, and resources symbols; Groups 2 and 3 (e.g. Switzerland) demonstrate the most balance (i.e. a continuous topographic range); Group 4 (e.g. Austria) is characteristically dominated by settlement, territory, and resources symbols; Groups 5 and 6 (e.g. Finland) are distinguished by the equal dominance of natural features and settlement, territory, and resources classes; and Group 7 (e.g. Ireland) is characterized by the dominance of the accessibility and transport class but also maintains the highest proportion of tourism, recreation, and conservation symbols.

In some cases, the most plausible explanation for group membership appears to be the influence of geographical proximity (e.g. Groups 3 and 7). This factor features prominently in the comparison of European transport maps by Morrison[62]:

> [...] *there is a striking similarity between the extent of the 'French style' of maps and the French language. This suggests that despite the supposed free flow of goods, labour, information etc. within the European Union, language still presents a substantial barrier even to the flow of graphical ideas! Since maps are themselves expressed in the universal language of cartography, the mechanism which restricts the diffusion must simply be the travel habits and professional contacts of the map designers. People who do not speak French will tend to avoid taking holidays in French-speaking areas, and will tend not to compete for business in such areas, so are less likely to be exposed to the 'French style'.*

Although this explanation relates to the style of thematic maps, it does seem relevant, especially when three Scandinavian countries (Denmark, Norway, and Sweden) happen to fall into one group. But it does not account for the membership of all groups, particularly Groups 1 and 4, which include countries that are not contiguous and whose populations do not speak the same language. Why, for example, does Austria merge with Spain and not Switzerland, which would appear to be more logical if proximity and its associated elements of language, climate, and terrain — all relevant to topographic map design — are considered so influential? Geographical proximity may therefore be just one contributing factor for group membership.

The similarities behind the most consistently distinctive grouping — of Great Britain and Ireland — probably result from a long historical association. Ordnance Survey conducted the topographical mapping of Ireland as proposed by the Spring-Rice report of 1824.[63] After the Ordnance Survey's methods of survey, landscape description, and portrayal were established in Ireland, it is likely that some of its legacy remains in the design of current Irish topographic maps. Of course, proximity has played a role in this particular historical association, whether the portrayal of landscape has been (or is) congruous with Irish culture and society or not, but the point is that historical and political associations between other European nation-states have not relied solely on their proximity with one another. Figures 11.16 present extracts from the 1:50 000 topographic maps within each cluster group and allow a visual comparison of similarities and differences in appearance between and within these seven groups. Whether these group characteristics extend to the appearance of the maps, or if other traits emerge more strongly, will be investigated using the qualitative analysis of the symbology.

62. Morrison, 1996, page 95.

63. Doherty, G. M. (2004). **The Irish Ordnance Survey: History, Culture and Memory**, Dublin, Four Courts Press. Page 14.

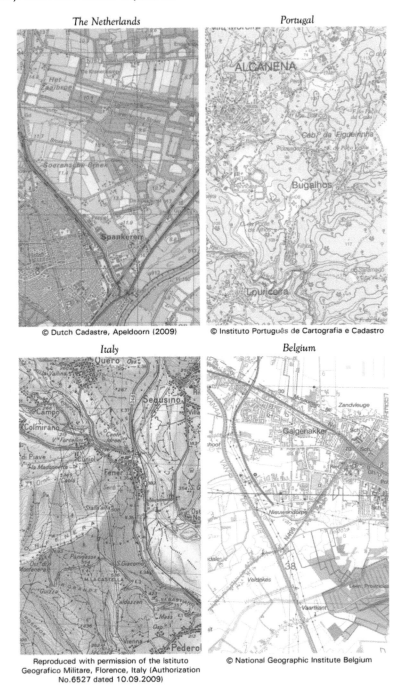

FIG. 11.   Extracts from maps comprising Group 1

Czech Republic   Switzerland

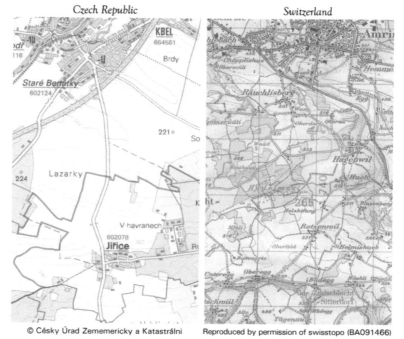

© Césky Úrad Zememericky a Katastrálni   Reproduced by permission of swisstopo (BA091466)

France

© IGN 2009 Authorization No.80-9050.

Fig. 12.   Extracts from maps comprising Group 2

Denmark   Norway

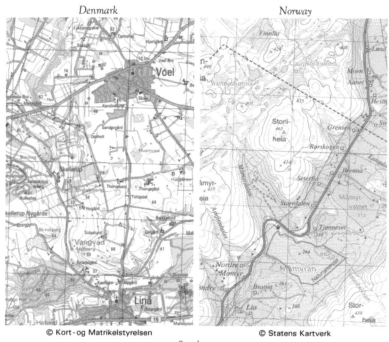

© Kort-og Matrikelstyrelsen   © Statens Kartverk

Sweden

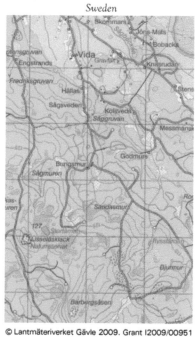

© Lantmäteriverket Gävle 2009. Grant I2009/00951

FIG. 13.   Extracts from maps comprising Group 3

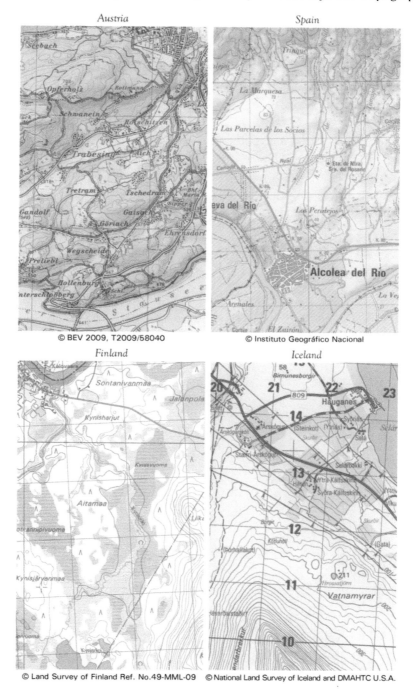

Austria

Spain

© BEV 2009, T2009/58040

© Instituto Geográfico Nacional

Finland

Iceland

© Land Survey of Finland Ref. No.49-MML-09   © National Land Survey of Iceland and DMAHTC U.S.A.

FIG. 14.   Extracts from maps comprising Groups 4 and 5

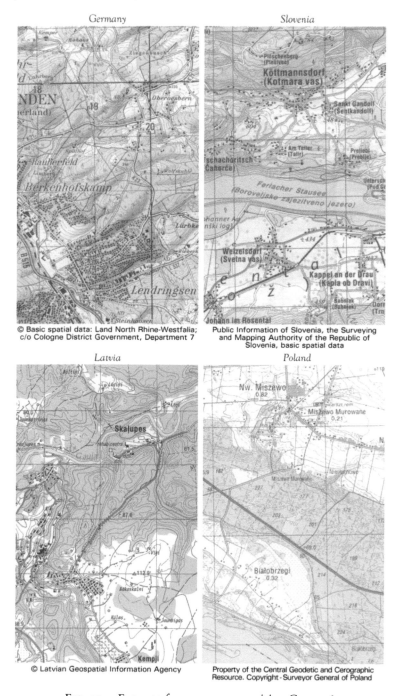

Germany

Slovenia

© Basic spatial data: Land North Rhine-Westfalia;
c/o Cologne District Government, Department 7

Public Information of Slovenia, the Surveying
and Mapping Authority of the Republic of
Slovenia, basic spatial data

Latvia

Poland

© Latvian Geospatial Information Agency

Property of the Central Geodetic and Cerographic
Resource. Copyright - Surveyor General of Poland

FIG. 15. Extracts from maps comprising Group 6

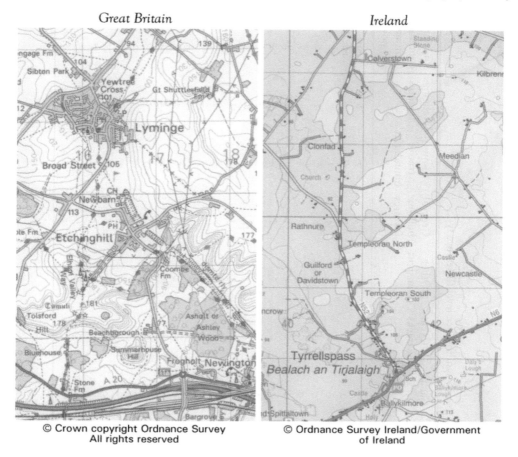

FIG. 16.   Extracts from maps comprising Group 7

Stylistic analysis of the classification of symbols — level III

The results of the level III symbol classification reveal a high degree of variation in how landscapes are classified by NMOs, as is clear from the relative symbol counts across the range of feature types (Tables 5 and 6). Some classes, i.e. roads; industry, communications, and power; water management and utilization; and hydrology, generally employ more symbols and exhibit more variation than others. For example, in comparing the number of symbols used for the class dealing with roads (which includes petrol stations, parking, and road bridges), the symbology of Great Britain adopts 40 symbols, whereas Norway uses only nine (Figure 17 below illustrates the comparative treatment of level III feature types). Additionally, for the industry, communications, and power class, Slovenia uses 32 symbols whereas the Czech Republic uses only two; and for the hydrology class, Slovenia uses 28 symbols while Italy uses just two. Countries with higher total symbol counts, however, may not necessarily

Table 5. Extract from the typology indicating the number of symbols per level III feature type

Country	Road	Rail	Paths	Canals	Cycle tracks	Other transport	General built-up features	Administrative boundaries	Religious features	Industry, communications, and power
Austria	12	8	3	2	0	4	3	7	8	19
Belgium	31	13	3	2	0	0	3	6	3	12
Czech Republic	15	8	1	0	0	3	4	12	1	2
Denmark	10	4	2	0	0	6	6	2	1	10
Finland	10	2	1	0	0	2	8	4	2	9
France	22	8	2	2	0	1	1	9	4	9
Germany	11	5	2	2	0	5	5	4	3	13
Great Britain	40	9	10	5	4	4	2	3	3	12
Iceland	10	0	1	0	0	1	11	2	1	8
Ireland	15	4	2	2	1	3	3	2	2	4
Italy	35	16	3	5	0	8	9	4	4	15
Latvia	18	9	0	0	0	3	9	3	2	20
The Netherlands	39	6	2	4	1	8	13	6	4	14
Norway	9	8	4	0	0	2	9	6	2	7
Poland	16	12	3	0	0	5	7	4	4	14
Portugal	25	15	1	2	0	1	7	5	2	10
Slovenia	22	8	3	0	1	10	14	6	4	32
Spain	12	10	3	2	0	0	3	5	1	21
Sweden	18	7	3	0	1	4	8	7	3	12
Switzerland	26	18	3	0	1	8	5	4	2	17
Mean	19.80	8.50	2.60	1.40	0.45	3.90	6.50	5.05	2.80	13.00
Standard deviation	9.95	4.59	2.01	1.70	0.95	2.90	3.65	2.48	1.64	6.68

Table 6. Extract from the typology indicating the number of symbols per level III feature type (continued from Table 5)

Country	Water management and utilization	Navigation and military features	Tourist and sport facilities	Historical features	Managed land	Hydrology	Vegetation	Terrain and relief	Non-landscape features
Austria	14	4	7	4	2	11	8	8	0
Belgium	5	5	2	2	2	13	5	10	0
Czech Republic	4	0	3	4	2	6	4	7	0
Denmark	4	4	4	3	2	9	6	7	0
Finland	3	11	2	1	2	17	8	9	0
France	9	4	3	7	5	8	9	6	0
Germany	10	6	3	5	4	12	9	9	0
Great Britain	1	4	14	7	9	7	4	10	0
Iceland	5	2	0	1	2	16	3	11	1
Ireland	0	4	11	3	4	5	3	4	0
Italy	19	5	6	2	2	2	21	5	1
Latvia	9	4	2	3	5	21	13	10	0
The Netherlands	10	15	3	1	2	10	15	6	0
Norway	2	4	8	0	3	11	5	6	0
Poland	10	4	3	2	5	13	13	17	0
Portugal	17	7	0	4	1	11	6	6	0
Slovenia	20	13	8	7	5	28	15	22	0
Spain	11	4	2	4	4	5	7	6	0
Sweden	4	3	14	5	4	13	8	11	0
Switzerland	14	2	11	3	3	13	8	17	0
Mean	8.55	5.25	5.30	3.40	3.40	11.55	8.50	9.35	0.10
Standard deviation	5.960	3.697	4.366	2.062	1.847	5.934	4.718	4.591	0.308

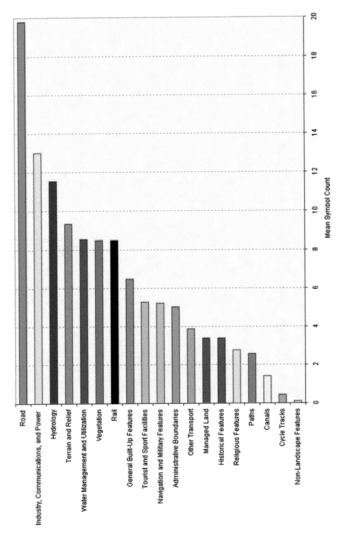

FIG. 17. Ranked mean symbol counts for each level III class

use the greatest number of symbols for each feature type; Italy uses 21 symbols for vegetation whereas Slovenia uses 15. Some classes demonstrate consistently small numbers, such as paths, canals, cycle tracks, religious features, historical features, and managed land, and consequently little variation. However, in terms of presence and absence, they demonstrate an important dichotomy, as symbols representing these types of feature are not included in all map symbologies. For example, Great Britain uses four symbols for cycle tracks, whereas 14 countries in the sample use none. Perhaps more surprisingly, The Netherlands uses only one symbol for this feature.

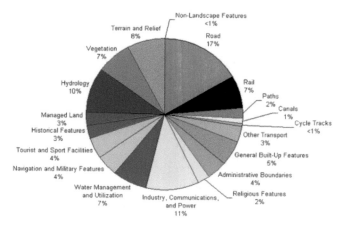

FIG. 18.   Pie chart indicating the mean level III symbol counts

If users of topographic maps have the same demands, as Arnberger[64] suggests, it would seem that these are not reflected in the range of symbols as counted on topographic maps from different countries. This demonstrates that the mapping concerns of these countries — as interpreted by their national mapping organizations — are by no means standardized.

As previously established, a comparison of percentage values is more useful for identifying potential stylistic similarities and differences than symbol counts. Figure 18 demonstrates that a simple pie chart is capable of providing a holistic impression of the mean relative proportions of symbol classes constituting the whole sample, again revealing the dominance of road symbols on the 1:50 000 maps of Europe.

With more classes, star plots provide an even more effective means of comparing the characters of country symbologies, especially at the most detailed level of symbol classification (Figures 19 and 20), and it is possible to discern three broad categories. These are: topographic ranges that are reasonably balanced (e.g. Slovenia); those with few feature types dominating (e.g. Iceland); and those with one feature type dominating (e.g. Great Britain). Moreover, a visual comparison of these star plots suggests that some countries are perhaps more 'articulate' than others in describing particular features, in using more symbols for these classes, such as 'road'. Of more relevance, however, is that plots of some pairs of countries have similar shapes (e.g. Great Britain and Ireland; Finland and Iceland; Germany and Denmark; and Latvia and Slovenia). This suggests that supranational styles might still exist as an expression of the extent of cartographic vocabulary at this level of classification. Comparing these observations with those resulting from the cluster analysis of the levels I and II data above, the uniqueness of the symbologies of Great Britain and Ireland again emerges strongly, although other similarities are not as clear.

64. Arnberger, E. (1974). 'Problems of an international standardization of a means of communication through cartographic symbols', **International Yearbook of Cartography**, 16, pp. 19–35. Page 22.

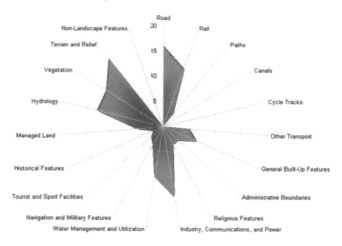

FIG. 19.   Star plot indicating the level III symbol count for Poland

For a more rigorous test, the same cluster analysis techniques used for the broader levels of classification were applied to the level III classes and the results of the two sets were compared. The main objective is to determine whether the same groupings resulting from the broader classes are consistent for a disaggregation of these data at level III. This provides a more thorough basis upon which to identify stylistic similarities.

In this case, the dendrograms which result from the three different clustering techniques applied previously (i.e. between-groups linkage, within-groups linkage, and Ward's method of linkage), exhibited much less resemblance to each other than to those derived from the levels I and II analysis. Furthermore, the structure resulting from the contrasting methods of clustering provides little basis for identifying consistent supranational groups. Indeed, the results from using Ward's method (Figure 21) show two major clusters (separated between Iceland and Italy) fusing together much later in the process.

A closer examination of the clustering, however, reveals that some pairs of individuals (countries) are consistently fused together more easily than others. Moreover, these match the visual pairing of countries described earlier: for example, Finland with Iceland and Latvia with Poland. Additionally, Germany and Slovenia are consistently joined at an early stage in the process, while Germany and the Czech Republic always remain far apart.

The dendrogram below (Figure 22) results from the same methods of clustering but where the data have been standardized, involving a transformation of the values so that the mean for each variable (level III class) equals 1. This allows the relationships between the clusters to be modified so that all variables have an equal impact on the computation of (Euclidean) distances, thus providing an even platform from which to form clusters. This results in a clearer demarcation between groups of countries, with a greater number of clusters merging at similar, initial stages. It therefore becomes easier to divide the sample into plausible groups of clusters, especially between those merging at the initial stages and those fusing later. For example, Figure 22 below

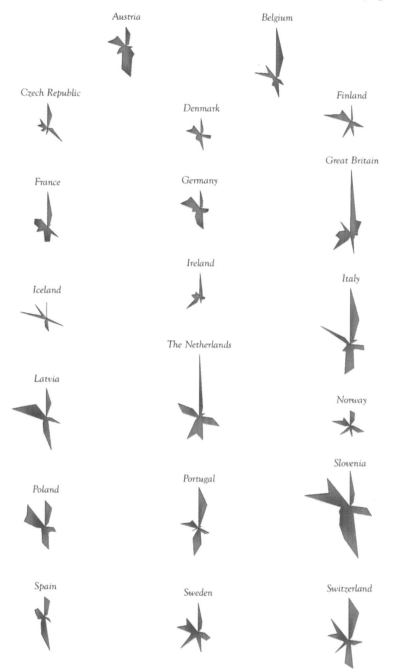

FIG. 20.   Star plots of each national symbology based on the symbol counts for each level III feature and plotted on identical axes (surface area of each plot reflects total symbology size)

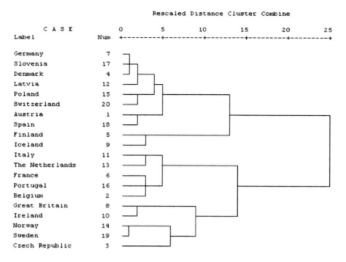

FIG. 21. Dendrogram from cluster analysis of countries based on percentage values for level III classes using Ward's method of linkage

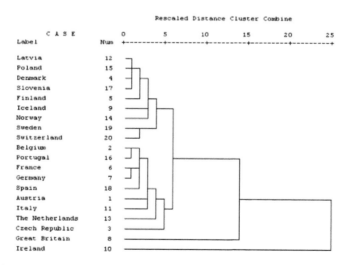

FIG. 22. Dendrogram from cluster analysis using the within-groups linkage method, based on percentage values for level III (data standardization: mean value for variables=1)

indicates that Great Britain and Ireland are clearly very different from the rest of the sample, with Ireland requiring more 'effort' to fuse to another cluster than any other country.

From the results of these cluster analyses, it is possible to divide the countries into three general groups, which begin to emerge at around the rescaled distance of 5 or further, as follows:

- Group 1: Denmark, Finland, Iceland, Latvia, Norway, Poland, Slovenia, Sweden, and Switzerland;
- Group 2: Austria, Belgium, Czech Republic, France, Germany, Italy, The Netherlands, Portugal, and Spain;
- Group 3: Great Britain and Ireland.

In terms of the level III classification, for example, Group 1 is characterized by the smallest overall symbol count and the most balanced topographic range, using similar proportions of symbols for the road and industry, communications, and power categories, and with a relatively high number of symbols allocated to hydrology, general built-up features, and terrain and relief. Group 2 countries allocate a relatively high proportion of symbols to the industry, communications, and power and water management and utilization classes, and, in contrast to Group 1, vegetation would seem to be more important than both hydrology and terrain and relief. Group 3 countries, however, are dominated by symbols belonging to the road class, but with a relatively high proportion dedicated to tourism and sport facilities, historical features, managed land, and paths. This group also has the least extensive selection of rail symbols. The deficiency of symbols from other classes (such as industry, communications, and power and water management and utilization) would suggest that the topographic maps of Group 3 present a landscape to be treated as a commodity for leisure and recreation rather than resources or territory, where users can escape to and access a sense of wilderness. In addition, Group 1 consists of mostly non-EU 15 countries and those towards the periphery of countries with lower total symbol counts. Group 2 is made up almost exclusively of contiguous EU 15 countries and Group 3 is made up of the two EU 15 countries representing the British Isles.

Summary of quantitative analyses

It is clear from the clustering process that countries merge into groups with more difficulty using the level III classes than with a smaller representative range of classes characterized by the broader classification afforded by levels I and II. The seven groups derived from the levels I and II clustering above might have also been determined more reliably because, as explained previously, they are less likely to result from the misclassification of symbols — a possible consequence of the detail of the level III classes. For example, classifying a marshland symbol into the natural features class offers more accuracy (if not precision) than its classification into the hydrology class at level III, because at the broader level, its distinction from other classes is clearer. While a high degree of individuality is demonstrated — as the cluster analysis of the level III data reveals — from a comparison of more detailed classifications of symbology, the similarity between countries (and also the contrast between groups) is emphasized through their comparison using the broader classes.

Instead of treating these two sets of results as offering conflicting cluster solutions, they should, however, be regarded as revealing different aspects of similarity and difference in cartographic vocabulary. Apart from France, Germany, Switzerland, and the Czech Republic (which, apart from Germany, later formed a group on their own), the seven groups resulting from the levels I and II cluster analysis remain homogeneous within the three classes resulting from the level III grouping, revealing a degree of consistency between the two sets of cluster groups. Each country's individual

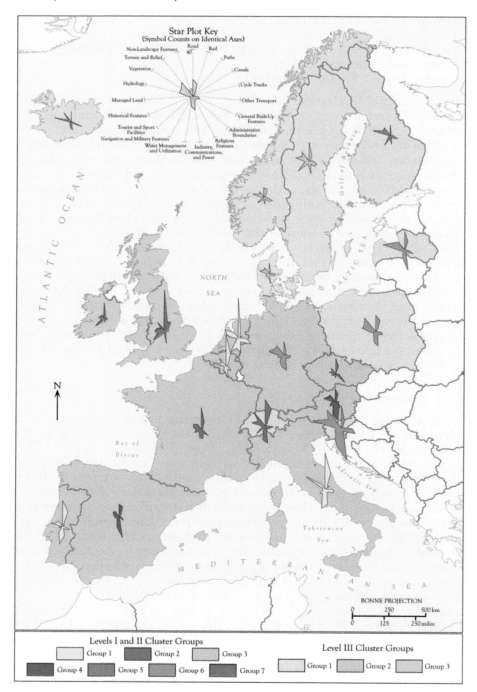

FIG. 23.   Geographical distribution of cluster groups derived from all classification data

expression of landscape is simply more apparent when symbologies are compared in more detail, as the simultaneous presentation of cluster groups in Figure 23 shows.

The next section concerns the analysis of the qualitative elements of the typology, which investigates similarities and differences in the graphical appearance of maps. As style is a case of how a feature is depicted in addition to whether or not it is depicted, the results of this analysis will determine whether the similarity between countries in the groups above extends to this aspect of the portrayal of their landscape.

QUALITATIVE ANALYSIS

This component of the typology concerns the visual appearance of the symbologies analysed above and as such, has a bearing on the immediate impression of the landscape presented in the maps. Countries possessing similar characteristics in their classification of landscape may not necessarily exhibit similarity in their choice of graphical variables and it is therefore possible that their topographic maps will vary greatly when it comes to appearance, as may be observed in Figures 11–16. Conversely, maps that appear to be similar in appearance might conceal differences arising from the classification of their landscape. So, while similarities may emerge from the ensemble effect of symbols in the above extracts, the methodology involves the analysis of individual aspects of their design in order to identify particular characteristics. The findings of the analysis for each subsequent element of the typology are presented and summarized in turn.

Colour

Advances in printing technology and the desire of national mapping organizations to exploit these has meant that the 'classical style' to which Keates[65] refers (i.e. black planimetry, blue water, brown contours, and green vegetation) is no longer as widespread as it once was (Tables 7–9). The most conservative countries seem to be Austria, Portugal, Slovenia, Italy, and France, which closely resemble this early supranational style in their use of colour. Most countries use either black or grey for symbols representing general built-up features. Departures from this scheme usually include some use of red, with countries using red (e.g. Spain), red brown (e.g. Poland), maroon (e.g. Sweden), light pink (e.g. Great Britain), or violet (e.g. Finland). The Alpine countries of France, Germany, Switzerland, Austria, and Italy demonstrate the sole use of black for portraying general built-up features, with the exception of Slovenia. The most complex colour system for general built-up features is that adopted by The Netherlands, with various shades used to discriminate between categories and the innovative use of maroon for high-rise buildings, which, using a darker figure, appear visibly closer to the user.

Where most country symbologies engage in a radical departure from the 'classical style' is through their depiction of roads. Almost gone are the early conventions of using a single colour (such as black or red) to depict this type of feature, which, as determined in the previous section, typically uses more associated symbols than any other feature. As it is likely that many NMOs would perceive road-users to be a key

---

65. Keates, 1996, page 256.

Table 7. Extract from the typology indicating the colour of symbols used for different types of feature

| Country | Land use | | | |
	General built-up features	Roads	Railways	Arable/ pastoral land
Austria	Black mix	Black mix, white with black mix outlines	Black mix and white	White
Belgium	Pure grey	Violet, full red, half yellow, white with pure grey outlines (except those under construction, which use full red outlines)	Black mix and white	White
Czech Republic	Pure grey, pure grey (dark)	Full yellow, white with black mix outlines	Black mix and white	White
Denmark	Pure grey, black mix	Half red, white with black mix outlines	Black mix and white	White
Finland	Violet, grey mix, black mix	Full magenta, black mix (no outlines)	Black mix	Orange, orange (light)
France	Black mix	Orange, white with black mix outlines	Black mix and white	White
Germany	Black mix	Orange, full yellow, and white with black mix outlines	Black mix and white	White
Great Britain	Half magenta (light) with black mix outlines	Full cyan, full green, full magenta, orange, full yellow, and white, all with black mix outlines	Black mix	White
Iceland	Red brown (light and very light) with black mix outlines	Red brown (dark) and white, both with black mix outlines	n/a	Half green (light)
Ireland	Pure grey, black mix	Full cyan, full green, full red, orange, full yellow, and white, all with black mix outlines	Black mix	Half green (light)
Italy	Black mix	Orange, white, both with black mix outlines	Black mix	White
Latvia	Pure grey, black mix	Red brown, red brown (light), and white, all with black mix outlines	Black mix	White
The Netherlands	Half magenta (light) with full red outlines; full red; full red (light) with full red outlines; maroon with black mix outlines	Violet, full red, orange, full yellow, white, full yellow and pure grey, pure grey, all with black mix outlines	Black mix and white	Half yellow (light)

Table 7. *Continued*

Country	General built-up features	Roads	Railways	Arable/ pastoral land
	Land use			

Country	General built-up features	Roads	Railways	Arable/ pastoral land
Norway	Black mix, white with black mix outlines	Full red (light), white, both with black mix outlines	Black mix	Half yellow (light)
Poland	Red brown, brown, brown (light), the latter two symbols with red brown outlines	Orange, white, both with red brown outlines	Black mix and white	White
Portugal	Full red (light)	White with full red or black mix outlines	Black mix and white	White
Slovenia	Black mix, orange, pure grey, the latter two symbols with black mix outlines	Orange, orange (light), and white, with pure grey outlines	Black mix	White
Spain	Full red (halftone) with full red outlines, full red with black mix outlines	Full yellow with full red outlines, white with full red outlines, and white, full red, orange, lime green, all with pure grey outlines	White with black mix outlines, black mix	Half yellow (light)
Sweden	Maroon, maroon (light), maroon (very light), all with pure grey outlines	Red brown, white, black mix (all as centrelines for various classes of road)	Black mix and white	Full yellow (light)
Switzerland	Black mix	Orange, half red, half yellow, white, all with black mix outlines	Black mix and white	White

user group for this scale of map, the benefits of using different colours as an aid to classification and navigation are clear and so road symbols exploit the greatest variation in the utilization of colour. Orange, red, yellow, and white are the most common colours for roads, usually in conjunction with a black casing. The colour scheme for the Alpine countries (especially Slovenia, France, and Italy) is similar, as is that between Belgium and The Netherlands, which both use violet for motorways. Great Britain and Ireland are also distinctive in this respect, particularly in their use of cyan for motorways; a colouring system designed to reflect road signage.[66] Railways are depicted universally in monochrome and appear either as a solid line or a dashed

66. Harley, J. B. (1975). **Ordnance Survey Maps: A Descriptive Manual**, Southampton, Ordnance Survey. Page 127.

Table 8. Extract from the typology indicating the colour of symbols used for different types of feature (continued from Table 7)

Country	Vegetation	Land use		
		Rivers and river outlines	General contours	Forests
Austria	Black mix point symbols over lime green (light) background	Full cyan	Brown	Lime green (light)
Belgium	Lime green, half green, dark green	Half cyan	Brown	Deciduous — lime green, coniferous — half green, mixed — full green
Czech Republic	Lime green (light), half yellow, dark green	Half cyan	Red brown	Lime green (light)
Denmark	Half green, heathland as half magenta (light)	Half cyan	Red brown	Half green
Finland	Orange, orange (light), white	Half cyan	Brown	White
France	Half green or white with full green point symbols	Half cyan	Orange	Half green, with full green point symbols
Germany	Lime green (light) with full green point symbols	Half cyan	Orange	Lime green (light)
Great Britain	Lime green (light) with full green point symbols	Half cyan	Red brown	Lime green (light) with full green point symbols
Iceland	Half green (light) with dark green point symbols for shrubland	Dark blue	Red brown (dark)	Half green (light) with label
Ireland	Half green (light) with dark green point symbols	Half cyan	Pure grey	Half green (light) with dark green point symbols
Italy	White and half green (light) with dark green and black mix point symbols	Full cyan	Orange	Half green (light) with black mix point symbols
Latvia	Half green and half green (light) with black mix and full green point symbols	Half cyan	Red brown	Half green with black mix point symbols
The Netherlands	Lime green (light), lime green, half yellow (light) all with black/turquoise point symbols for cultivated/uncultivated land, heathland as half magenta (light)	Half cyan	Red brown	Lime green with turquoise point symbols
Norway	Half yellow, lime green (light) with full green dot symbols	Half cyan	Red brown	Lime green (light)

Table 8.   *Continued*

| Country | Land use | | | |
	Vegetation	Rivers and river outlines	General contours	Forests
Poland	Half green with full green point symbols	Full cyan	Orange	Half green with dark green point symbols
Portugal	Lime green with black mix point symbols	Full cyan	Brown	Lime green with black mix point symbols
Slovenia	Lime green (light) with black mix and lime green point symbols	Full cyan	Red brown	Lime green with black mix point symbols
Spain	Lime green (light and very light), half yellow (light), and half green, with full green point symbols	Full cyan	Brown	Lime green with full green point symbols
Sweden	Lime green (light), white, or full yellow (light), with lime green point symbols	Half cyan	Red brown	Lime green (light) with full green point symbols
Switzerland	Half green (light) with black mix point symbols	Half cyan	Red brown	Half green (light) (open forest with only black mix point symbols)

black line within a black casing. While countries differ in the number of symbols devoted to representing this type of feature and therefore offer varying amounts of detail, the classification of railways is restricted to the variable of line thickness and style (e.g. dashed).

Vegetation is almost universally shown using shades of green, although Finland characteristically uses orange and white area symbols. In most countries, different point symbols are placed over a green background to indicate the type of vegetation. These symbols normally use a darker shade of green, but appear as black in some Alpine countries and also Portugal. Countries do not easily group together regarding the background colour. Although it might be assumed that lime green would perhaps be used by Mediterranean countries and shades of green possessing a higher cyan content by northern European countries (perhaps to mimic the colour of vegetation — coniferous trees for the latter, for example), this is not in fact the case.

Rivers and general contours both tend to be depicted in fairly consistent colours, again drawing parallels with the 'classical style', with various forms of cyan in use for the former and orange or brown for the latter. Departures from this include Iceland, which adopts a darker blue[67] for hydrology, and Ireland, which uses grey for

67. This map has been designed to be red-light readable and so this affects the gamut of colours chosen.

Table 9. Extract from the typology indicating the colour of symbols used for different types of feature (continued from Table 8)

Country	Tourist point symbols	International borders	Any other significant features/ observations
		Other	
Austria	Black mix	Black mix with half magenta (halftone) highlight ribbon	Peat cutting point symbol in black mix
Belgium	n/a	Black mix with violet (halftone) highlight ribbon	Sports complex area symbol in full red line pattern over full yellow background
Czech Republic	Black mix	Full magenta with half magenta highlight ribbon	Arable or other land is shown in the legend using white
Denmark	Full magenta	Black mix	Yachting harbour tourist symbol is shown in the legend using half cyan
Finland	Black mix	Black mix with full magenta diagonal line pattern highlight ribbon	Marshes of an easy passability are shown as a blue line pattern over a full yellow area, giving the effect of a yellow green area; isobaths in brown
France	Black mix	Black mix with orange highlight ribbon	Printed in fewest colours in the sample
Germany	Black mix	Black mix in centre of half magenta highlight ribbon	Gardens in half green give urban areas a more natural association
Great Britain	Half cyan	Black mix	Park or ornamental ground is in half grey (light) which gives them an urban association
Iceland	n/a	Black mix (first order administrative)	Colours are designed to be red light readable, which imposes limitations on the gamut of colours used
Ireland	Full red	Black mix	Land appears in half green (light) with a lighter tint for areas 100 m or higher, blending in with built-up areas
Italy	Black mix	Black mix in centre of orange diagonal line pattern highlight ribbon	Printed in minimal colours as per France, but with full magenta overlay
Latvia	Black mix	Black mix in centre of red brown (half tone) highlight ribbon	Bogs have a depth measurement with arrow
The Netherlands	Black mix	Black mix with full yellow highlight ribbon	Buildings symbolized using various shades of red
Norway	Pink	Black mix with half magenta (halftone) highlight ribbon	Cultivated areas in full yellow

Table 9. *Continued*

Country	Other		
	Tourist point symbols	International borders	Any other significant features/observations
Poland	Brown	Black mix with half magenta highlight ribbon	Predominance of brown
Portugal	n/a	Black mix	Buildings and roads in full red (light) have an almost neon effect over background lime green (light)
Slovenia	Black mix	Black mix with orange (light) highlight ribbon	High variety of symbols but all in full or light versions of full cyan, orange, black mix, lime green, or brown
Spain	Black mix	Black mix	Irrigated land is shown using half green background with full green diagonal line pattern
Sweden	Maroon	Black mix	All buildings maroon, except industrial zones in pure grey (light)
Switzerland	Black mix	Black mix with half magenta highlight ribbon	Legend printed solely in black mix

contours.[68] Irish maps also exercise a unique hypsometric tinting scheme. The depiction of forests usually follows two main trends, i.e. that of other forms of vegetation, with point symbols depicting the particular type of forest, or simply a plain area symbol. Forests on Finnish maps appear as white, which curiously associates their appearance with orienteering maps, especially given the high number of navigational symbols, but a practical reason for this particular depiction would be that it saves ink.

Of those countries choosing to create a set of point symbols to denote tourist features, most apply black, with the result that tourist symbols exhibit little distinction from other built-up features. This, in itself, points to the 'classical style', with all built-up features appearing in black. Sweden uses maroon for its tourist symbols, although this is the same colour as all built-up features in the symbology (with the exception of industrial zones), while Poland uses brown in a similar fashion. However, countries incorporating more tourist symbols within their symbology generally use the most distinctive colours: Great Britain uses cyan symbols, while Denmark and Norway respectively use red and pink. Apart from indicating a consciousness to meet the needs of a specific user group, this use of colour perhaps also reflects the market status of the topographic map within a country. Such topographic maps may be designed to target a sector that may otherwise be bereft of other cartographic products at that scale.

68. Grey and blue are used for rocky and glacial terrain respectively in other symbologies, e.g. Switzerland.

The last category to be considered is that of international boundaries. This is particularly significant because it may provide clues as to how a country considers its national status and possibly how it wishes to express its identity among its neighbours. Most borders at this level follow a similar appearance, i.e. a dashed black line (usually dot –dash –dot) surrounded by a band (or ribbon) in light pink or orange. The countries of Denmark, Iceland, Great Britain, Ireland, Portugal, Spain, and Sweden do not employ this coloured ribbon and this gives the impression of a continuous landscape devoid of national boundaries. For some of these countries, such boundaries have not been the subject of such recent contention. Conversely, France utilizes a bold orange ribbon and the Czech Republic a relatively striking light magenta line surrounded by a lighter ribbon, as well as devoting more symbols to representing boundaries of any kind than any other country. It might be expected that Schengen Agreement countries would utilize the most discreet border symbols, but curiously, this is not the case.

From this description of the use of colour, it is clear that there are both individuality and conformity in the symbologies of Europe, with some features exhibiting more variation than others. There is, however, some consistency among countries in their choice of colour for different types of feature, with the Alpine countries of Austria, France, Germany, Italy and Slovenia exhibiting most similarity across the range of features and remaining most faithful to the 'classical style' described by Keates.

'White' space

Most countries leave 'white' space unclassified (Table 10), letting it remain as white and forcing the reader to assume its many possible connotations. Some countries offer a classification similar to 'other' land (e.g. Belgium and Sweden), while others denote something more specific, e.g. 'dry and clear' land (Spain). In The Netherlands, however, 'white' space is simply eliminated altogether — there is neither a 'background' nor an ambiguous 'other' category in the drive by Dutch mapmakers to classify space and offer a 'complete' national landscape. As a substantial part of the background of most maps, 'white' space can also contribute to the ensemble effect of the colour combination of hill shading, forests, and contour lines. A more striking departure is found in the map series of Ireland, where a light green has been used instead of white. This forms part of the hypsometric tinting, but this particular choice of colour and its omnipresence is perhaps intended to evoke connotations associated with the myth of the 'Emerald Isle' by reinforcing this as the essence of the Irish landscape.

Visual hierarchy

As shown in Table 11, roads frequently appear in the top three most visually dominant features, along with buildings. The use of bold colours, which are often encased in black outlines, for the road networks, is likely to comprise the strongest 'figure' in the design and attract the most attention, while buildings (of any sort) also stand out with their dark, filled rectangles. The prominence of roads is perhaps justified both on the grounds that they form major landmarks (particularly motorways) and the importance of the road system as a whole. Given that roads appear to be so dominant in the majority of sample maps, it would appear that a similar approach is adopted

Table 10. Extract from the typology indicating how 'white' space is treated by each national mapping organization

Country	Treatment of 'white' space
Austria	White
Belgium	White
Czech Republic	Incorporated in the legend as arable and other land
Denmark	White
Finland	Incorporated in the legend as wooded area (white)
France	White
Germany	White
Great Britain	White
Iceland	White
Ireland	Hypsometric tinting: half green (light) through half yellow (light) to red brown (light)
Italy	White
Latvia	White
The Netherlands	Eliminated: all land is classified (no category appears in the legend as white; arable land is shown using a light half yellow)
Norway	White
Poland	White
Portugal	White
Slovenia	White
Spain	Incorporated in the legend as dry and clear land
Sweden	Incorporated in the legend as other open land
Switzerland	White

Table 11. Extract from the typology indicating features at the top of the visual hierarchy for each country

Country	Most dominant features		
Austria	Buildings	Railways	Main roads
Belgium	Main roads	Forests/vegetation	Canals
Czech Republic	Main roads	General buildings	Boundaries
Denmark	Main roads	General buildings	Grid
Finland	Lakes	Main roads	Fields
France	Main roads	Other roads	Forests
Germany	Main roads	Buildings	Forests
Great Britain	Main roads	Buildings	Forests
Iceland	Main roads	Buildings	Relief
Ireland	Main roads	Other roads	Land over 100 m
Italy	Main roads	Other roads	Buildings
Latvia	Main roads	Railways	Forests
The Netherlands	Main roads	Railways	Forests
Norway	Main roads	Cultivated land	Schools
Poland	Main roads	Other roads	Buildings
Portugal	Railways	Buildings	Main roads
Slovenia	Buildings	Railways	Main roads
Spain	Buildings	Irrigated land	Railways
Sweden	Main roads	Other roads	Cultivated land
Switzerland	Buildings	Main roads	Railways

across Europe. While railway lines might, perhaps, be thought to be more dominant through their use of black, they frequently use a thinner lineweight than main roads or appear as a dashed symbol, both of which are likely to recede in comparison with the qualities of road symbols. Of course, indicating the presence of railway stations is likely to be more relevant to the needs of map users as navigation is not usually required for rail travel. In Austria, however, where black is used for both roads and railways, the reverse is apparent because road symbols are often represented by thin parallel lines rather than thick, single lines, which have more dominance.

Lettering

From the range of typographic styles adopted for the features included in Table 12, most countries would seem to subscribe to certain conventions. These include the use of black for all the examined feature labels except those for rivers (in cyan), italicized fonts for hydrographic labels, plain (upright) fonts for settlement labels, and the use of sans serif fonts for legend explanations. Additionally, all countries use increasing sizes and capitalization to suggest importance for the upper stages of certain types of feature, such as towns and large bodies of water. The choice of font for relief features is the most varied in the sample, with roughly half the countries adopting a sans serif plain font and the other half a serif italic font. But distinctive similarities are also present. For example, the italicized lettering used for labelling hydrological features leans backwards in Finland, Germany, and Norway. Similarly, Austria, Portugal, and Switzerland all use serif italic fonts throughout the range of features described in the typology.

The mean cumulative score in the table, which is intended to serve as an indicator of the overall character of type in use for the national map series, is 9.25. This low score suggests that most countries tend to adopt the use of uniform, sans serif fonts, perhaps to support the view of a current, modern landscape on an up-to-date map, as opposed to a traditional, artistic, and possibly antiquated impression that may be obtained from the use of many serif italic fonts. In performing a Spearman's rank correlation coefficient test between the cumulative lettering scores and all the feature types in the typology classification, a correlation of 0.568 (significant at the 0.01 level) was found with the percentage values for the water management and utilization class (a level III feature type), implying that countries adopting more serif fonts devote a larger proportion of symbols to this type of feature. At first, this appears to be a completely arbitrary correlation with little explanation. However, considering the apparent convention above that hydrological features tend to be labelled in italics, this correlation might have an aesthetic rationale. If this particular type of feature is common, as suggested by the symbol classification, a wider application of italicized fonts for other features would provide a more continuous surface and unified appearance. In effect, this would mean that the presence of this one type of feature has a substantial aesthetic influence over the labelling of others.

Summary of qualitative analyses

There is, on the whole, a great variety of colour in use among maps in the sample, although some 'classical style' conventions persist — Alpine countries are generally

Table 12. Extract from the typology showing the styles of lettering used for particular features*

Country	Settlement text (excluding large urban areas)	Relief feature text	River text	Legend explanation text	Lettering style cumulative total
Austria	Black mix, SI (4) and SP (3)	Black mix, SSP (1), SI (4)	Full cyan, SI (4)	Black mix, SSI (2)	18
Belgium	Black mix, SSP (1) and SSI (2)	Black mix, SSP (1)	Half cyan, SSI (2)	Dark green, SSP (1) with English in SSI (2)	9
Czech Republic	Black mix, SSP (1) and SSI (2)	Black mix, SSP (1)	Full cyan, SSI (2)	Black mix, SSP (1)	7
Denmark	Black mix, SSP (1)	Grey mix, SSP (1)	Half cyan, SSI (2)	Black mix, SSP (1)	5
Finland	Black mix, SSP (1)	Black mix, SSI (2)	Black mix, SSI (2)	Black mix, SSP (1)	6
France	Black mix, SSP (1)	Black mix, SSI (2)	Black mix, SSI (2)	Black mix, SSP (1)	6
Germany	Black mix, SI (4)	Black mix, SP (3)	Half cyan, SI (4)	Black mix, SSP (1)	12
Great Britain	Black mix, SSP (1)	Black mix, SSP (1)	Half cyan, SSP (1)	Black mix, SSP (1)	4
Iceland	Black mix, SSP (1)	Black mix, SSI (2)	Dark blue, SI (4)	Black mix, SSP (1)	8
Ireland	Black mix, SSP (1) with Gaelic in SSI (2)	Black mix, SSP (1)	Half cyan, SSP (1) with Gaelic in SSI (2)	Black mix, SSP (1)	8
Italy	Black mix, SSP (1)	Black mix, SSP (1), SSI (2)	Full cyan, SSI (2) and SI (4)	Black mix with English in full magenta, SSI (2)	12
Latvia	Black mix, SSI (2) and SSP (1)	Black mix, SSI (2)	Half cyan, SSI (2)	Black mix, SSP (1)	8
The Netherlands	Black mix, SSP (1)	Black mix, SSI (2)	Half cyan, SSI (2)	Black mix, SSP (1) with English in SSI (2)	8
Norway	Black mix, SI (4)	Black mix, SSP (1)	Half cyan, SSI (2)	Black mix, SSP (1) with English in SSI (2)	10
Poland	Black mix, SSP (1)	Black mix, SSP (1)	Full cyan, SSI (2)	Black mix, SSP (1) with English in SSI (2)	7

Table 12. *Continued*

Country	Settlement text (excluding large urban areas)	Relief feature text	River text	Legend explanation text	Lettering style cumulative total
Portugal	Black mix, SSP (1) and SSI (2)	Black mix, SSP (1)	Full cyan, SI (4)	Black mix, SSI (2)	10
Slovenia	Black mix, SSP (1)	Black mix, SI (4)	Full cyan, SSI (2)	Black mix, SSP (1)	8
Spain	Black mix, SSP (1)	Black mix, SI (4)	Full cyan, SI (4)	Black mix, SSP (1)	10
Sweden	Black mix, SSP (1)	Black mix, SSI (2)	Half cyan, SI (4)	Black mix, SSP (1)	8
Switzerland	Black mix, SI (4) and SP (3)	Black mix, SI (4), SP (3)	Half cyan, SI (4)	Black mix, SSP (1) and SSI (2)	21

*SSP=sans serif plain (1); SSI=sans serif italic (2); SP=serif plain (3); SI=serif italic (4); H=handwritten (5) — unused (number in brackets indicates points allocated).

the most conservative in this respect, although aspects are clearly present in most countries. Road symbols employ the greatest variation in colour (only Austria retains a monochrome depiction), while railways are universally monochrome in design. The depiction of built-up features offers most potential for making stylistic associations, with similarities found within three distinct groups: the Iberian countries, the Alpine countries, and EU enlargement countries — the latter seeming to follow a legacy of Soviet design (Figure 24). 'White' space is present on most maps but is eliminated on Dutch maps and possibly used to reinforce national identity as a dominant green hypsometric tint in Irish maps. In terms of visual hierarchy, features that are classified in more detail tend to be symbolized to exhibit more prominence and road symbols are generally the most dominant type of feature on European 1:50 000 topographic maps. There are also some general conventions in lettering: all type in black except for hydrological features (which use cyan); sans serif fonts for settlements and legend text; and italicized fonts for hydrology. In addition, countries with a higher proportion of symbols devoted to the water management and utilization class tend to use more varied or italicized lettering on their maps in general.

There was not scope in this investigation to undertake a detailed comparison of cover designs, but it was observed that only the maps of Great Britain and Ireland have photographs relating to the area covered by the map on their front covers, as others usually have a location map of some sort. While this could be taken as evidence to support the stylistic similarity between the maps of these two countries, a comparison with other maps in the sample would be inconsistent as some maps were received in sheet format and therefore lacked covers.

CONCLUSIONS

Far from being standardized, European 1:50 000 state topographic mapping exhibits a rich diversity of cartographic styles. Although it was possible to group countries

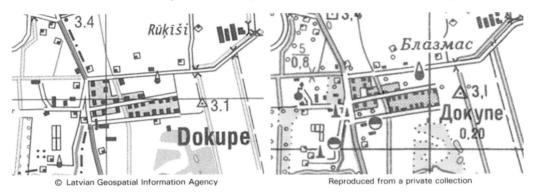

© Latvian Geospatial Information Agency   Reproduced from a private collection

FIG. 24.   An example of the legacy of Soviet design in an EU enlargement country, comparing an extract of Latvian sheet no. 4141: Ventspils[69] (left) with an extract of Soviet sheet O-34-92-B: Ventspils[70] (right) (both enlarged to 270%)

using a cluster analysis based on the proportion of symbols within each class, the findings of this study reveal much stylistic individuality in European state cartography, which is demonstrated through the classification and graphical appearance of each national symbology. Greater similarity, however, is apparent when symbologies are compared in less detail, perhaps indicating that an assortment of traits is blended harmoniously with the stylistic input of the native country.

Stylistic associations were either made on the level of similar classification of landscape (e.g. Great Britain and Ireland) or similar appearance (e.g. the Alpine countries), but not both. It would therefore seem that particular features and their representation are gradually — and selectively — incorporated from other national map series. The populations of contiguous countries may be more familiar with the neighbouring language as well as each other's maps (and thus 'cartographic language') and may well have surveyed each other's land at some point. The maps of Great Britain and Ireland possess the fullest expression of a supranational style in the sample, although the degree of similarity in their classification of landscape does not extend to their appearance.

The application of the typology used here highlights its effectiveness in identifying stylistic traits and it is designed to allow further comparisons of topographic maps within other regions. Furthermore, the clear and consistent grouping of Great Britain and Ireland demonstrates that the state classification of landscape is perhaps more likely to leave a lasting legacy, and even without strong visual similarities, it is possible to detect the stylistic association. The method may therefore have particular relevance for investigating the evolution of cartographic style and perhaps for detecting and identifying the legacy of colonial styles in current topographical mapping.

There appears to be a deeper cultural reason for the persistence of stylistic diversity, despite the apparent pragmatic advantages of a standardized series of

69. Latvian State Land Service. (2001). '4141: Ventspils' (2nd ed.) from **Topogrâfiskâ Karte 1:50 000**, Rîga, Latvian State Land Service.

70. General Staff of the USSR Armed Forces. (1988). 'O-34-92-B: Ventspils' **1:50 000**, Moscow, General Staff of the USSR Armed Forces.

international topographic maps. Indeed, the eventual failure of the *International Map of the World*, proposed by Albrecht Penck in 1891, deftly illustrates the problems of international collaboration over the portrayal of landscape and it is no coincidence that recent initiatives for collaborative mapping in Europe have concentrated on the standardization of the referencing and coding of information rather than the design of cartographic symbols. State topographic maps therefore maintain a cartographic legacy that can penetrate deep into the national consciousness. Regarding their symbologies as 'vocabularies' that have gradually evolved to express the national landscape more articulately may therefore offer a route to understanding why national differences in symbology persist, and perhaps how the resulting styles may be understood as 'dialects' of the cartographic language of European state 1:50 000 topographic maps.

ACKNOWLEDGEMENTS

This investigation could not have been completed without the generous cooperation of the European national mapping organizations, particularly those who graciously to send sample maps. While every reasonable effort has been made to contact copyright holders, the authors apologize for any outstanding inaccuracy or omission. The research was undertaken as part of a PhD studentship at Canterbury Christ Church University, whose financial assistance is gratefully acknowledged. The authors should particularly like to thank Dr Peter Thomas for his helpful comments. Dr Kent would like to express sincere gratitude for the bestowment of an ICA Travel Award, through which the presentation of some aspects of this research at ICC Moscow was made possible.

BIBLIOGRPAHY

Arnberger, E. (1974). 'Problems of an international standardization of a means of communication through cartographic symbols', **International Yearbook of Cartography**, 16, pp. 19–35.

Board, C. (1981). 'Cartographic communication', **Cartographica Monograph**, 27, pp. 42–78.

Böhme, R. (1989). **Inventory of World Topographic Mapping**, Vol. 1, Western Europe, North America and Australasia, Barking, Elsevier Applied Science Publishers.

Böhme, R. (1991). **Inventory of World Topographic Mapping**, Vol. 2, South America, Central America and Africa, Barking, Elsevier Applied Science Publishers.

Böhme, R. (1993). **Inventory of World Topographic Mapping**, Vol. 3, Eastern Europe, Asia, Oceania and Antarctica, Oxford, Elsevier Applied Science Publishers.

Bundesamt Für Eich- und Vermessungswesen. (1998). '202: Klagenfurt' from **Öesterreichische Karte 1:50 000**, Vienna, Bundesamt für Eich- und Vermessungswesen.

Bundesamt Für Landestopographie (1999). '217: Arbon' from **Landeskarte der Schweiz 1:50 000**, Wabern: Bundesamt für Landestopographie.

Césky Úrad Zememericky a Katastrálni. (2003). '13-11: Benátky Nad Jizerou' **1:50 000**, Prague, Césky Úrad Zememericky a Katastrálni.

Collier, P., Fontana, D., Pearson, A. and Ryder, A. (1996). 'The state of mapping in the former satellite countries of Eastern Europe', **The Cartographic Journal**, 33, pp. 131–139.

Collier, P., Forrest, D. and Pearson, A. (2003). 'The representation of topographic information on maps: the depiction of relief', **The Cartographic Journal**, 40, pp. 17–26.

Collier, P., Pearson, A. and Forrest, D. (1998a). 'The representation of topographic information on maps: vegetation and rural land use', **The Cartographic Journal**, 35, pp. 191–197.

Collier, P., Pearson, A., Fontana, D. and Ryder, A. (1998b). 'The state of mapping in the European republics of the former Soviet Union', **The Cartographic Journal**, 35, pp. 165–168.

Doherty, G. M. (2004). **The Irish Ordnance Survey: History, Culture and Memory**, Dublin, Four Courts Press.

Dorling, D. and Fairbairn, D. (1997). **Mapping: Ways of Representing the World**, Harlow, Longman.

EuroGeographics. (2003). Members (NMO Statistics), **http://www.eurogeographics.org/eng/01_members. asp** (accessed 2003–2006).

Everitt, B. S. (1993). **Cluster Analysis**, 3rd ed., London, Arnold.

Forrest, D., Pearson, A. and Collier, P. (1996). 'The representation of topographic information on maps — a new series', The Cartographic Journal, 33, pp. 57–58.

Forrest, D., Pearson, A. and Collier, P. (1997). 'The representation of topographic information on maps — the coastal environment', The Cartographic Journal, 34, pp. 77–85.

General Staff of the USSR Armed Forces. (1988). 'O-34-92-B: Ventspils' 1:50 000, Moscow, General Staff of the USSR Armed Forces.

Geodetska Uprava Republike Slovenije. (2003). '12: Jesenice' from **Državna Topografska Karta 1:50 000**, Ljubljana, Geodetska Uprava Republike Slovenije.

Główny Geodeta Kraju. (1995). 'N-34-124-D: Słubice' from **Mapa Topograficzna Polski 1:50 000**, Warsaw, Główny Geodeta Kraju.

Hansen, K. M. and Tukey, J. W. (1992). 'Tuning a major part of the clustering algorithm', **International Statistical Review**, 60, pp. 21–44.

Harley, J. B. (1975). **Ordnance Survey Maps: A Descriptive Manual**, Southampton, Ordnance Survey.

Harvey, P. D. A. (1980). **The History of Topographical Maps: Symbols, Pictures and Surveys**, London, Thames & Hudson.

Imhof, E. (1982). **Cartographic Relief Presentation**, trans. By Steward, H. J., Berlin, Walter de Gruyter.

Institut Géographique National Belgique. (2002). '13: Brugge' (Edition 2-M737) 1:50 000, Brussels, Institut Géographique National Belgique.

Institut Géographique National. (1982). '1422: Chalonnes-Sur-Loire' (2nd ed.) from **Serie Orange 1:50 000**, Paris, Institut Géographique National.

Instituto Geográfico Nacional (2003). '963: Lora Del Río' from **Mapa Topográfico de España 1:50 000**, Madrid, Instituto Geográfico Nacional.

Instituto Português de Cartografia e Cadastro. (2004). '27-C: Torres Novas' (3rd ed.) 1:50 000, Lisbon, Instituto Português de Cartografia e Cadastro.

Istituto Geografico Militare. (1972). '083: Monte Grappa' 1:50 000, Firenze, Istituto Geografico Militare.

Jervis, W. W. (1936). **The World in Maps: A Study in Map Evolution**, London, George Philip & Son.

Keates, J. S. (1972). 'Symbols and meaning in topographic maps', **International Yearbook of Cartography**, 12, pp. 168–181.

Keates, J. S. (1996). **Understanding Maps**, 2nd ed., Harlow: Longman.

Kent, A. J. (2008). 'Cartographic blandscapes and the new noise: finding the good view in a topographical mashup', The Bulletin of the Society of Cartographers, 42, pp. 29–37.

Kent, A. J. (2009). 'Topographic maps: methodological approaches for analyzing cartographic style', **Journal of Map and Geography Libraries**, 5, pp. 131–156.

Knowles, R. and Stowe, P. W. E. (1982). **Western Europe in Maps: Topographical Map Studies**, Harlow: Longman.

Kort- og Matrikelstyrelsen. (2003). '1214-I: Silkeborg' 1:50 000, Copenhagen, Kort- og Matrikelstyrelsen.

Kraak, M. and Ormeling, F. (1996). **Cartography: Visualization of Spatial Data**, Harlow: Longman.

Landesvermessungsamt Nordrhein-Westfalen. (2004). 'L4512: Unna' (Editon 11-DGID) from **Topographische Karte 1:50 000**, Bonn: Landesvermessungsamt Nordrhein-Westfalen.

Landmælingar Íslands. (n.d.). '1916-I: þorvaldsdalur' (Edition 1- DMA) from **Staðfrædikort 1:50 000**, Akranes, Landmælingar Íslands.

Lantmäteriet. (2001). '13-H-SV: Gävle' (5th ed.) from **Terrängkarta 1:50 000**, Gävle, Lantmäteriet.

Latvian State Land Service. (2001). '4141: Ventspils' (2nd ed.) from **Topogräfiskä Karte 1:50 000**, Rīga, Latvian State Land Service.

Latvijas Ģeotelpiskās Informācijas Aģentūra (LGIA). (2006). '4323: Sigulda' (2nd ed.) from **Topogräfiskä Karte 1:50 000**, Rīga, LGIA.

Lawrence, V. (2004). 'The role of national mapping organizations', The Cartographic Journal, 41, pp. 117–122.

Maastokartta. (2001). '2724: Palojoensuu' from **Topografinen Kartta 1:50 000**, Espo, Maastokartta.

Morrison, A. (1994). 'Why are French transport maps so distinctive compared with those of Germany and Spain?', The Cartographic Journal, 31, pp. 113–122.

Morrison, A. (1996). 'Public transport maps in Western European cities', The Cartographic Journal, 33, pp. 93–110.

Munro, T. (1946). 'Style in the arts: a method of stylistic analysis', **The Journal of Aesthetics and Art Criticism**, 5, pp. 128–158.

Nicholson, T. (2004). 'Cycling and motoring maps in Western Europe 1885–1960', **The Cartographic Journal**, 41, pp. 181–215.

Olson, E. C. and Whitmarsh, A. (1944). **Foreign Maps**, New York, Harper & Brothers.

Ordnance Survey. (2004). '189: Ashford & Romney Marsh' (Edition D1) from **Landranger** 1:50 000, Southampton, Ordnance Survey.

Parry, R. B. and Perkins, C. R. (2000). **World Mapping Today**, 2nd ed., London, Bowker-Saur.

Piket, J. J. C. (1972). 'Five European topographic maps: a contribution to the classification of topographic maps and their relation to other map types', **Geografisch Tijdschrift**, 6, pp. 266–276.

Rhind, D. (2000). 'Current shortcomings of global mapping and the creation of a new geographical framework for the world', **The Geographical Journal**, 166, pp. 295–305.

Robinson, A. H., Morrison, J. L., Muehrcke, P. C., Kimerling, A. J. and Guptill, S. C. (1995). **Elements of Cartography**, 6th ed., New York, John Wiley & Sons.

Rogerson, P. A. (2001). **Statistical Methods for Geography**, London, Sage.

Simpson, J. A. and Weiner, E. S. C. (Eds.) (1989). **The Oxford English Dictionary**, 2nd ed., Oxford, Oxford University Press.

Statens Kartwerk (1997) '1623 III: Roan' (Edition 4-NOR) from **Topografisk Hovedkartserie M-711** 1:50 000, Hønefoss, Statens Kartwerk.

Suibhéireacht Ordanáis (2003). '48: Offaly, Westmeath' (2nd ed.) from **Discovery Series** 1:50 000, Dublin, Suibhéireacht Ordanáis.

Sylvester, D. (1952). **Map and Landscape**, London, George Philip & Son.

Taylor, D. R. F. (1989). 'Foreword', in **Inventory of World Topographic Mapping**, Vol. 1, Western Europe, North America and Australasia, ed. by Böhme, R., Barking, Elsevier Applied Science Publishers.

Topografische Dienst. (1999). '33-O: Oost Apeldoorn' from **Topografische Kaart** 1:50 000, AC Emmen, Topografische Dienst.

Tyner, J. A. (1987). 'Interactions of culture and cartography', **The History Teacher**, 20, pp. 455–464.

Wood, M. (1968). **Foreign Maps and Landscapes**, London, George G. Harrap & Co.

Reflections on 'Stylistic Diversity in European State 1:50 000 Topographic Maps'

CHRIS PERKINS

University of Manchester

Until the emergence of Internet-served global mapping portals, National Mapping Organizations (NMOs) maintained local monopolies over the depiction of landscapes in topographic mapping (Parry and Perkins, 2000). Arguably, the level of detail and quality of these surveys remains unchallenged by Google, OSM or Bing. In 1991, Nick Chrisman observed that the national cartographic styles, which emerged from these organizations reflected their different cultural contexts, tracing, for example, the cartographic landscape of the USA to historical precedents as diverse as Roman notions of individuality, medieval property relations, enlightenment ideas of equality and more recent environmental discourse. He argued that graphical expression and inclusion criteria reflected particular local influences on cartographic forms. Chrisman made large generalizations, which are closely related to Benedict Anderson's (1991) influential consideration of the role of the map in the emergence of shared national imaginaries. However, he failed to back these up with evidence sourced from the detail of specific cartographic designs.

Stylistic diversity in European State 1:50 000 topographic maps addresses exactly how this variation is reflected in mapping and provides the evidence. It is one of the longer papers published in the history of *The Cartographic Journal*, comprising 35 pages, 12 tables and 24 figures. The paper is based on research carried out by Alex Kent for his PhD, and was first submitted as a paper at the 2007 International Cartographic Conference in Moscow. Kent and Vujakovic anchor their analysis to the recognition that style reflects local, pragmatic needs, and that a typological approach can unpack similarities and differences. They use the legends deployed in printed 1:50 000 scale civilian maps from a sample of 20 European national mapping agencies as a source for classifying map symbologies. Interestingly, the sample reported in this research ended up reflecting the real-politic of different nation states, since access to mapping is itself still variable even in the Internet age. Only 20 from an original target of 38 NMOs provided material and post-Soviet national mapping in Eastern Europe is strikingly absent. A three-level classification explores differences in land cover, land use and specific features included in the map specifications. Colour is analysed in relation to pre-determined categories in a CMYK colour chart; visual hierarchy is deployed in order to assess relative importance of different features; the presence of white space or an image base is recorded, while lettering is described according to its position on a subjective–objective hierarchy.

Together, 2388 symbols in 17 different legends were analysed with a mix of quantitative and qualitative techniques. Slovenian maps deployed most symbols, Irish the least. Most symbols related to human or artificial features. Star plots and pie charts are deployed to explore relative proportions of different feature types. Hierarchical cluster analysis sets up the identification of seven different 'groups' of countries with similar mixes of symbols. For example, Austrian and Spanish maps are dominated by settlement, territory and resources symbols, whereas maps from Great Britain and Ireland are dominated by accessibility and transport symbols, and include the highest proportion of tourism, recreation and conservation symbols. Explanations for these patterns are advanced and relate to geographical proximity, or shared cartographic histories. A qualitative analysis builds on the quantitative analysis of landscape classification and numerous descriptive generalizations are made about the patterns in these results. For example, it is found that classical conservative approaches to design persist in particular in Alpine countries; that road symbols display the greatest variation in colour use and were usually the most visually prominent features; and that lettering is almost always in black. Kent and Vujakovic reflect that similarities emerge, but that these in particular depend upon the level of analysis. They conclude that topographic mapping in their sample 'exhibits a rich diversity of cartographic styles'. Apparent similarities and standards start to dissolve as soon as a detailed aesthetic and typological approach is deployed.

So in an era of increasing standardization, when national mapping agencies are arguably in decline in the face of globally-served commercial products such as those offered by Google, the design of different national European hard copy topographic survey maps has been shown to be both complex and strongly local. This finding is significant for a number of reasons. It suggests that an aesthetic approach to cartography is still strongly relevant: even in an age when tools for standardized displays have supposedly encouraged a convergence in practice, the 'tramlines' of cartographic history still matter (Rhind, 1997). The other reason that this paper is significant lies with the rigorous application of systematic analysis of the sample of mapping, and the use of innovative visualizations to portray results. Relating the broad-brush cultural generalizations advanced by Chrisman (1991), to a detailed analysis of the role that maps play in the construction of imagined national communities is of course beyond the scope of Kent and Vujakovic. They speculate, but their study unpacks an analysis of the forms, instead of tracing the genealogy of practices through which particular aesthetic forms emerged. However, as a necessary first step, and as part of the resurgence of interest in map design evidenced elsewhere in this volume, this paper makes a really significant contribution and offers us hope that the cartographic bland-scape critiqued elsewhere by Alex Kent (2008), may indeed be less dominant than feared. Design still matters!

REFERENCES

Anderson, B. R. O'G. (1991). **Imagined Communities: Reflections on the Origin and Spread of Nationalism** (**Revised and Extended. ed.**), Verso, London.
Chrisman, N. R. (1991). 'Building a Geography of Cartography: Cartographic Institutions in Cultural Context', in **15th Conference of International Cartographic Association**, pp. 83–92, Bournemouth, Sep 23–Oct 1.
Kent, A. J. (2008). 'Cartographic blandscapes and the new noise: finding the good view in a topographical mashup', **The Bulletin of the Society of Cartographers**, 42, pp. 29–37.
Parry, R. B. and Perkins, C. R. (2000). **World Mapping Today**, 2nd ed., Bowker Saur, London.
Rhind, D. (Ed.) (1997). **Framework for the World**, Geo-Information International, Cambridge.

Visualization of Origins, Destinations and Flows with OD Maps

JO WOOD, JASON DYKES AND AIDAN SLINGSBY

Originally published in *The Cartographic Journal* (2010) 47, pp. 117–129.

We present a new technique for the visual exploration of origins (O) and destinations (D) arranged in geographic space. Previous attempts to map the flows between origins and destinations have suffered from problems of occlusion usually requiring some form of generalization, such as aggregation or flow density estimation before they can be visualized. This can lead to loss of detail or the introduction of arbitrary artefacts in the visual representation. Here, we propose mapping OD vectors as cells rather than lines, comparable with the process of constructing OD matrices, but unlike the OD matrix, we preserve the spatial layout of all origin and destination locations by constructing a gridded two-level spatial treemap. The result is a set of spatially ordered small multiples upon which any arbitrary geographic data may be projected. Using a hash grid spatial data structure, we explore the characteristics of the technique through a software prototype that allows interactive query and visualization of 105–106 simulated and recorded OD vectors. The technique is illustrated using US county to county migration and commuting statistics.

INTRODUCTION

There are many applications for which associations between pairs of known geographic locations provide an important data source. These associations might involve direct movements of people, for example, migration,[1] commuting behaviour,[2] GPS tracklog analysis[3] or interaction between actors in social networks and service use.[4]

1. Tobler, W. (1987). 'Experiments in migration mapping by computer', **The American Cartographer**, 14, pp. 155–163.

2. Chiricota, Y., Melançon, G., Quang, T. T. P. and Tissandier, P. (2008). 'Visual Exploration of (French) Commuter Networks', in **Geovisualization of Dynamics, Movement, and Change**, AGILE'08 Satellite Workshop, Girona, May 5.

3. Andrienko, G. and Andrienko, N. (2008). 'Spatio-temporal Aggregation for Visual Analysis of Movements', in **IEEE Symposium on Visual Analytics Science and Technology** (VAST 2008), pp. 51–58, Columbus, OH, Oct 19–24.

4. Radburn, R., Dykes, J. and Wood, J. (2009). 'vizLib: Developing Capacity for Exploratory Data Analysis in Local Government –Visualization of Library Customer Behaviour', in **GIS Research UK 17th Annual Conference**, pp. 381–387, Durham, Apr 1–3.

Alternatively, movements of goods, knowledge,[5] disease and animals[6,7] may be explored. A range of techniques have traditionally been used to represent such associations, including direct mapping of geographic flow vectors,[8] flow density maps,[9] origin–destination (OD) matrices[10] and statistical summaries of spatial association.[11]

We present the OD map — a new method of visualizing associations between datasets of this type that overcomes some of the problems traditionally associated with mapping geographic vectors. Our aim is to develop visualization techniques that enable trends in OD relationships to be identified and their sensitivities explored while maintaining maximum cognitive plausibility of the representation.[12] In so doing, we contribute a spatially embedded view of trajectories that may be combined effectively with other multiply-coordinated views.

REQUIREMENTS AND PRIOR WORK

This work was motivated by the need to analyse and understand patterns in a number of large datasets in which paired origins and destinations were key characteristics. These included collections of mobile search enquiry locations and result destinations,[13] spatio-temporal records of borrowing behaviour among library users relating home origin to library destination[14] and GPS records of traffic across London recorded by all vehicles from a courier company over a 1 month period that included pick-up (origin) and drop-off (destination) points.[15]

In each case, our objective was to gain insight into the nature of the journeys defined between pairs of locations. Developing an environment for the visual exploration of movement and other interactions in time and space has potential for building knowledge and understanding, but doing so has been recognized as one of the major research challenges associated with visual analytics.[16,17] When dealing with such large

5. Paci, R. and Usai, S. (2009). 'Knowledge flows across European regions', **The Annals of Regional Science**, 43, pp. 669–690.

6. Gilbert, M., Mitchell, A., Bourn, D., Mawdsley, J., Clifton-Hadley, R. and Wint, W. (2005). 'Cattle movements and bovine tuberculosis in Great Britain', **Nature**, 435, pp. 491–496.

7. Guo, D. (2007). 'Visual analytics of spatial interaction patterns for pandemic decision support', **International Journal of Geographic Information Science**, 21, pp. 859–877.

8. Tobler, 1987.

9. Rae, A. (2009). 'From spatial interaction data to spatial interaction information? geovisualisation and spatial structures of migration from the 2001 UK census', **Computers, Environment and Urban Systems**, 33, pp. 161–178.

10. Voorhees, A. (1955). 'A general theory of traffic movement', in **Institute of Traffic Engineers Past Presidents' Award Paper**, ITE, New Haven, Proceedings, Institute of Traffic Engineers, pp. 46–56.

11. Gilbert et al., 2005.

12. Skupin, A. and Fabrikant, S. (2003). 'Spatialization methods: a cartographic research agenda for non-geographic information visualization', **Cartography and Geographic Information Science**, 30, pp. 99–119.

13. Wood, J., Dykes, J., Slingsby, A. and Clarke, K. (2007). 'Interactive visual exploration of a large spatio-temporal dataset: reflections on a geovisualization mashup', **IEEE Transactions on Visualization and Computer Graphics**, 13, pp. 1176–1183.

14. Radburn et al., 2009.

15. Slingsby, A., Dykes, J. and Wood, J. (2008). 'Using treemaps for variable selection in spatio-temporal visualization', **Information Visualization**, 7, pp. 210–224.

16. Hernandez, T. (2007). 'Enhancing retail location decision support: The development and application of geovisualization', **Journal of Retailing and Consumer Services**, 14, pp. 249–258.

17. Andrienko, G., Andrienko, N., Dykes, J., Fabrikant, S. and Wachowicz, M. (2008). 'Geovisualization of dynamics, movement and change: key issues and developing approaches in visualization research', **Information Visualization**, 7, pp. 173–180.

data volumes, the transformation into a meaningful visual representation of the spatio-temporal structure requires data reduction though selection or aggregation in a manner that suits the need of the analyst.[18] This led to the following requirements:

1. to be able to visually represent large numbers of origin-destination vectors (of order 10^5–10^6);

2. to create a visual environment that provides both overview and detail on demand;

3. to emphasize representation of the origin and destination locations over the geometry of the paths that link them;

4. to use a projection that preserves the spatial configuration of the study area to allow integration of supplementary geographic data;

5. to provide a visual representation that can show both long and short trajectories with minimal occlusion;

6. to be able to distinguish origin from destination without visual clutter and so infer direction of flows;

7. to be able to compare origin–destination vectors with destination–origin vectors;

8. to be able to distinguish artefacts of the visualization from characteristics of the data under investigation.

Existing techniques for flow mapping meet some of our requirements. Tobler[19] provides some early examples of computer generated flow maps, Rae[20] in his review of flow mapping methods showed how some of these techniques could be implemented in geographic information systems. The principle behind direct flow mapping is to project geographic space onto a plane and plot each trajectory as a line from origin to destination. While this approach has a long history and can produce maps that are familiar to many users, it does not scale well to large numbers of trajectories. As the number of links increases, it also becomes increasingly difficult to indicate the direction of flow visually without clutter. Occlusion of trajectories by others sharing the same space produces maps that are difficult to interpret unless some form of generalization is applied. This problem increases as data sizes grow to orders of 105 or more and has resulted in a trend identified in Andrienko *et al.*[21] towards the 'derivation, depiction and visualization of abstract data summaries [such as] — aggregates, generalization, samples'. For example, Cui *et al.*[22] and Holten and van Wijk[23] have proposed aggregation of flows of high local density to overcome this problem; Guo[24] proposed aggregatation of flows according to space and attribute. However, the spatial distortion involved in combining flows may result in unacceptable loss of detail,

18. Rae, 2009.
19. Tobler, 1987.
20. Rae, 2009.
21. Andrienko *et al.*, 2008.
22. Cui, W., Zhou, H., Qu, H., Wong, P. C. and Li, X. (2008). 'Geometry-based edge clustering for graph visualization', **Transactions on Visualization and Computer Graphics**, 14, pp. 1227–1284.
23. Holten, D. and van Wijk, J. (2009). 'Force-directed edge bundling for graph visualization', **Computer Graphics Forum**, 28, pp. 983–990.
24. Guo, D. (2009). 'Flow mapping and multivariate visualization of large spatial interaction data', **IEEE Transactions on Visualization and Computer Graphics**, 15, pp. 1041–1048.

and perhaps more significantly, still retains the problem of long trajectories occluding shorter ones that occupy the same graphic space. Even with an acceptable level of generalization, such as the flow density surfaces of Rae,[25] density of flow lines does not necessarily indicate the density of origin and destination locations and can result in arbitrary patterns of flow density. The suitability for flow line density will depend on the importance attached to the path of the trajectory connecting origin and destination. In our case, we are more interested in the topology of connections between origin and destination rather than their geometry (requirement 3 above).

An alternative solution to the occlusion problem is to filter selected flows.[26] This may include only filtering to a single limited set of origins or destinations.[27] While this can declutter the display of flow maps, it requires some intelligent selection of the origins and destinations to filter, or dynamic interaction to vary these on demand. It is also difficult to provide a visual overview using this approach.

Instead of plotting trajectories as vectors, another commonly used approach is to construct some form of origin–destination matrix (OD matrix) where matrix rows represent the locations of flow origins and columns the locations of destinations. Ghoniem et al.[28] compared the usability of node-link views and matrix views and found that for more than 20 nodes, the matrix view was superior for most tasks, although notably, 'path finding' was the only exception. Path finding or similar geographic interpretation of flows may be important for geographically arranged trajectories when, for example, considering routes taken between mobile search locations and result destinations, home location and library or traffic flows in a large city to evaluate and plan the delivery of services.

Visualizing the OD matrix directly also has a long history[29] where the number of flows between an origin and destination is used to colour the appropriate matrix cell. Some attempts have been made to view the matrix as a three-dimensional surface,[30] although the cognitive plausibility of the results is highly questionable. To enhance the utility of OD matrices, some form of sorting and aggregation is frequently applied. However, there are many ways to aggregate and reorder.[31,32] This approach has been applied for computational efficiency, especially for sparse matrices, but for visualization, the main benefit of matrix sorting is to make clusters more apparent, famously illustrated by Bertin.[33] Reordering can be achieved in a number of ways — for

25. Rae, 2009.

26. Tobler, 1987.

27. Phan, D., Ling, X., Yeh, R. and Hanrahan, P. (2005). 'Flow Map Layout', in IEEE Symposium on Information Visualization (Infovis 2005), pp. 219–224, Minneapolis, MN, Oct 23–25.

28. Ghoniem, M., Fekete, J. and Castagliola, P. (2004). 'A Comparison of the Readability of Graphs Using Node-link and Matrix-based Representations', in 10th IEEE Symposium on Information Visualization (Infovis 2004), pp. 17–24, Austin, TX, Oct 19–21.

29. Wilkinson, L. and Friendly, M. (2009). 'The history of the cluster heat map', The American Statistician, 63, pp. 179–184.

30. Marble, D., Gou, Z., Liu, L. and Saunders, J. (1997). 'Recent advances in the exploratory analysis of interregional flows in space and time', in Innovations in GIS 4, pp. 75–88, Taylor & Francis, London.

31. Wilkinson, L. (1979). 'Permuting a matrix to a simple pattern', in Proceedings of the Statistical Computing Section, pp. 409–412, American Statistical Association, Washington, DC.

32. Guo, D. and Gahegan, M. (2006). 'Spatial ordering and encoding for geographic data mining and visualization', Journal of Intelligent Information Systems, 27, pp. 243–266.

33. Bertin, J. (1983). Semiology of Graphics, University of Wisconsin Press, Madison, WI.

example, by clustering strongly connected nodes,[34,35] to reveal spatial clusters[36,37,38] or to preserve spatial proximity relations.[39]

While cluster detection has obvious benefits when analysing matrices, the reordering required fails to meet requirement 4 outlined above as the original spatial configuration of the matrix cells is lost during reordering. Guo[40] and Guo *et al.*[41] suggest overcoming this problem by providing dynamic linking between the transformed matrix space and a conventional geographic view. Alternative strategies for preserving some of the spatial properties of the matrix cells include reprojecting two-dimensional space into one-dimensional sequences using space-filling Morton ordering[42] and aggregation into identifiable spatial regions.[43] These approaches, which we might term 'quasi space-preserving', allow some meaningful geographical interpretation of results, but are limited in their ability to integrate with other geographic data in the same projected space.

The approach we propose here attempts to retain as much of the geographic space as possible by dividing it into a regular grid nested at two levels. As with the work of Andrienko and Andrienko,[44] this is a form of S6S aggregation, but unlike their regional clustering, the regular grid is a property of the geographic space, not the clustering algorithm or dataset. This has a significant benefit in interpretation and integration with supplementary spatial information which may support spatial tasks (such as the analysis of spatio-temporal pattern across a city) more effectively.[45] Because of the spatial autocorrelation of most geographic phenomena, retaining spatial structure can also reveal clustering in the data.

THE OD MAP

To overcome problems of occlusion inherent in node-link representations of origin–destination vectors, we propose a transformation of graphical space whereby each two-dimensional vector $\vec{v}$ from origin (p_o) to destination (p_d) is represented by a cell p_{od} in a two-dimensional matrix. But unlike a conventional OD matrix, we order cells to reflect their original two-dimensional geographic location. This is achieved by dividing geographic space into a regular coarse grid. Each trajectory can therefore be referenced by two grid cell locations — the grid cell in which the trajectory's origin

34. Jarvis, R. and Patrick, E. (1973). 'Clustering using a similarity measure based on shared near neighbours', **IEEE Transactions on Computers**, 22, pp. 1025–1034.

35. Karypis, G. and Kumar, V. (2000). 'Multilevel k-way hypergraph partitioning', **VLSI Design**, 11, pp. 285–300.

36. Andrienko and Andrienko, 2008.

37. Guo, D., Chen, J., MacEachren, A. and Liao, K. (2006). 'A visualization system for space-time and multivariate patterns (VISSTAMP)', **IEEE Transactions on Visualization and Computer Graphics**, 12, pp. 1461–1474.

38. Guo, 2007.

39. Marble *et al.*, 1997.

40. Guo, 2007.

41. Guo *et al.*, 2006.

42. Marble *et al.*, 1997.

43. Andrienko and Andrienko, 2008.

44. Andrienko and Andrienko, 2008.

45. Ghoniem *et al.*, 2004.

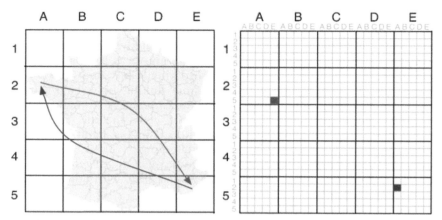

FIG. 1.   Left: geographic space partitioned into a regular grid. Right: OD map space. Origin space uses identical gridding of geographic space (dark grid lines); destination space consists of nested small multiples of the geographic space (light grid lines). The red trajectory from geographic location (A,2) to (E,5) is represented by the single OD map cell with coodinates (AE,25). The reverse blue trajectory is shown as a cell with coordinates (EA,52)

lies, and the grid cell in which its destination lies. By nesting the destination grid cells within the origin grid cells, we can preserve the spatial relationships between origin cells and the relationships between destination cells in a single matrix (Figure 1).

Assuming that both OD map columns and rows are numbered from 0, the transformation from OD map to geographic space involves a simple rounding and modulus arithmetic operation

$$p_o = \mathrm{floor}(p_{od}/n) \tag{1}$$
$$p_d = p_{od} \bmod n \tag{2}$$

where n is the number of cells along one side of the geographic space. The transformation from geographic to OD space simply involves adding the two locations, scaling origin space by n

$$p_{od} = np_o + p_d \tag{3}$$

The result is a set of small multiples of destination space each arranged in their geographic position in origin space. This is a special case of a treemap[46] in which the top two levels of the hierarchy are an identical set of spatial nodes. These may be consistently sized cells that partition space as a regular raster or the result of a spatial treemap.[47] Indeed, the arrangement is a spatial treemap, denoted as sHier(/,$origin, $destination); sLayout(/,SP,SP);sSize(/,FIXED,FIXED) using the hierarchical visualization expression language (HiVE) to describe the representation.[48]

46. Shneiderman, B. (1992). 'Tree visualization with tree-maps: a 2-d space-filling approach', **ACM Transactions on Graphics**, 11, pp. 92–99.

47. Wood, J. and Dykes, J. (2008). 'Spatially ordered treemaps', **IEEE Transactions on Visualization and Computer Graphics**, 14, pp. 1348–1355.

48. Slingsby, A., Dykes, J. and Wood, J. (2009). 'Configuring hierarchical layouts to address research questions', **IEEE Transactions on Visualization and Computer Graphics**, 15, pp. 977–984.

The recursive layout has similarities to the Map2 arrangement of trajectories proposed by Guo *et al.* (2006) in their VIS-STAMP system, but unlike their system, we guarantee that the projection of origin space is an identical, but scaled, projection of destination space. This consistent spatial arrangement between original and destination maps is likely to reduce cognitive load in comparison to arrangements that use a different spatial projection for origin and destination spaces.

It also has the advantage that asymmetry of flows can be shown by swapping O space and D space in the tree, thus meeting requirements 6 and 7 above. In HiVE, this is a move to sHier(/,$destination,$origin) through a swap operation denoted as oSwap(/,1,2). Dynamically swapping O space and D space provides a visual indication of flow asymmetry that can be identified without the need for extra symbolization to represent flow direction (e.g. Rae[49]).

In addition to the removal of the line occlusion problem, a cell-based representation of OD vectors allows us to colour each OD cell according to some attribute of the aggregated trajectories between a pair of origin and destination locations. The obvious symbolization is to colour according to total flows, but other attributes may be represented such as difference maps or the signed chi statistic[50,51,52] when comparing actual flow magnitudes with those expected based on some underlying model.

Simulated OD vectors

To test the ability of the OD map to discriminate between different structures of geographic vectors, various simulated flow sets were created. In particular, the origin location distribution, the destination location distribution, the length of and direction of flow were randomly generated under a range of assumptions. The aim was to see whether the OD map was adequately discriminating between different forms of flow distribution in large datasets (requirements 1 and 2 above); to identify any visual artefacts produced by the OD map or conventional flow map (requirement 8).

In the first set of simulations, 100,000 vectors were generated each with a uniformly distributed random origin and destination (Figure 2 top row). They were represented as a conventional flow map (Figure 2 left) using alpha blending to reveal the density of flow vectors. The same vectors were shown as an OD map (Figure 2 right). As expected, the OD map indicates an approximately uniform density of origin–destination cells. In contrast, the flow map suggests an apparent increase in density away from the edges. This is purely an artefact of the probability of line overlap, and not a true variation in OD density. For studies where the geometric path between origin and destination is not considered, it is important to appreciate that a conventional flow map or flow density surface (e.g. Rae[53]) can produce such arbitrary variations in line density that do not reflect the true distributions of OD locations.

49. Rae, 2009.
50. Census Research Unit (1980). **People in Britain: A Census Atlas**, HMSO, London.
51. Dykes, J. and Brunsdon, C. (2007). 'Geographically weighted visualization — interactive graphics for scale-varying exploratory analysis', **IEEE Transactions on Visualization and Computer Graphics**, 13, pp. 1161–1168.
52. Wood *et al.*, 2007.
53. Rae, 2009.

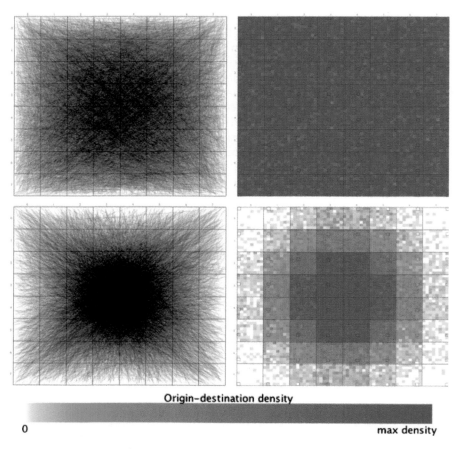

Origin–destination density

0   max density

FIG. 2. 100,000 simulated trajectories shown as a vector flow map (left) and an OD map (right). Destination locations are uniformly randomly distributed. In the upper example, origin locations are also uniformly random, while in the lower example, origin locations have a Gaussian random distribution about the centre. In all cases, the OD map colour uses a Brewer 'YlOrBr' colour scheme[54] exponentially scaled between 0 and the maximum OD density

To distinguish arbitrary edge effects from true variations in OD density, a second set of 100,000 trajectories was generated, again with a uniformly random set of destination locations, but containing a Gaussian distribution of origin locations around a centre point and standard deviation of 20% of the map width (Figure 2 bottom row). In this case, the concentration towards the centre is clearly evident in both the flow map and OD map. The OD map further reveals the uniform random nature of the destination cells as the approximately homogeneous density in each large grid square is apparent in this representation. This would not be detectable from the flow map directly.

54. Brewer, C. (2002). **Selecting Good Color Schemes for Maps**, http://www.colorbrewer.org.

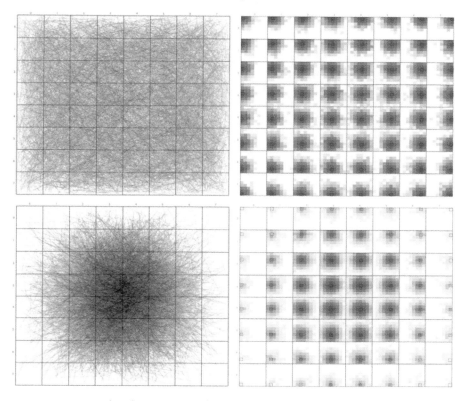

FIG. 3.   100,000 simulated trajectories shown as a vector flow map (left) and an OD map (right). In the upper example, origin locations are uniformly random, while in the lower example origin locations have a Gaussian random distribution about the centre. In both cases, destination locations are a random direction with Gaussian random distance from the origin

Many real geographic flows have a tendency to show spatial autocorrelation,[55] i.e. shorter flows between origins and destinations are more likely than longer flows. To simulate this effect, a further set of simulations was created where each destination was generated in a uniformly random direction and Gaussian distance (with standard deviation of 20% of the map width) from each origin location (Figure 3). For both a uniform (Figure 3 top row) and Gaussian (Figure 3 bottom row) distribution of origins, the OD map representation shows the Gaussian flow length distribution clearly. The flow maps shown on the left of the figure, while distinguishable from each other, do not clearly reveal the positive spatial autocorrelation of the trajectories, nor do they strongly distinguish themselves from their uncorrelated equivalents in Figure 2.

55. Cliff, A. and Ord, J. (1973). **Spatial Autocorrelation**, Pion, London.

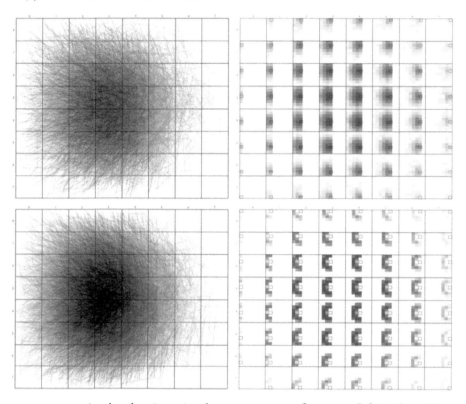

FIG. 4.   100,000 simulated trajectories shown as a vector flow map (left) and an OD map (right). Origin locations have a Gaussian random distribution about the centre and a directional bias that prevents any left-to-right flows. In the upper example, length of the trajectory is Gaussian, while in the lower example, length of the flow is fixed at 25% of the map width (two grid cells)

To investigate the ability to reveal directional and distance bias, a final set of simulations was created where flows from left to right were excluded (Figure 4). This represents an extreme case of bias, but provides a useful benchmark with which to compare output. In Figure 4 upper row, the length of each flow is Gaussian (with standard deviation of 20% of the map width). Here the directional bias is visible only around the margins of the flow map where densities are sufficiently low. In contrast, the OD map shows the bias as a clear edge in each origin grid cell. In Figure 4 lower row, the length of each flow is fixed at 25% of the map width, thus precluding short flows. Visually, this flow map is indistinguishable from the Gaussian distribution, but the OD map clearly shows 'hollow' OD densities where close origin–destination flows are not permitted. It is important that the visualization is able to reveal these types of structure as many real geographic vectors will possess a tendency to have a minimum length (e.g. vacation trips or courier deliveries) or directional bias (e.g. animal migration).

It is recognized that there are many other structures that might exist in real geographic OD interactions (e.g. polycentric commuting patterns, impact of geographic barriers to movement, etc.) that are not tested with these simulations. The case study (see the section on 'CASE STUDY') demonstrates how such structures may be identified, and the interaction enabled by our OD map software (see the section on 'Interaction') allows their characterization to be explored.

Rasterization artefacts

Aggregation is often necessary when analysing large datasets and geography provides a meaningful framework for doing so. This enables the exploration of large datasets while identifying trends and structure in the data.[56] However, any process that involves geographic aggregation can lose information that may have some value. Equally it is important to recognize that any trends or structure revealed could be an artefact of the aggregation undertaken rather than a property of the data. In particular, if the spacing between origin or destination locations is at approximately the same scale as the OD grid spacing, aliasing effects can be introduced. Figure 5 shows a flow map for travel-to-work flows for the US counties of Ohio. In this dataset, all home and work locations are aggregated to the county/counties in which they occur. In turn, these locations are mapped to the county centroid. Dividing the state into an 868 grid yields cells of approximately the same size as the counties. In most cases, the distribution of counties results in one centroid per grid cell, but in some cases, two centroids can coincide with a cell (e.g. Union and Logan in cell [3,2]), or even three (e.g. Geauga, Portage and Summit in cell [1,6]). The resulting OD map may therefore show higher OD densities not necessarily due to a regional control, but simply the coincidence of data points and arbitrary grid boundaries. This is one type of example of the modifiable area unit problem (MAUP).[57] Dynamic OD maps allow us to explore the sensitivities and effects of the MAUP and we have implemented two interactive approaches to do this.

The first is to permit dynamic variation of the number and location of the grid cells used to aggregate the data. This allows a visual spatial sensitivity analysis to be undertaken where the user interactively changes the grid aggregation to see if this has any significant effect on the trends evident in the OD map. A sample set of OD map excerpts with differing grid aggregations is shown in Figure 6. Offsetting the grid location (Figure 6 top row) has relatively little effect on the flow trends in this example, but changing the grid resolution (Figure 6 bottom row) has a more significant impact. In particular, increasing the number of grid cells also increases the number of blank origin and destination cells as fewer county centroids coincide with the smaller grid cells. This suggests that selecting an appropriate grid size is important if origins or destinations are clustered in space.

The choice of appropriate grid size can be facilitated through interactive control over grid size and location. In our optimized implementation (see the section on 'Optimization with hash grids' below), immediate visual feedback is given on the effect of grid changes for OD maps with up to order 10^6 OD vectors.

56. Andrienko and Andrienko, 2008.
57. Openshaw, S. (1984). 'The modifiable area unit problem', **Concepts and Techniques in Modern Geography**, 38, p. 41.

FIG. 5. Ohio travel-to-work flows showing county aliasing. Flows are located at each county's centroid. Where the scale of the county is approximately equal to the scale of the OD grid, aliasing effects can occur

An alternative solution to the aliasing problem can be applied when grid cells are approximately the same size as the spatial units under investigation. Here we tessellate the county centroids to form a grid arrangement. This is in effect a special case of the spatial treemap,[58] where all centroids are forced to their nearest unique grid location. Where spatial units have some meaning, this can produce a more interpretable OD map (Figure 7). This quasi space-preserving solution may be more spatially consistent than others proposed and thus has advantages over existing alternatives.

A spatial treemap may not result in a regular grid tessellation, and while this would still allow an OD map to be created, its cognitive plausibility would be reduced as each coarse grid cell could potentially be a rectangle of a different aspect ratio. As a result, the destination cells within each origin cell would be subject to an inconsistent

58. Wood and Dykes, 2008.

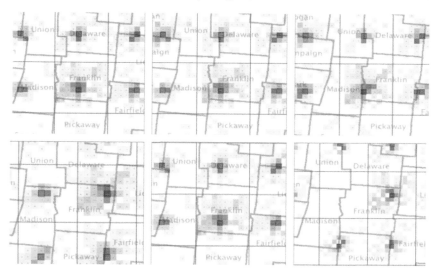

FIG. 6.   Effect of changing grid resolution and position on OD map densities of travel to work
flows from Franklin county Ohio. Top row shows effect of moving the 10×10 grid in an
east–west direction. Bottom row shows the effect of changing grid resolution from 9×9 (left)
through 10×10 (middle) and 11×11 (right)

scaling. To overcome this problem, we ensure that the number of tessellated grid cells is a perfect square, by if necessary, adding some blank 'dummy' grid cells where no flows occur. The location of these cells is selected at the points furthest away from any known origin or destination cells. Typically this will be around the edge of non-rectangular study regions (see bottom corners and top central portion of the spatial treemap shown in Figure 7).

The disadvantage of regular grid tessellation is that for geographic regions with elongated aspect ratios, the nested cells become even more elongated (the aspect ratio is squared). Geographic integrity is maintained, but at the cost of very thin cells that can be difficult to interpret when coloured. This can be overcome by either reprojecting geographical space to give a squarer aspect ratio or by inserting dummy cells along the 'thin' edges of each cell. The former approach has the cost of transforming geographic space to something that may be unfamiliar to users (e.g. Chile projected to a square), while the latter can result in inefficient use of graphical space (dummy cells repeated for each nested cell). A balance between geographic familiarity, cell interpretability and space efficiency must be struck by the analyst when constructing an OD map space.

Optimization with Hash Grids

The OD map provides a visual interface for filtering of a set of vectors or trajectories. By selecting any given cell in the OD map, a query may be made of just those trajectories that originate from the given origin cell and end in the given destination cell implied by the OD cell location. Since the OD map itself retains the geographic

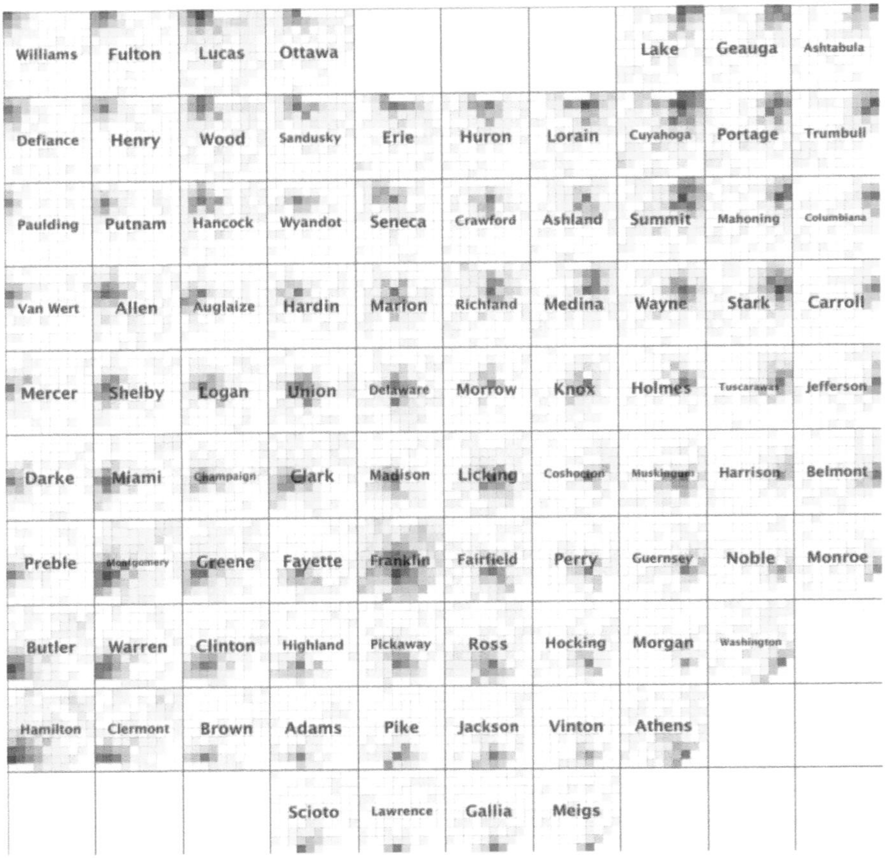

FIG. 7. Ohio travel-to-work OD map. Here counties are tessellated into a grid using a spatial treemap (one county per grid cell plus 12 'dummy' grid cells)

projection of the original data, such a query could be used to project the full geographic path of the selected trajectories over the OD map by plotting lines or polylines. To allow this to happen interactively, and to facilitate rapid brushing over the OD map, an efficient data structure is required to store the set of 105–106 OD trajectories. Here we use the spatial hash grid, more commonly deployed for rapid collision detection.[59]

The hash grid divides space into a regular grid, where each grid cell is accessible via a one-dimensional hash code. Each cell then stores a collection of references to spatial objects associated with the region within the grid cell boundaries. This has an obvious mapping to the OD map that also uses a regular gridding of space. In this case, we construct two hash grids, one that stores references to all the trajectories that

59. Eitz, M. and Lixu. G. (2007). 'Hierarchical Spatial Hashing for Realtime Collision Detection', in **IEEE International Conference on Shape Modeling and Applications** (SMI 2007), pp. 61–70, Lyon, Jun 14–16.

originate from any given cell, and the other that stores references to the trajectories that end in each cell. The cell hash value is easily constructed in such a coarse two-dimensional grid as

$$hash = row \times n + col \tag{4}$$

where *row* and *col* are the OD map coarse row and column values and *n* is the number of rows or columns in the OD map. Given that n is likely to be of order 10, there is no danger of overflow in the hash value if stored using at least 16 bit integers.

To select the set of trajectories between given origin and destination cells, the origin hash grid is first queried to retrieve only those trajectories that originate from the given cell. The destination hash grid is likewise queried to find only those trajectories that end in the given cell. Finally, the intersection of these two sets filters all but those vectors between the queried origin and destination cells. If the collections within each hash grid cell in turn are sorted using an optimization structure (implemented here using Java's TreeSet structure), queries are sufficiently fast to allow interactive brushing of trajectory sets of order 10^5–10^6 trajectories.

CASE STUDY

To explore the validity of OD mapping, we created a prototype environment for visual exploration that could show flow maps, OD maps and OD matrices, and provide interactive control and linking between these views. It was developed in Processing,[60,61,62] an extension to Java for rapid visually orientated application development.

We selected county to county migration flows for the conterminous USA from the US 2000 Census.[63] This dataset recorded the numbers of changes in home address between 1995 and 2000 aggregated to the county of origin and destination. Locations were added by combining the dataset with the 2000 census county gazetteer and projecting from latitude/longitude to an Albers Equal Area projection. In total 721,432 separate migration vectors representing the intercounty movements of 46.6 million people were recorded. These vectors are shown as a conventional flow map in Figure 8. Evident in this view of the data is the 'population footprint' showing some of the major population centres, most clearly where they contrast with regions with fewer migration paths. However, from the experiments shown in the section on 'Simulated OD vectors' above, caution must be exercised in interpretation of artefacts. For example, it is not always evident whether the higher flow line densities in the central states are due to migrants at those locations or simply because they happen to be placed on the path between origins and destinations to the west and east.

In contrast, Figure 9 shows the OD map of the same data. As in the previous OD maps, each OD cell is coloured according to the absolute numbers of movements

60. http://www.processing.org.
61. Reas, C. and Fry, B. (2007). **Processing: A Programming Handbook for Visual Designers and Artists**, MIT Press, Cambridge.
62. Fry, B. (2007). **Visualizing Data**, O'Reilly, Cambridge.
63. US Census Bureau. (2008). **County to County Migration Flow Files**, http://www.census.gov/population/www/cen2000/ctytoctyflow.

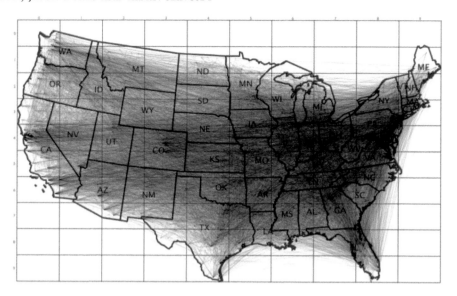

FIG. 8.   20,000 US county–county migration vectors (3% random sample). Vectors rendered using transparency and anti-aliasing to allow 'occlusion density' to be seen

between the origin and destination using an exponentially scaled colour scheme, and the 'home cell' for each origin is highlighted. The spatial autocorrelation inherent in migration is evident in that the highest densities of migration are where origins and destinations are close (darker red cells clustered around the home cell). There appears to be significant migration along the west coast of the US, both within California (e.g. LA and Orange County) and further north in Oregon and Washington. Some grid cells appear to show a much more homogeneous set of destination cells than others. For example, Southern California (0,6) and Colorado (3,5) have destinations evenly spread, while New York and Florida show much more concentrated (and reciprocal) flows. The non-coastal northwest states show some inter-state migration between them, but relatively little to other large parts of the USA.

The grid cells containing large urban populations (e.g. Southern California, Chicago and New York) inevitably show higher migration across the USA, but caution should be exercised when interpreting these patterns, especially with an exponential colour scale.

The signed chi statistic

Analysis of Figure 9 demonstrates that it is not always possible to separate high frequency of migration from the underlying population footprint. It is inevitable that on the whole areas of higher population density will have more migration, simply because there are more people available to move. The cell-based symbolization of the OD map allows us to substitute more discriminating measures than absolute counts, such as the signed chi statistic for comparing observed with expected values

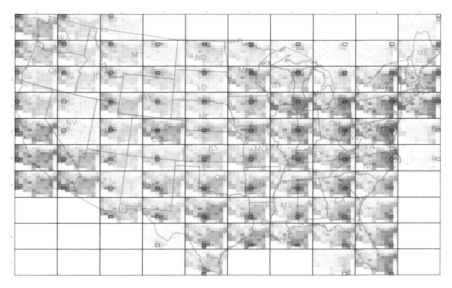

FIG. 9. All 721,432 US county–county migration vectors shown as an OD map. Each large grid cell represents origin location, within which the map of destination densities using the same grid is shown

$$\chi = \frac{obs - exp}{\sqrt{(exp)}} \tag{5}$$

In this case, we define the observed value as the numbers of migrations from any given origin to a given destination and the expected value is weighted according to the mean population of the observed and expected counties

$$exp_{od} = \frac{\sum m}{\sum pop} \frac{pop_o + pop_d}{2(n-1)} \tag{6}$$

where $\sum m$ is the total number of people who have moved from one county to another, $\sum pop$ is the total population of all counties, pop_o and pop_d are the populations of origin and destination counties, and n is the total number of counties. In other words, this particular measure of expectation assumes that people migrate to all other counties in proportion to their respective populations. This simple model makes one of a number of different possible assumptions about expectation, but serves to illustrate how the OD map can be used to summarize such statistical measures. Other more sophisticated measures that could be quantified using the chi statistic might incorporate socioeconomic status, geographical permeability or trends over time. The results of mapping the population-based chi value is shown in Figure 10.

The chi OD map confirms that there is indeed greater than expected migration along the west coast (dark red destination cells). Cells that contain significant

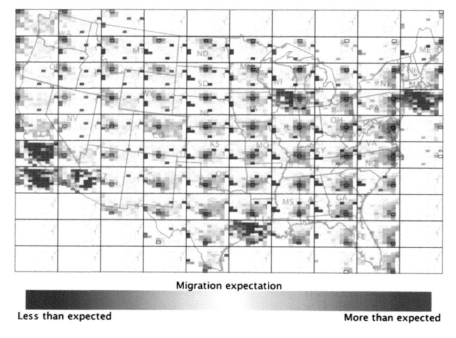

Migration expectation

Less than expected   More than expected

FIG. 10. Chi statistic OD map showing the difference between observed county–county migration and that expected based on the populations of origin and destination counties (100% sample). Geographical context is provided by both a large origin map of US states and small multiples of destination maps. Chi values are coloured using an exponentially scaled diverging Brewer 'RdBuɪɪ' scheme

numbers of darker blue cells show greater 'selectivity' in migration destination given the size of the origin populations. The OD resolution and spatial origin were varied interactively in order to examine the persistence of patterns with scale and aggregation. It is apparent that there is less migration from Southern California to most of the USA, with the exception of the Pacific coast, Chicago, the large cities of Texas and New York. Similar patches of blue can be seen at these cities, suggesting a population less inclined to migrate away from big cities than our simple model suggests. New York, Chicago and Southern California show a relative lack of east–west migration, whereas Houston shows a resistance for north–south movement. Dark red cells that are not close to the home cell indicate larger than expected movements between geographically distant locations. For example, there are greater than expected flows from the NE coast to Florida and from Florida to Colorado.

Interaction

Viewing static OD maps provides some insight into the structure of the geographic vectors under investigation, particularly in overview. Adding interactive features to the prototype supports the exploratory process in a number of ways. Various interactions are possible[64] and provide access to alternative layouts,[65] transformations between them[66] and details to be accessed on demand.[67]

We found the following interactions useful in visually exploring spatial interactions in a number of datasets and have implemented them in our flowMappa environment for visual exploration[68]:

- zooming and panning;
- toggling of numeric indicators of OD densities;
- toggling of context maps in both O space and D space;
- dynamic changing of grid cell size and offset;
- ability to swap O space for D space and *vice versa*;
- brushing to overlay selected OD vectors;
- linked views between flow map, OD map and ODmatrix;
- varying colour scheme between ColorBrewer alternatives;
- varying colour scaling between log and linear scales.

Some selected examples of the effects of this interactive control are shown in Figure 11. While in this example, selected OD vectors are shown as straight lines, the projection into geographic space would allow the full geometry of the trajectories between selected origins and destinations to be overplotted.

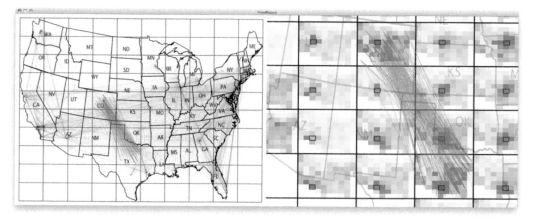

FIG. 11.   Example of interaction query and linked views. The application shows a flow map view on the left hand side and an OD map view on the right. Both can be independently zoomed and panned allowing both overview and detail-on-demand. By brushing over the various destination cells in the North Texas origin cell, the trajectories between the locations are overlaid on both views. Trajectories are stored at the resolution of the county centroids

64. Yi, J. S. and Stasko, J. (2007). 'Toward a deeper understanding of the role of interaction in information visualization', **IEEE Transactions on Visualization and Computer Graphics**, 13, pp. 1224–1231.

65. Slingsby *et al.*, 2009.

66. Heer, J. and Robertson, G. (2007). 'Animated transitions in statistical data graphics', **IEEE Transactions on Visualization and Computer Graphics**, 13, pp. 1240–1247.

67. Shneiderman, B. (1996). 'The Eyes Have it: A Task by Data Type Taxonomy for Information Visualizations', in **IEEE Symposium on Visual Languages**, pp. 336–343, Boulder, CO, Sep 3–6.

68. http://www.flowmappa.com.

CONCLUSIONS

Revisiting our list of requirements (see the section on 'Requirements and Prior Work'), it is apparent that none of the existing techniques for showing large sets of trajectories are able to meet all eight of these. The OD map offers several advantages over the more commonly used flow mapping and OD matrix representations. Owing to aggregation of OD vectors into a regular grid, the OD map is scalable to large collections of vectors (requirement 1). The cost of this aggregation is twofold. First, like any form of aggregation, a potential loss of detail in origin and destination location results as the geographic grid resolution of the OD map is limited to geographic grids 20×20 or so (each cell having to contain a further 20×20 cells). This limit can be partially overcome by interactive zooming to reveal detail on demand (requirement 2). Second, aggregation into grid cells that do not reflect the underlying geographic structure can give rise to aliasing effects. These can be partially overcome through the use of a spatial treemap to aggregate into meaningful geographic units along with dummy cells to preserve key geographic properties.

For analysis where the vector between a pair of two locations has greater importance than the geometric path between them, the OD map provides this detail with minimal loss of information and little visual clutter (requirements 3 and 5). Where the geometry of the path is important, such as transportation infrastructure management, the geographic projection of the OD map allows this geometry to be overlaid (requirement 4). This works well for spaces with relatively square aspect ratios, but would be more problematic when nesting long thin regions. Methods for transforming geographic spaces, such as the spatial treemap algorithm used in Figure 7, provide some solutions but with some loss of spatial coherence and possible impacts on cognitive load.

All forms of trajectory mapping are liable to visualization artefacts that do not reflect true properties of the data. Despite their common use, flow maps of large collections of vectors seem particularly vulnerable to this. Occlusion effects are removed by the OD map, although the partitioning of space into a grid introduces possible aliasing and MAUP artefacts. The effect of these on the stability of the visualization can be explored though dynamic change in gridding parameters (requirement 8). If the resolution of the vector data is much finer than the grid resolution, this effect is minimal, but aggregation is greater. Where grid resolution is closer to the data resolution, the spatial treemap provides one way of removing aliasing effects, but at the cost of some spatial distortion.

The asymmetry between flows in both directions between pairs of points is explicit in the OD map (requirement 6). It can be further explored by interactive swapping of O space and D space as well as through brushing over OD cells (requirement 7).

We do not propose the OD map as replacement for other forms of trajectory exploration, but suggest that it offers a new spatial view of large collections of geographic vectors that may be integrated with existing systems to help reveal and consider the geography of associations between pairs of locations. As the cells in the OD map are identical to those of the OD matrix, it provides a supplementary spatial ordering of this much-used aspatial representation of geographic information to which spatial cognition can be applied as we endeavour to explore geographic interactions and processes.

Reflections on 'Visualization of Origins, Destinations and Flow with OD Maps'

MENNO-JAN KRAAK

ITC, University of Twente

Geographers have been interested in movements for over a more than a century. These could be migration of people, commuters in an urban environment, the transport of good between harbours or airports, the behaviour of animals, the spread of diseases and more recently, also the dynamic relations in social networks or Internet traffic. In the past, it was not easy to obtain the necessary data to get a proper insight in these movements. However, technological developments during the last decades have made this less of a problem, so that the problem has actually shifted into massive data abundance (big data) and currently, many scientists are studying this new problem. An example of this effort is the European MOVE project, a COST action, where people work together to 'develop improved methods for knowledge extraction from massive amounts of data about moving objects'.

Maps play an import role in revealing spatio-temporal patterns in movement data. The simplest map depicting movement would consist of an arrow indicating the start and end of the movement as well as the path followed. In the middle of the nineteenth century, these maps evolved into flow line maps. In addition to the start, the end and the path, the quantitative (how much is moving) and qualitative (what is moving) characteristics of the movement were also indicated. Minard was one of the first to creatively develop this type of map, whose design is not yet matched by current automated methods (Figure 1), although some get close (Verbeek *et al.*, 2011). An alternative to the flow map was introduced in the 1970s: the Space–Time cube, which deals with time more explicitly. As soon as more data became available, alternative map representations were used to summarize data; an example being density surfaces.

A non-spatial solution used to indicate movement is the OD Matrix. It indicates the start (Origin) and end (Destination) of the movement. The cell in the matrix can indicate the nature of the movement, whether such movement exists, how much is transported between O and D or how long does it take to go from O to D. The OD matrices are often used as the basis for graph algorithms in network analysis. For large matrices, summarization techniques have been applied as well (Wilkinson and Friendly, 2009).

Cartographers are not unfamiliar with the matrix and it is often found on road maps to indicate distances between selected cities. Examples can be found from as early as 1750, as Figure 2 shows. In a different guise, this is known as Bertin's re-orderable matrix, to organize data for thematic mapping (Bertin, 1967). However, the OD matrix lacks easy answers to typical geographic questions.

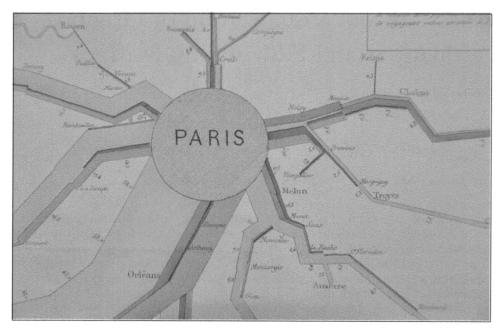

Fig. 1.   Minard's flow maps: detail carte figurative et approximative des poids des bestiaux venus Paris sur les chemins de fer en 1862 (Library Lasage, Collection Ecole Nationale des Ponts et Chaussées)

Fig. 2.   OD matrix as distance chart by R. and J. Ottens (Amsterdam) around 1750

It difficult to apply the above mapping and matrix solution within today's 'big data' environment. Researchers are using all their geocomputational and design skills to identify solutions. Figure 3 shows a set of examples derived from the Visual Complexity website which collects examples of innovative network graphics dealing with large (movement) data sets. In our cartographic domain, solutions are characterized by keywords such as interactivity, aggregation, design and coordinated multiple views in the context of geovisualization and geovisual analytics.

This is where Wood, Dykes and Slingsby's paper fits in. The authors begin to combine the best of the flow map and the OD matrix, and 'propose mapping OD vectors as cells rather than lines, comparable with the process of constructing OD matrices, but unlike the OD matrix, we preserve the spatial layout of all origin and destination locations by constructing a gridded two-level spatial treemap'. Basically what happens is that geographic space (the map) is partitioned using a regular grid, which forms the base for the OD map. However, each OD grid cell contains a nested geographic grid.

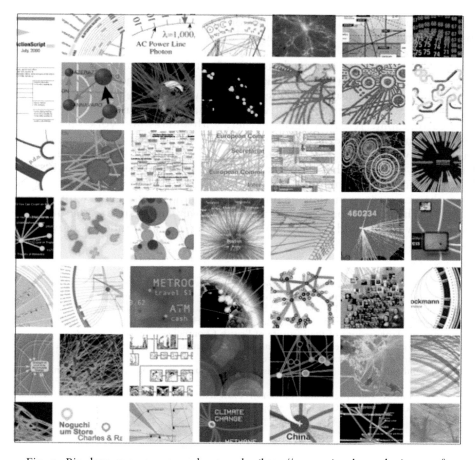

Fig. 3. Big data: movements and networks (http://www.visualcomplexity.com/)

These nested grid cells are filled based on the destination of movements. Figure 1 in the paper is essential for understanding how the OD maps work. The resulting OD maps represent the origin–destination density, since each nested grid cell is coloured based on the amount of links using the visual variable value. The authors discuss the strengths and weaknesses of their approach using synthetic datasets with a case study using real world data 5 years of county-to-county migration in the USA. In this example, they also combine the number of movements with existing population data in the area to see if the trend revealed in the OD map is as one might expect. The OD map has been implemented in the processing software environment which allows for all kind of interaction with the OD map content as well as with the linked flow map and the accompanying website (http://www.flowmappa.com) gives some additional static examples. In their conclusion, the authors discuss the problem of selecting the grid size in relation to the ability to recognize the geographic area. It would be interesting to see the results of usability research with regard to the understanding of the OD map by potential users.

The authors have been active in our cartographic domain for over 20 years, including participation in the ICA's Commission on Geovisualisation. Their institutional website currently states: 'Our research looks at the way interactive information visualization, cartography and visual analytics help us make sense of the world around us. We take a creative approach to visualization and interaction design, which combines best practice, the latest visualization research and our own novel ideas, often testing and demonstrating these with highly interactive prototypes. We develop theory, best practice, prototypes, software and evaluative work' (giCentre, 2013). This line of cartographic research began earlier, with interesting work on CDV (Dykes, 1998) and LandSerf (Fisher et al., 2004) and each combining novel visualization ideas with interactive software developments to demonstrate the value of their ideas.

Part of the challenge that lies ahead is to see who will be the new Minard and is able to extend the OD map — an alternative for the simple movement map with arrows that can accommodate many different movements, each with a beginning, end and a schematized path — but to also answer and visualize the questions that Minard's flow map could handle: what is moving and how much is moving?

REFERENCES

Bertin, J. (1967). **Semiology Graphique**, Mouton, Den Haag.
Dykes, J. A. (1998). 'Cartographic visualization: exploratory spatial data analysis with local indicators of spatial association using Tcl/Tk and cdv', **The Statistician**, 47, pp. 485–497.
Fisher, P., Wood, J. and Cheng, T. (2004). 'Where is Helvellyn? Fuzziness of multiscale landscape morphometry', **Transactions of the Institute of British Geographers**, 29, pp. 106–128.
giCentre. (2013). **giCentre — Department of Information Science**, City University, London, http://www.soi.city.ac.uk/organisation/is/research/giCentre/research.html.
Wilkinson, L. and Friendly, M. (2009). 'The history of the cluster heat map', **The American Statistician**, 63, pp. 179–184.
Verbeek, K, Buchin, K. and Speckmann, B. (2011). 'Flow map layout via spiral trees', **IEEE Transactions on Visualization and Computer Graphics**, 17, pp. 2536–2544.

Reasserting Design Relevance in Cartography: Some Concepts

KENNETH FIELD AND DAMIEN DEMAJ

Originally published in *The Cartographic Journal* (2012) 49, pp. 70–76.

Cartographers have always been concerned about the appearance of maps and how the display marries form with function. An appreciation of map design and the aesthetic underpins our fascination with how each and every mark works to create a display with a specific purpose. Yet debates about what constitutes design and what value it has in map-making persist. This is particularly acute in the modern map-making era as new tools, technology, data and approaches make map-making a simpler process in some respects, yet make designing high-quality maps difficult to master in others. In the first part of a two-part paper, we explore what we mean by map design and how we might evaluate it and apply it in a practical sense. We consider the value of aesthetics and also discuss the role of art in cartography taking account of some recent debates that we feel bring meaning to how we think about design. Our intent here is to reassert some of the key ideas about map design in cartography and to provide a reference for the second part of the paper where we present the results of a survey of cartographers. The survey was used to identify a collection of maps that exhibit excellence in design which we will showcase as examplars.

INTRODUCTION

Cartographers have always been concerned about the appearance of their maps and also have a tendency for being critical about the appearance of other people's maps.[1] This trend seems to have increased as the number of professional cartographers has declined, GIS has become prevalent as a map-making tool and online web maps are now relatively easy to make by pretty much anyone. We face a situation where the pool of cartographic expertize has undoubtedly become smaller; it is possible that the need for high-quality maps has decreased and that ubiquitous map-making has lowered people's expectations. Criticism of map quality is prevalent in blogs and other un-policed platforms used to pass comment though, so clearly there is a group for which sub-par cartography causes deep concern.

In this two-part paper, we aim to explore a little of why it is that cartographers are so passionate about the design and appearance of maps. This might go some way to explaining why some feel that they should act as map police and point out where

---

1. Kent, A. J. (2005). 'Aesthetics: a lost cause in cartographic theory?', **The Cartographic Journal**, 42, pp. 182–188.

someone else's work is not performing as an optimum communication device. Is this helpful or does it serve to isolate the profession whether they have fair comments or not? And why might people be reluctant to listen to expert advice or constructive criticism of their work? In short, we intend to shed light on some of the concepts that cartographers know and understand and which drives their passion for design. We deal with this in part 1 of the paper.

In part 2 of the paper, we back up the concepts here by providing the results of a survey of cartographic experts that have led us to select a range of maps that we feel exhibit the best of cartographic design.[2] Many lists have been complied of the 'top ten' movies, albums, restaurants or any number of other facets of daily life, but rarely do cartographers state what they believe to be indicators of excellence in design. An exception is provided by Forrest[3] though by self-admission his examples are very much a personal choice. Are they too busy criticising to tackle the issue in a more positive manner? Perhaps the problem has more to do with deciding on a top ten in a world that contains so many different types and styles of maps which are hard to assess side by side? Perhaps it may be that, ultimately, good design is partly subjective and every map will be seen in different ways by different people, so there is bound to be difficulty in creating an objective assessment? Perhaps it is because design is misunderstood and people mistake aesthetics and beauty as being a proxy for good design? We intend our surveyed examples to provide a point from which others might be able to reflect on their own work. Figure 1 provides a taster of the collection of surveyed maps that provide the focus for part 2 of this paper.

DEFINING DESIGN IN CARTOGRAPHY

No generally accepted formal definition of design exists.[4] Here, then, is the first problem cartographers face when trying to explain that a map is well designed or poorly designed. Without a clear definition of what cartographic design is or how it might be conceived, it is no wonder it is such a contentious issue. By definition, or lack of it, we face difficulties trying to describe how a map might exhibit good design. Informally, design as a noun refers to some sort of plan or set of conventions used in the construction of an object or a system. We apply design to the creation of furniture, cars, architecture, business processes and most other things. As a verb, 'to design' refers to the implementation of the plan. Ralph and Wand[5] do offer a formal definition as '(noun) a specification of an *object*, manifested by an *agent*, intended to accomplish *goals*, in a particular *environment*, using a set of primitive *components*, satisfying a set of *requirements*, subject to *constraints*; (verb, transitive) to create a design, in an *environment* (where the designer operates) [original emphasis]'. That seems to cover it!

2. Demaj, D. and Field, K. (2012). 'Reasserting design relevance in cartography: some examples', **The Cartographic Journal**, 49, pp. 77–93.

3. Forrest, D. (2003). 'The top ten maps of the twentieth Century: a personal view', **The Cartographic Journal**, 40, pp. 5–15.

4. Ralph, P. and Wand, Y. (2009). 'A proposal for a formal definition of the design concept', in **Design Requirements Workshop, Notes on Business Information Processing**, ed. by Lyytinen, K., Loucopoulos, P., Mylopoulos, J. and Robinson, W., pp. 103–136, Springer-Verlag, Berlin.

5. Ralph and Wand, 2009, p.109.

FIG. I.   The 39 maps: examples of excellence in map design drawn from a survey of
cartographic experts

So design is some sort of roadmap or approach that achieves a particular outcome and given that the process of making a map can be matched to these descriptors, then we are all, by definition, designers. Nowhere, though, is there a description of what makes a designer good or what makes one bad. Is it experience, intuition, skill, luck or a combination of those and many other dimensions? Can a novice map-maker design a great map. Yes. Can an experienced map maker following a sequential design process turn out a poor or uninspiring map...again, yes. Of course, when we reflect on many other areas of life, we exert our preferences in taste by exclaiming something 'looks good' or 'works really well'. Here then, we are making an emotional judgement on the aesthetic or functional dimensions of the object designed, the trade-offs and compromises that have been made and, by inference, attributing some sort of credit to the designer. Their design may have involved considerable effort and painstaking systematic application of principles or it may have been an accident of serendipity. Designers may follow a rational model which involves optimizing their work given known constraints and objectives as part of a clear plan[6] though in the absence of unknown goals or changing requirements and constraints, they may alternatively improvise the design process and rely on creativity to generate alternate designs. If the resulting object looks good and works well, then it constitutes good design on some level and that is when form and function work in harmony.

Design philosophies therefore exist either explicitly or implicitly to guide the principles that underpin how someone approaches a specific design task. Some philosophies guide the overall goal of a design such as KISS (Keep it Simple Stupid). A use- or user-centred design might focus on the use of the object or the needs, wants and limitations of the end user. Translating overarching philosophies into methods allow us to focus on exploring possibilities, redefining specifications and prototyping alternative solutions to improve the final object. Cartographic methods have often encouraged us to consider design and production (planning and executing) together, but design involves identifying the cartographic problem and solution(s) and is in part a creative process. Production on the other hand is largely pre-planned, though its own processes may have a major impact on constraining the creative process.

Cartography, then, is a creative professional career where problem-solving is part of the production process. For nonprofessional map-makers, there is no expectation that they should know of the multifarious design philosophies or approaches to design and production, yet they are still able to produce a map given the available technology, data and the spark of an idea. Maps are every-day objects and, as such, the realm of cartography is being seriously challenged because so many modern-day maps (and particularly web maps) break accepted cartographic practice. Maps are being designed outside of formal boundaries and by new patterns of production. Is this a problem? One view might be that it is a healthy development that will undoubtedly yield new ways of illustrating information using map objects. Alternatively, one might question why people seem so intent on ignoring decades of work that has gone before which is there precisely to guide map makers. Is it laziness or ignorance not to acknowledge what has gone before or perhaps it is simply a lack of awareness and uncertainty?

6. Newell, A. and Simon, H. (1972). **Human Problem Solving**, Prentice Hall, London.

The dominant theme in cartography for a number of decades was the communication of information.[7,8] This remains relevant at its basic level since the cartographer's intent is to communicate something about the world by using a map and the language of cartographic symbology.[9] In many ways, information theory focused attention on the key ideas behind cartographic communication since it sought to isolate the key elements of a map and establish how they served the key function which is transmission of information.

The argument goes that an awareness and control of the factors that inhibit transmission of information serves to give the map-maker a way to focus activities to meet the overall aim.[10] Ethics also plays a part in the transmission of information since a cartographer is entrusted with the choices of what to show and how to depict it. The message of a map can be fundamentally altered depending on who makes it and how it is made. Harley[11] stressed that cartographers have a professional responsibility to question the world-view that they present on their maps and be cognisant of the impact of their choices. He went further by urging them to think of the consequences of their maps and prompt their social conscience to the point of accusing them of complacency.[12] Monmonier[13] suggests that a cartographic solution is a highly selective, authored view that might equally be consciously manipulated or based on 'ill-conceived design decisions about many factors, such as map scale, geographic scope, feature content, map title, classification of data, and the crispness or fuzziness of symbols representing uncertain features'.[14] Continuing the discussion, Monmonier suggests that it is impossible for a map reader to know whether the map-maker had a biased view when making the map or whether they were too lazy to explore alternative designs to offer a more coherent or complete picture of reality. Of course, this presupposes the map reader is aware that the map in front of them is in any way sub-par which, of course, many will be unaware of so we are back to the assertion of Harley that the map-maker has an ethical responsibility given the possible shortcomings in the reader. It is worth repeating the basic tenets of ethics in cartography as Dent *et al.*[15] outlines them because they are important values to hold when designing a map:

- always have a straightforward agenda, and have a defining purpose or goal for each map;
- always strive to know your audience (the map reader);

7. Keates, J. S. (1989). **Cartographic Design and Production**, 2nd ed., Longman Scientific and Technical, Harlow.

8. Kent, 2005.

9. Board, C. (1981). 'Cartographic communication', **Cartographica**, 27, pp. 42–78.

10. Robinson, A. H. and Petchenik, B. B. (1975). 'The map as a communication system', **The Cartographic Journal**, 12, pp. 7–15.

11. Harley, J. B. (1989). 'Deconstructing the map', **Cartographica**, 26, pp. 1–20.

12. Harley, J. B. (1991). 'Can there be a cartographic ethics', **Cartographic Perspectives**, 10, pp. 9–16.

13. Monmonier, M. (1991). 'Ethics and map design: six strategies for confronting the traditional one-map solution', **Cartographic Perspectives**, 10, pp. 3–8.

14. Monmonier, 1991, p. 3.

15. Dent, B., Torguson, J. S. and Hodler, T. W. (2009). **Cartography: Thematic Map Design**, McGraw Hill, New York.

- do not intentionally lie with data;
- always show all relevant data whenever possible;
- data should not be discarded simply because they are contrary to the position held by the cartographer;
- at a given scale, strive for an accurate portrayal of the data;
- the cartographer should avoid plagiarizing: report all data sources;
- symbolization should not be selected to bias the interpretation of the map;
- the mapped result should be able to be repeated by other cartographers; and
- attention be given to differing cultural values and principles.

EVALUATING CARTOGRAPHIC DESIGN

Evaluating a map is often difficult because the wide range of impacts on the design process can never be fully understood by the map reader. There may be perfectly good reasons why something is depicted in a particular way that might at first not seem particularly logical. Of course, it is up to the map-maker to keep this confusion to a minimum, but the map as an object should be the end product of the design process and the trials and tribulations of its construction hidden. Southworth and Southworth[16] suggest the following range of guidelines that list the design characteristics of a successful map:

- a map should be suited to the needs of its users;
- a map should be easy to use;
- maps should be accurate, presenting information without error, distortions or misrepresentation;
- the language of the map should relate to the elements or qualities represented;
- a map should be clear, legible and attractive; and
- many maps would ideally permit interaction with the user permitting change, updating or personalization.

While useful, many map-makers would be unaware of this informal cartographic code of practice. These basic tenets drive the design process and have formed the cornerstone of cartographic training for decades, but in modern, democratized map-making, the guidelines are perhaps relevant but less visible or understood.

CREATIVITY IN CARTOGRAPHIC DESIGN

There is potential for lists of rigid rules to be seen as constraining the creative process in map design which might counter the desire for aesthetic appeal. Successful maps are not made by a single recipe and experience teaches a cartographer that some recipes work well for them while others do not. Though following some basic rules can help with the process of design, true creativity is often associated with going beyond the boundaries of convention. This is perhaps where many modern maps cause cartographers such alarm, though we would contend that where some argue

16. Southworth, M. and Southworth, S. (1982). **Maps: An Illustrated Survey and Design Guide.** Boston: New York Graphic Society/Little Brown.

that they are being creative and pushing the boundaries to challenge assumptions, actually, it was more by accident than design because they were poorly equipped to realize what it was they were trying to show in the first place. Creativity might challenge assumptions, but in a cartographic sense, it is more commonly seen when someone sees patterns in data that were only revealed through their map. Creativity here, then, is the art of seeing something worth mapping and making it visible to a readership. It goes beyond simply making a map and expecting the reader to make sense of the patterns; it reveals the pattern and by so doing, illuminates some aspect of the data in a new or interesting way. Another way of applying creativity is to tackle a familiar subject in a different way. By taking a new direction or perception, we often reveal new insights and this, in turn, might help to clarify spatial patterns and relationships or reveal new connections between data. Certainly, some of the most evocative maps were designed by risk-takers who dared to try a completely new approach to a mapping task. These are more often than not maps that evoke a strong emotional response in people and in their own right set a new precedent in design. This sort of work is rare but makes cartography such an exciting area of creative design when you happen upon a new or visually stunning and effective representation of data. Whether a function of innovative, challenging design or following rules, a map will not be successful unless it meets user requirements. While pushing the boundaries in design often creates the most inspiring maps, they all share a clear sense of having been designed for a particular purpose and user group.

MAP AESTHETICS

Kent[17] focuses on the role of aesthetics in cartography. He suggests that aesthetics has been largely ignored because of the assumptions that it neither influences the process of cartographic design (it remains a by-product) and that it exists independently from geographical information. Aesthetics is concerned with the nature and appreciation of beauty and has roots in the branch of philosophy that deals with artistic taste. Debate surrounds whether objects, or groups of objects, possess aesthetic properties or whether they are a function of subjective perception. Cartographers have tended to assert thatmaps can contain aesthetic properties and that this in part is why they succeed.[18] This may be one area that has led to a marginalization of cartography since if design is difficult to define and aesthetics even more so, then how can amap-maker hope to create a well-designed map that meets the needs of the users? It is true that beautiful maps are often those that aremost appealing but at their basic level they are tools.[19] There is, of course, no reason why we cannot make tools that are also beautiful to use[20] and Norman[21] notes that attractive things make people feel good.

17. Kent, 2005.
18. Robinson, A. H., Morrison, J. L., Muehrcke, P. C., Kimerling, A. J. and Guptill, S. C. (1995). **Elements of Cartography**, 6th ed., John Wiley & Sons, New York.
19. Petchenik, B. B. (1985). 'Value and values in cartography', **Cartographica**, 22, pp. 1–59.
20. Kent, 2005.
21. Norman, D. A. (2004). **Emotional Design: Why We Love (or Hate) Everyday Things**, Basic Books, New York.

Technology has aggravated the issue of aesthetics and led to sterile designs. Powerful mapping software allows anyone to easily create convincing maps and graphics with little understanding of their data, design philosophies or principles of mapping. It is certainly questionable how many of today's map-makers have even heard of, let alone consulted, Bertin's 'Semiology of Graphics'[22,23] or Tufte's 'The Visual Display of Quantitative Information'.[24] Of course, the counter is that we now have greater openness and inclusivity and a greater range of map-making opportunities given the prevalence of freely available datasets. The words of John Wright (first published in 1944) still hold true though, regardless of technology:

> *An ugly map, with crude colours, careless line work, and disagreeable, poorly arranged lettering may be intrinsically as accurate as a beautiful map, but it is less likely to inspire confidence.*[25]

Karssen[26] defines three main components of aesthetics: harmony, composition and clarity. Harmony relates to the extent to which the different map elements look good together (as a whole). Composition is the way in which different map elements are positioned and the different emphasis each has been given. Clarity deals with the ease of recognition of the map elements. When we enter the realm of what looks 'good' we are, in part entering the realm of subjectivity though if map-makers at least appreciate these basic ideas and use them as a guide then their maps should reflect them. How well these principles of aesthetics are applied is part intuition, part experience and part training. Map critique is actually the vocalization of the application of aesthetics to a design solution with the goal to make it more harmonious, composed or clear.

Successful maps have at their core a clear, identifiable purpose, show correct information and are 'correct' graphically. These are fundamental cornerstones we can trace back to Robinson's seminal 'The look of maps'.[27,28] Additionally, Both Keates[29] and Robinson *et al.*[30] are clear that a successful map should be aesthetically pleasing and that visual efficiency creates clarity. If the goal in making a map is to create clarity of communication, then clarity itself will lead to an aesthetically pleasing object and a successful map will therefore require some aesthetic appeal to function.

APPLYING DESIGN IN A PRACTICAL SENSE

How we apply design through a process is most easily achieved by following a rational approach through sequential steps. Learning these steps is useful in shaping an understanding of the overall design process. There are, essentially, six main stages in

22. Bertin, J. (1983). **Semiology of Graphics**, The University of Wisconsin Press, Madison, WI.

23. Bertin, J. (2011). **Semiology of Graphics**, Esri Press, Redlands, CA.

24. Tufte, E. (2006). **The Visual Display of Quantitative Information**, Graphics Press, Cheshire, CT.

25. Wright, 1977, p. 23.

26. Karssen, A. (1980). 'The artistic elements in design', **The Cartographic Journal**, 17, pp. 124–127.

27. Robinson, A. H. (1952). **The Look of Maps: An Examination of Cartographic Design**, The University of Wisconsin Press, Madison, WI.

28. Robinson, A. H. (2010). **The Look of Maps: An Examination of Cartographic Design**, Esri Press, Redlands, CA.

29. Keates, 1989.

30. Robinson *et al.*, 1995.

designing anything: problem identification, preliminary ideas, design refinement, analysis, decision and implementation,[31] though, of course, they might not act or be applied independently of one another. In the first stage, need and the criteria to guide the overall design are identified. Here, we identify the map purpose and the intended audience and set limitations and deadlines. In the second step, we explore a range of ideas by brainstorming, developing alternative ideas and supporting creative thinking. In the third stage, we refine these ideas into a clear plan where ideas are selected or rejected. We refine ideas to sharpen them and it is this from here on that drives the creation of models (prototypes) which will involve sketching right through to the development of a detailed version of the map. At this stage, we work out problems and test alternatives and where we might solicit independent views of the map to determine if it works. Changes are made throughout the decision and final stages and implementation before work is committed to a final product. Evaluation and modification is continuous and, crucially, each time this process is followed, our collective knowledge about what works or what does not work grows. Of course, this is equally true for experienced cartographers or amateur map-makers and helps us refine our own work.

Design is both indefinable and critical to the effectiveness of a map. This is not particularly helpful particularly when trying to encourage novice map-makers (or even lazy experienced cartographers!) to consider the importance of design in their own work. Design is certainly a dynamic activity[32] which in part explains the problems of defining it. Perhaps we should adopt the simple definition that 'map design is the aggregate of all the thought processes that cartographers go through'.[33] It involves all of the major decision-making objectives such as choice of projection, scale, typography, colour, symbology and so on, and as a designer, we seek synthesis: all features of a product must combine to satisfy all the characteristics we expect it to possess with an acceptable relative importance for as long as we wish, bearing in mind the resources available to make and use it.[34] This last point is particularly intriguing given the predilection of rapid publication supported by large datasets and easy-to-use mapping tools. The idea that a 'quick and dirty' map is useful and revealing is often cited as a reason for spending minimal time on a map. However, they are usually less than optimally designed, so we would contend that by developing an appreciation for and applying some ideas from map design, it would leave the map-makers in a better position to make effective use of limited time. Being under pressure of a deadline or facing technological limitations should not be an excuse for the inappropriate application of basic design principles. For a number of years, GIS software was poorly equipped to support good design, though this is no longer an issue. It is also true that online mapping tools have taken time to mature and this technological progress has only recently provided a medium for authoring high-quality products. Technology, then, no longer limits opportunity in design terms.

Map design revolves around the need to satisfy a particular communication goal to someone. By first knowing the nature of the problem, what the map is going to show

31. Hanks, K., Bellistan, L. and Edwards, D. (1978). **Design Yourself**, William Kaufman, Los Altos, CA.
32. Dent *et al.*, 2009.
33. Dent *et al.*, 2009, p. 19.
34. Mayhall, W. (1979). **Principles of Design**, Van Nostrand Reinhold, New York. See p.90.

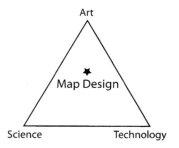

FIG. 2.   Map design in cartography

and who it serves is crucial to how we proceed. The precision with which this is considered is also important as a clear, tight focus will help eliminate noise from the design process. Ultimately, every manipulation of the marks on the map is planned so that the end result will yield a structured visual whole that serves the map's purpose.[35] Map design is a complex intellectual and visual activity. A cartographer makes good use of the sciences of communication, geography and psychology when creating a map and applies them through an appreciation of the use of the language of graphics.

THE ART IN OR OF CARTOGRAPHY

Cartography has always been described as an art, science and technology. For decades, this seems to have been a suitable catch all and cartographers have often placed themselves firmly at one of the points of that particular equilateral triangle (Figure 2). It would seem sensible for those whose pursuit is in the study and appreciation of map projections, for instance, to place themselves somewhere along the bottom axis. And what of map design? Many would tend towards placing design near to the 'art' apex but is that really where it should go? As we have tried to expound in this paper, we would contend it should be placed in the centre. It is not a purely artistic dimension of cartography — neither is it entirely to do with aesthetics and the look and feel of the map (what we might call the affective objective). It is as much about the science of cognition and the effective use of technology at your disposal. Actually, being playful, a designer would place map design at the *visual centre* of the diagram (Figure 2).

The debate about art in cartography has come to the fore recently as a number of people have questioned the value of seeing art as a component of cartography. Huffman[36] suggests that cartography is a form of art rather than being a component of it and this, we feel, has some merit when we consider design. Art, then, is what cartography is made of and it follows that design is something that encompasses the entire gamut of the mapping process. It is impossible to divorce design from either the science of map-making or the technology you use. Huffman contends that while

35. Dent *et al.*, 2009.

36. Huffman, D. (2011). **On human cartography**, http://somethingaboutmaps.wordpress.com/2011/04/20/on-human-cartography/ (accessed 10 January 2012).

a lot of science goes into map-making (through the use of digital tools, algorithms, projections, colour spaces, etc.), it does not necessarily follow that cartography is a science as well as an art. As an example, many of what we might see as purely artistic endeavours (painting, sculpture) also use tools based in science...yet they are not in any way regarded as such. So why is cartography still seen as much as a science as an art? It is true that cartographers use tools and data developed through scientific experimentation and research but so do other arts. So the art of cartography is in the way in which the cartographer applies the scientific tools of the trade and this is what we assert is the process of design towards the creation of an object for someone. The art of cartography is in the doing.

In the same way that this paper was largely prompted by a perception on our part of reluctance by some to consider design as a crucial part of good cartography, Huffman suggest that what is missing from a lot of cartography is humanity. Because cartography (and design) is fundamentally a human activity, it is incumbent on us to apply design. Machines cannot do this job for us and what we see in so many of the poorer maps that surround us is a failure to apply design by relying on the defaults offered by the machine in front of us. For most of human history, maps have been in most or in part made by hand. The marks made on a map have been deliberate and made by choice, not by a computer, which means that thought and choice has been applied. Map-making, then, is about decision-making in order to evoke a response from the map reader, not simply the loading of data into a piece of software and hitting the 'make map' button. Maps were made to build understanding, to offer a judgement or, perhaps, to create an emotional response. Machines do not do this. Humans do, yet so much of the cartographic process is now either fully or semi-automated that it acts as a design bypass. So now we have some sort of human–machine hybrid; we tend to see more and more unappealing maps or maps that suffer basic errors of design and construction. Worse, they offer very little in the way of aesthetic appeal or possess a graphical hook for our emotional response. Even worse, as Carl Steinitz once said 'far too much of what we see today goes round and round or up and down without actually saying very much and, worse, it's often accompanied by music'.[37] Far too many maps do not make us think about anything because the humanity has been removed from the design and production process as more maps are made without humans or made without humans who know or who are willing to explicitly exert some design influence. Huffman ends by noting that 'there is no art without creative intention...there is no cartography without a human creator'.

The art of cartography and the art of the cartographer are therefore about purposeful design.[38] Cartography is much more than simply visuals and aesthetics, it is about every step of the design process that goes into creating a map and as Woodruff[39] asserts, 'It's about careful thought behind the design of a map, not just any work (automated or otherwise) that results in a map'. The ability to understand and apply design effectively is part of what distinguishes a cartographer from a map-maker and

---

37. Steinitz, C. (2010). Panel discussion, in **CASA Conference on Advances in Spatial Analysis and e-Social Science**, London, Apr 13.

38. Woodruff, A. (2011). **Web cartography, that's like Google Maps, right?** http://www.axismaps.com/blog/2011/12/web-cartography-thatslike-google-maps-right/ (accessed 10 January 2012). Wright, J. K. (1977). 'Map-makers are human: comments on the subjective in maps', **Cartographica**, 19, pp. 8–25.

39. Woodruff, 2011.

with that knowledge and expertize, comes a responsibility to enthuse others of the fundamental value of design.

SUMMARY

In this paper, we have reasserted the relevance of design in map-making and focused on the concepts that underpin the application of design in a practical sense. We have also explored a little of the debates that exist in terms of defining design, the scope and place of aesthetics and the notion of humanity in map-making. To summarize, we would like to reassert the work of William Balchin and Alice Coleman who, in their classic paper[40] coined the term 'graphicacy' as an intellectual skill necessary for the communication of relationships which cannot be successfully communicated by words or mathematical notation alone. As human beings, we learn the skills or articulacy, numeracy and literacy at school, but rarely do we learn the medium of visual communication. Cartographers do learn this in a formal sense and learn how to apply it effectively. Most contemporary map-makers do not, so it is no surprise that, visually, many maps fail in some way or other and as Huffman[41] notes, cartography is a very human activity that requires our intellectual input. Balchin and Coleman's ideas about graphicacy have as much (if not more) relevance in the modern mapping landscape than ever before given the plethora of map-makers and the rise of map-making. Learning by example, it seems to us, is a good way to appreciate the importance of design as a component of graphicacy in map-making. This is the focus of part two of this paper.[42]

40. Balchin, W. G. V. and Coleman, A. (1966). 'Graphicacy should be the fourth ace in the pack', Cartographica, 3, pp. 23–28.
41. Huffman, 2011.
42. Demaj and Field, 2012.

Reasserting Design Relevance in Cartography: Some Examples

DAMIEN DEMAJ AND KENNETH FIELD

Originally published in *The Cartographic Journal* (2012) 49, pp. 77–93.

While concepts and theories about design underpin the work of the professional cartographer, it is unrealistic for most map-makers to be cognisant of the plethora of techniques available to support excellence in design. In the second part of our paper to reassert the relevance of design in cartography, we present the results of a survey of cartographic experts drawn from the academic and professional world. The survey asked participants to present their top ten most expertly designed maps from throughout history. Here, we share the most frequently cited maps that emerged from that survey and describe why they exhibit design excellence. By showing map-makers a range of high-quality cartographic work, we aim to provide exemplars that demonstrate how design affects a map and expertly marries form with function. The techniques on display are well executed and create products that are both well suited to their purpose and have an aesthetic quality that invites people to take notice. They are all, in their own ways, beautiful examples of the art of design in cartography. We have deliberately avoided a 'top ten' approach and, instead, offer three examples in a range of map categories. The examples are neither definitive, nor exhaustive and should act as a starting point to explore design in cartography from those who have managed to set the bar high.

INTRODUCTION

The first part of this paper outlined the detailed and rich conceptual setting that has, for decades, underpinned cartography and helped shape what it is that we do and how we strive to achieve excellence in design.[1] Cartographers are aware of a multitude of design related debates and have learnt to adopt what works and reject what does not often through innovation but with a focus to communicate something specific to a defined audience. Cartographic language, involving grammar and syntax, enables us to be able to judge when one symbol will work more effectively than another, or where a particular colour has associations that might hinder the message in a map. In short, it is what makes us able to speak graphically to reduce the ambiguity in our maps.

---

1. Field, K. and Demaj, D. (2012). 'Reasserting design relevance in cartography: some concepts', The Cartographic Journal, 49, pp. 70–76.

Rather than encouraging map-makers to get to grips with all of this conceptual stuff (though some is always worth learning!), let's accept that it is the domain of the cartographic profession and for many, they just want to make a map. Understanding concepts is part of what describes professional cartographers as not simply makers of maps. So how do we bridge the gap? There exists a need for cartographers to encourage, cajole and educate the new wave of map-makers to infuse their work with something of what we know to avoid the design vacuum and lack of humanity that Huffman[2] has identified. One way of doing this is to provide examples that demonstrate excellence in design. Our intention here is not to search for, and disassemble, why a map works in minute detail but to provide a comment of some unique, innovative aspect that simply makes the map work. In essence, the process of design is about problem-solving and continually asking what works, what does not work, what are the alternatives and would they work any better. If we manage to answer the small questions during the design process, then the bigger picture will emerge, richer and more harmonious. This is where the maps we have selected here have succeeded. They each have a particular purpose and the map-maker (they were not all made by cartographers!) has struck upon the magic formula that harmoniously combines form with function, clarity and an aesthetically pleasing result that encourages readers to look at and interact with the map.

The maps we present here are the summary of a survey of some 20 acknowledged cartographic experts from across the globe who were simply asked to provide their top ten maps that exhibit what they consider to be an excellent design ethos. We have summarized the diverse list to show those that were most often cited and which illustrate a clear user-led design principle in tandem with an innovative, engaging, clear and harmonious result. They are all visually engaging though the examples illustrate a huge variety in approach and depiction showing how wide the palette of cartographic design extends. While some maps are included based on widespread agreement, the fact that over 100 separate maps were suggested shows that an appreciation of good design is inherently subjective and cannot easily be quantified objectively. The maps herein solve a unique problem or display a very specific dataset. Yes, they are (we hope you agree) aesthetically pleasing as objects, but what they do successfully is take a map-making problem and apply it well to create a coherent product that exhibits clarity for a specific purpose and user group. Clearly, we cannot include every map that exhibits good design and we may have excluded some obvious examples, but we believe that the purpose of displaying extracts acts as a set of exemplars which will provide a starting point for debate about design as well as an end-point for map-makers to look at and appreciate good quality design and to learn from such examples.

Ultimately, even poor maps are designed so design is whatever the designer makes of the task at hand. When it comes to artistic endeavour, differences in opinion begin to appear and objectivity is replaced by a splash of subjectivity and what might be perceived as well designed by one person could equally be criticized by another. But our argument here is that by looking at examples of what we consider to be good

2. Huffman, D. (2011). On human cartography, http://somethingaboutmaps.wordpress.com/2011/04/20/on-human-cartography/ (accessed 10 January 2012).

cartographic design we might learn something from what has gone before and endeavour to reflect on them as part of our own design process. This list is a 'highlights of cartography' as put forward by the cartographic profession and not a definitive model to emulate. They serve as ideas and models of success from which inspiration can be drawn.

The maps we have chosen that reflect our survey are by no means definitive or exhaustive. In listing the maps we have specifically avoided the 'top ten' approach (see, for example, Forrest[3]), because it seems invidious to assess the qualities of one map against another when they vary so much from one another. Instead, we present three examples for each of a series of general map types. This allows us to show maps in a broad context and avoids the inevitable issue with a 'top ten' of naming the overall 'best map'. It also allows us to explore similarities in form relative to function or, how different maps take completely different approaches. We have tried to select maps that display a wide range of styles and purposes as well as including both historical and contemporary examples. There are 39 maps representing 13 separate map type categories. One of the fascinating aspects of the survey is that only 9 of the maps most often cited are actually made by someone whose profession and training classifies them as a cartographer. This is interesting in itself since so many of today's maps are criticized on the basis that the author has little understanding of cartographic principles. Well, neither did the authors of 30 of the maps presented here, as chosen by our survey of cartographic experts. This shows clearly that great maps do not need a cartographer at the helm. They require someone with passion, insight and a story to tell. They also require someone with a keen eye for telling the story well using a graphical language. Whether they learnt anything of cartography to inform their own map-making process is hard to identify but their results certainly inform professional cartographers and map-makers through their own drive for excellence in communication design.

Finally, it is impossible to present the maps here at their true size so we include only a thumbnail image in most instances that gives a flavour of the map's appearance. We have included URLs to online resources where they exist and are grateful for the ICA Commission on Map Design for hosting a digital archive of these extracts to encourage people to add their own comments and propose alternatives. See mapdesign.icaci.org for further details. So, in no particular order...

---

3. Forrest, D. (2003). 'The top ten maps of the twentieth century: a personal view', *The Cartographic Journal*, 40, pp. 5–15.

ARTISTIC MAPS

FIG. 1. *Yellowstone National Park* by Heinrich Berann, part of a set published by the US National Parks Service between 1986 and 1995

Heinrich Berann's work is predominantly in the panoramic style of mapping. As a painter, he used his artistic talent to invent a new way of painting landscapes for the purposes of tourist mapping. Berann's work is meticulous in its attention to detail, uses highly saturated colours and a unique curved projection that mimics what might be seen (though exaggerated) from an aeroplane. The foreground of the map is almost planimetric which curves across away from the point of view to a horizon depicting the mountains in profile. The map is immediately pleasing to the eye and creates a unique sense of place that, for tourism mapping, is well suited to the need to attract visitors. Berann also developed a trademark way of rendering cloudscapes which again, represent the sky in a way that is unlikely to be seen in the natural environment. This map of Yellowstone national park was one of four he prepared for the United States Park Service and inspired numerous others to work in a similar style.[4]

FIG. 2. *The Island* by Stephen Walter published by TAG Fine Arts, 2008

The Island is a satirical map which takes the view of London, UK being an isolated island floating somewhere amidst its various commuter towns.[5] It appears independent from the rest of the country, emphasized by the border of Greater London being depicted as a coastline. Walter's map is entirely hand drawn using pictorial sketches and text and instead of the known landmarks you might find on a traditional topographic map, he fills the space with a vast array of local information based on his personal knowledge, feelings and impressions of a place. He details the interesting and mundane and the map becomes a social commentary that invites others to create an emotional bond with the work through a shared lens. Walter uses a large format (101×153 cm) to give himself enough space to contain the intricate detail and builds visual hierarchy in the map through the density of ink. Central London, for instance, contains reverse white type on a black shaded background to emphasize the density of the centre of the city.

4. Troyer, M. (2002). The world of H. C. Berann, http://www.berann.com/ (accessed 18 January 2012).

5. Walter, S. (2008). The Island, http://www.stephenwalter.co.uk/drawings/drawa1.php (accessed 18 January 2012).

FIG. 3. *Angling in Troubled Waters: a Serio-Comic map of Europe* by Fred W Rose published by GW Bacon, 1899

Rose's pictorial illustration depicts the threat posed to British interests by Russian territorial ambitions during the Balkan crisis in late Imperial Europe. The use of maps in this way features heavily in cartoons and other satirical works where familiar shapes of countries are coupled with images of people or events to make a geopolitical point. Rose makes good use of the fishing metaphor to illustrate which countries are fishing and what their catches (colonial possessions) are. It is a highly illustrative and engaging form that draws the eye in to explore the interplay between figures and parts of the map. As a means of stirring debate and controversy, these types of map are a particularly provocative way of capturing the imagination. The use of map shapes and images as a basis for artistic impression is a good approach to communicate such messages since the outlines of countries are familiar shapes and artists such as Rose successfully play on the familiarity to evoke a response.[6]

DEPICTING RELIEF

FIG. 4. *Mount Everest* published by the National Geographic Society, 1988

Mount Everest has been mapped extensively using a plethora of relief representations. Possibly the most frequently cited example of excellence in design is by Eduard Imhof[7] for his impressive use of colour. Here, though, Bradford Washburn used Swissair Survey aerial photos and Space Shuttle infrared photos to plot Mount Everest at 1 : 50 000.[8] Possibly the last example of hand-drawn Swiss relief representation makes clear the most detailed and accurate map ever made of Mount Everest. The digital age has yet to provide ways of matching such exquisite artistry. The peaks, glaciers, rocks and hydrography are particularly clear with scree slopes depicted in astonishing detail. Blue contours sit well in the overall design and take on the appearance of layers of ice. The typography is beautifully set and the map has a soft, photo-realistic feel that adds visual impact. The border separating China and India is so subtle it looks like it is actually painted on the ridgelines. It is a masterpiece of terrain representation showing natural beauty and scientific information in the most vivid possible way.[9]

6. BibliOdyssey. (2009). Satirical maps: an incomplete evolution of the cartoon political map, http://bibliodyssey.blogspot.com/2009/06/satirical-maps.html (accessed 18 January 2012).

7. Imhof, E. (1962). Schweizerischer Mittelschulatlas, 13th ed., Konferenz der kantonalen Erziehungsdirektoren, Zurich.

8. Washburn, B. (1988). 'Mount everest — surveying the third pole', National Geographic, 174, pp. 653–659.

9. National Geographic. (2012). Bradford and Barbara Washburn, http://www.nationalgeographic.com/explorers/bios/washburns/ (accessed 18 January 2012).

FIG. 5. *The Heart of the Grand Canyon* published by the National Geographic Society, 1978

Bradford and Barbara Washburn were mountaineers, explorers and cartographers. During an impressive career, they were strongly supported by the National Geographic Society and many of Bradford Washburn's maps are unrivalled in the realm of mountain cartography (see Figure 4). It took 7 years and numerous skilled individuals to survey and map the Grand Canyon at 1 : 24 000.[10] Washburn's original maps were combined by Lockwood Mapping, cliff-drawing was by Rudi Dauwalder and Alois Flury in Switzerland and relief-shading crafted by Tibor Toth at National Geographic. Browns bear similarity with natural wood and the textures help immerse readers in the landscape. The colour transition of the landscape from the vivid green plateaus to the ocher red canyon arms to the deep brown-grey valleys and turquoise waters creates a stunning contrast. At such a scale, Washburn was able to represent the Canyon in a way never before seen and as a large format poster the map remains a classic National Geographic product for which the Washburns received the Alexander Graham Bell Medal for 'unique and notable contributions to geography and cartography'.[11]

FIG. 6. *Atlantic Ocean Floor* by Heinrich Berann, 1977

In 1967, Berann painted the first in a series of plan oblique physiographic maps of the ocean floor for Marie Tharp and Bruce Heezan and their collaboration culminated in the 1977 World Ocean Floor map. The spectacular 1977 map revolutionized the theories of plate tectonics and continental drift as well as demonstrating the plan oblique technique of relief representation effectively. Berann skilfully combined blue-greys to create a topologicallly accurate, though hugely exaggerated, picture of the ocean floor that leads readers to want to explore. The Mid-Atlantic Ridge and fracture zones appear so life-like with a rippling effect and intricate detail that it draws your eye in and captures your attention. This creates a strong figural component positioned central to the map page that suggests the page (and fracture zone itself) splitting down the middle. The yellow land and deep grey-blue ocean floor provides a strong contrast between land, shoreline and oceans.[12]

10. Garrett, W. E. (1978). 'Grand Canyon', National Geographic, 154, pp. 2–5.
11. National Geographic. (2012). Bradford and Barbara Washburn, http://www.nationalgeographic.com/explorers/bios/washburns/ (accessed 18 January 2012).
12. Troyer, 2002.

GEOLOGICAL MAPPING

FIG. 7. *Alluvial map of the Lower Mississippi Valley* by Harold N. Fisk, 1944

Improving the navigability of the Mississippi River has been ongoing for decades due to the constantly changing morphology. In 1941, the Mississippi River Commission appointed Harold Fisk to undertake a geological survey of the Lower Mississippi Valley. His detailed and exhaustive report contained numerous maps that illustrated the historical courses of the river, colour coded for different ages of point bar migration, chute cut-offs and avulsions. As a collection, they succinctly present the complicated story of channel evolution of the river and are archived by the US Army Corps of Engineers.[13] Rather than attempting to fit all detail on one map, Fisk let the geography drive the size and scale necessary to show detail clearly. The maps exhibit a perfect blend of neutral basemap to provide a context for the coloured detail of the river morphology though almost every colour has a percentage of black to allow it to tone harmoniously with the grey background. The organic historical stream flow patterns make an intriguing visual and despite the fluidity of the mapped phenomena the maps appear very structured.

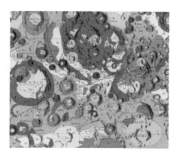

FIG. 8. *Geologic map of the central far side of the Moon* by Desiree E. Stuart-Alexander, 1978

Prepared for the National Aeronautics and Space Administration by US Department of the Interior and US Geological Survey as part of the Geologic Atlas of the Moon, 1 : 5 000 000, this map was the first of its kind. It was compiled from NASA Lunar Orbiter and Apollo photographs and Soviet Zond photographs as well as geochemical and geophysical data obtained from orbiting spacecraft to show the detailed geological character of the Moon in glorious detail. The map illustrates the topography as a technicolour mosaic that is almost Jackson Pollockesque in design. The engaging palette of colours immediately attracts interest in the map which accentuates the strange form of the Lunar landscape. What might appear to be a small design element, the thin black line outlining each feature helps to accentuate the image and delineate one feature from another as distinct forms in contrast to the monotonous appearance of the real landscape. The map is in two versions, one that includes geological notation and grids and a version without.[14,15]

13. Army Corps of Engineers. (2004). Lower and Middle Mississippi Valley Engineering Geology Mapping Program, http://lmvmapping.erdc.usace.army.mil/index.htm (accessed 18 January 2012).

14. Stuart-Alexander, D. E. (1978a). Geologic Map of the Central Far Side of the Moon I-1047, http://www.lpi.usra.edu/resources/mapcatalog/usgs/I1047/150dpi.jpg (accessed 18 January 2012).

15. Stuart-Alexander, D. E. (1978b). Geologic Map of the Central Far Side of the Moon, http://astrogeology.usgs.gov/Projects/PlanetaryMapping/DIGGEOL/moon/1047/lfar.htm (accessed 18 January 2012).

FIG. 9. *Geological Map of London and its Environs* by R. W. Mylne, 1871

First published in 1856 in a period of great change in the understanding of public health and disease in cities, this was an important map in its day. Robert Mylne was a Civil Engineer and Architect and knew that a detailed geological map was essential for informing major public works such as improved water supply and sewerage systems for London. The original version contained only contours to show differences in elevation and though proposals to modernize the sewage system were neglected at the time due to a lack of funds, the Great Stink of 1858 persuaded Parliament of the urgency of the problem. The map informed the design of an extensive underground sewerage system that drained downstream of the centre of population. By 1871, the engraved map had been hand coloured to show the underlying geological structure and informed the construction of deep artesian wells and bore holes to supply the city with clean water. The combination of plan view and cross-section help to tell the story of London's topography and geology.

MAPPING THE Z DIMENSION

FIG. 10. *View and Map of New York City* by Herman Bollmann, 1962

Published for the 1964 New York World's Fair, Bollman's map maintains scale equally throughout by an axonometric projection, a technique developed as early as the 15th Century. Bollmann, a woodcarver and engraver, drew this spectacularly detailed map by hand from 50,000 ground and 17,000 aerial photographs to allow readers to view all parts of the map at the same scale.[16] The map exaggerates widths of streets to create a perfect amount of white space in which buildings sit. The dense fabric of the city is represented at the same time as giving clarity to individual buildings. Vertical exaggeration is used to give a sense of the skyscrapers soaring. The street numbering is consistently placed and beautifully letter-spaced. The rich detail invites closer inspection and the colouring, predominantly in pastel shades (to identify building function) with deep grey roof-tops mimics the grey skyline of Manhattan. Other versions exist such as Constantine Anderson's 1985 map and Tadashi Ishihara's version from 2000.[17] Bollman himself went on to create similar 'Bildkarten' or picture plans of over 60 European cities.

16. Hodgkiss, A. G. (1973). 'The Bildkarten of Hermann Bollmann', Cartographica, 10, pp. 133–145.
17. Codex99. (2011). The streets of New York, http://www.codex99.com/cartography/110.html (accessed 19 January 2012).

FIG. 11. *Ascent from Eskdale in a Pictorial Guide to the Lakeland Fells* by Alfred Wainwright, 1957

Alfred Wainwright is probably best known for his seven Pictorial Guides to the Lakeland Fells.[18] The example here illustrates the hand-drawn style used throughout his 59 walking guides and publications, each one painstakingly detailed using pen and ink based on his own surveys. He drew all the maps, wrote the accompanying text and produced countless illustrations in the hand-lettered publications. The ascent maps are planimetric in the foreground and morph to become perspective in the distance showing natural features and the climb ahead along the route. Contours not only provide useful information but add to the representation of the third dimension.[19] The hand-drawn approach lends itself to giving a sense that the maps are somehow more real and match their in situ use perfectly. The maps are not just landscape sketches though. Planimetric detail is marked and pictorial symbols (e.g. trees) are also used to good effect. As a small format book, the publications and the maps they contain are perfectly suited to their purpose.

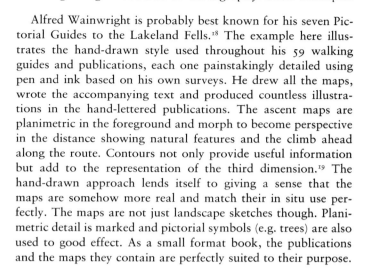

FIG. 12. *Here and there: a horizonless projection in Manhattan* by Jack Shulze and Matt Webb, 2009

Created as a pair of maps of Manhattan, one looking uptown from third and seventh and one looking downtown from third and thirty-fifth, BERG[20] explain how the maps are an exploration of the speculative projections of dense cities. Conceptually, they are an inverse of the approach used by Wainwright (Figure 11) and Berann (Figure 3). The map shows the viewer in the third person standing at the base of the map surrounded by large-scale local detail in perspective and bends to reveal the city stretched out ahead in plain view. It is an intriguing and innovative way of representing an environment that would normally be out of sight and Shulze and Webb cleverly take design cues from Google's map (grey buildings, yellow roads, haloed text) to give the map a sense of familiarity that juxtaposes the unfamiliarity of the view. The gridded streetscape of New York lends itself well to this type of representation, as does the length of the island of Manhattan which creates a tall, narrow, large format poster that emphasizes the approach.

18. Wainwright, A. (1955–1966). The Pictorial Guide to the Lakeland Fells, Vols. 1–7, Westmorland gazette, Kendal.

19. Garland, K. (1991). Lead, kindly light: the design and production of illustrated walkers' guides', Information Design Journal, 7, pp. 47–66.

20. BERG. (2009). Here and there: a horizonless projection inManhattan, http://berglondon.com/projects/hat/ (accessed 19 January 2012).

STREET MAPPING

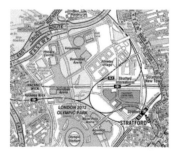

FIG. 13. *Melway Street Atlas*, first published 1966

The first edition of Melway was released in 1966 after 5 years of production and contained 106 original hand drawn maps. Now in its thirty-nineth edition (http://www.melway.com.au/), the map was created in response to shortcomings in available directories at the time. By the 1980s, Melway was the most popular street directory in Melbourne. The maps are designed with a rich and diverse palette of colours, from the blue suburb names to the bright orange secondary streets to the black major roads giving clarity to distinct features. The publication was awarded the International Cartographic Association award in 1982 and the inaugural award for cartographic excellence by the Australian Institute of Cartographers. Street labelling is positioned above the roads, instead of being placed within the road which was against the market trend at the time. Type hierarchy, positioning and colour provided space in which to label a wide range of contextual information. The maps maintain an often imitated 'house style' and it has become so ubiquitous that it's not unusual for people to give a Melway grid reference as directions.

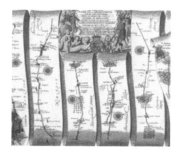

FIG. 14. *Britannia* by John Ogilby, 1675

In 1674, translator and publisher John Ogilby was appointed as His Majesty's Cosmographer and Geographic Printer and published Britannia, a road atlas of Great Britain, in 1675 which set the standard for many years to come. The atlas contained 100 strip maps accompanied by text at a scale of one inch to one mile.[21] The scale was innovative for the time and later adopted by Ordnance Survey in its first map series. Ogilby's maps are a linear cartogram and north varies between strips. People can orient themselves in the direction of travel regardless of the true direction. The scroll effect suggests their use for navigation as if they were to be opened and used on the journey itself. Features are artistically represented, but all have a practical value and a great deal of extraneous detail is omitted. The maps, marginalia and cartouches are particularly ornate and typography also includes flowing ascenders and descenders. Ingenious for its time and a style still used today to show the linearity of route networks (e.g. motorway networks) in many street atlases.

FIG. 15. *Atlas and Guide to London* by AtoZ (the Geographer's Map Company), first published in 1938

The Geographer's Map Company was started in 1936 and is now the largest independent map publishing company in the UK (http://www.az.co.uk/). It still produces the iconic London A–Z street atlas in addition to over 340 other mapping products. Phyllis Pearsall, a painter and writer, founded the company after discovering that the Ordnance Survey map she was following to get to a party was not up to the task and she became lost. She conceived the idea of mapping London which involved walking over 3000 miles and over 23,000 streets mapping each as she went. Pearsall proofread, designed and drew the map with the help of a single draftsman. Although a map containing hundreds of combinations of type form: bold, italics, spacing of characters, colour, san serif, reversed type, size, rotation, upper and lower case, the design and placement of the typography is meticulous. The use of orange primary routes, yellow secondary and white local was unique and possibly the inspiration for Google Maps at street level. The pocket book size of the original was a perfect form for navigation and despite the atlas being crammed with detail it is extremely well structured in graphical terms.

GENERAL REFERENCE ATLAS

FIG. 16. *The Times Survey Atlas of the World* by *The Times*, extract from second edition, 1920

The Times Atlas of the World (latterly with the addition of Comprehensive Edition) was first published in 1895 and is currently in its thirteenth edition (published in 2011 http://www.timesatlas.com). Originally containing 117 pages and over 130,000 names, it has grown to a 544 page publication with 125 map plates and over 200,000 indexed names and is marketed as The Greatest Book on Earth. The tenth edition, published in 1999, was the first to be produced entirely using computer cartography, but until that time much of the map drawing was by hand. Unsurpassed global coverage of the world's physical and political features in a single volume, the atlas is both prestigious and authoritative. It has a classic style and traditional appearance and is meticulously presented. Crucially, the intricate design has stood the test of time and the maps are beautifully laid out, easy to read and fascinating. In an age of querying the internet for answers, when questions of authority and accuracy remain, the Times Atlas is unparalleled and remains a reference publication of the highest quality.

FIG. 17. *National Geographic Atlas of the World* by the National Geographic Society, extract from the sixth revised edition, 1992 published by the National Geographic Society

First published in 1963, the National Geographic Atlas of the World is now in its 9th Edition (published in 2010 http://www.nationalgeographic.com/atlas/) and contains 300 maps and nearly as many illustrations. The maps are beautiful and engaging and the overall design has become a hallmark of National Geographic publications. One of the most notable design features involves colourfully marking the country boundaries yet keeping the interior of the geographic areas as black text on a white background. This helps establish a clear figure-ground between mapped features and the labels and harks back to the use of hand-painted tint bands on early historical maps. The atlas is also notable for its use of the Winkel Tripel Projection, a standard since 1998, and for its custom proprietary fonts originally designed in the 1930s and named after staff cartographers of the era (Darley, Bumstead, Riddiford, etc.). The fonts and extensive labelling alone sets the atlas apart from others and gives it an unmistakable style. Touches, such as the curved lettering from point features around a coastline, are also a signature national geographic style.

FIG. 18. *Great World Atlas* by Dorling Kindersley, 2008

Dorling Kindersley first published their Millennium World Atlas in 1999, inspired by success of their Eyewitness Travel Guide series. The latest version (published in 2008 http://www.dk.co.uk/nf/Book/BookDisplay/0,,9781405329859,00.html) contains 528 pages with rich, vibrant cartography, a wide range of cloud-free satellite images, high-quality terrain models and fold-out pages. Dorling Kindersley was one of the first to use satellite imagery linked to maps and as hybrid map/satellite image illustrations. Each page is beautifully presented with many using a unique approach of clipping map areas from their surroundings as opposed to allowing map detail to bleed off the edge of the page. This particular treatment allows the page to be filled with images, facts, illustrations and text which gives supporting information rarely found elsewhere in atlas mapping. Each map features its own legend rather than relying on one in the preliminary pages. Given the amount of content and irregular shapes, the balance, structure and harmony of each page is remarkable and the detail gives readers an opportunity to explore geography as a traveller rather than using the atlas merely as a reference tool.

CARTOGRAMS

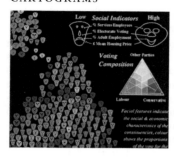

FIG. 19. *The distribution of voting, housing, employment and industrial composition in the 1983 General Election* by Danny Dorling, 1991

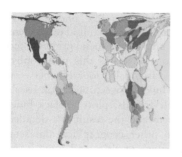

FIG. 20. *Total Population* by Worldmapper.org, 2006

As Tufte[22] said, 'The world is complex, dynamic, multidimensional; the paper is static, flat. How are we to represent the rich visual world of experience and measurement on mere flatland?'. This example of a multivariate cartogram by Dorling[23,24] does just that. The Dorling cartogram creates a social landscape so circles are proportional to the population of the area they represent. Here, though, Chernoff Faces[25] replace circles to ascribe additional information. Sizes of faces are proportional to the electorate and shape, eyes, nose and mouth each display additional socio-economic variables allowing a theoretical maximum of 625 different faces. In reality, only a fraction of these permutations exist, each coloured in one of 36 trivariate colours. The strength of the image is its overall impact as well as the ability to mine detail. Faces evoke emotional reactions and show social differences we can easily interpret. Sharp local divisions or gradual changes emerge. While such glyphs can often overload a map image, Dorling combines them masterfully and the strong colours on a black background create additional contrast and impact.

Gastner and Newman[26] outlined the holy grail of population density equalizing cartograms: a method to account for differences between areas of different population sizes while retaining the general shape of individual regions and their contiguity with neighbouring regions. Their method overcomes the stylized cartograms use of geometric shapes to represent areas and produces useful, elegant and easily readable maps which have been widely adopted, particularly to illustrate political differences or socio-economic conditions. The example here is one of a collection of over 700 maps of the world by Worldmapper.org (http://www.worldmapper.org; Dorling et al.[27]). An ambitious project, the maps paint a vivid picture of the world in a way never before seen from transport to poverty, pollution to religion and every conceivable human, social, economic and environmental condition in between. The maps are deliberately stark which emphasizes the single mapped variable. The colours are vivid against a light background and represent 12 geographical regions for visual comparison across different maps. The beauty of the maps lies in their effectiveness at portraying the discomfort of a wholly unequal world.

22. Tufte, E. (1990). Envisioning Information, Graphic Press, Cheshire, p. 9.

23. Dorling, D. (1991). 'The visualization of spatial social structure', PhD thesis, University of Newcastle Upon Tyne, http://www.sasi.group.shef.ac.uk/thesis/index.html (accessed 19 January 2012).

24. Dorling, D. (1994). 'Cartograms for visualizing human geography', in Visualization in GIS, ed. by Hearnshaw, H. and Unwin, D., pp. 85–102, Wiley, Chichester.

25. Chernoff, H. (1973). 'The use of faces to represent points in kdimensional space graphically', Journal of the American Statistical Association, 68, pp. 361–368.

26. Gastner, M. T. and Newman, M. E. J. (2004). 'Diffusion-based method for producing density equalizing maps', Proceedings of the National Academy of Sciences of the USA, 101, pp. 7499–7504.

27. Dorling, D., Newman, M. E. J. and Barford, A. (2010). The Atlas of the Real World: Mapping the Way we Live, Thames and Hudson, London.

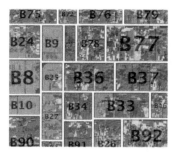

FIG. 21. *Egalitarian Mapping of People in Britain* by giCentre, City University, 2010

ABSTRACT MAPS

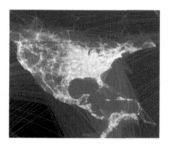

FIG. 22. *Visualizing friendships* by Paul Butler, 2010

Successful and elegant display of large multivariate datasets is rarely achieved because many attempt to fit their data into geographical space. Spatial treemaps provide an alternative way of mapping a large number of geographical units by modifying space.[28,29] Here, Wood et al.[30,31] show 1,526,404 postcode units in Great Britain, sized by population and arranged so that geographical relationships and postcode geography hierarchy are maintained. The map is richly coloured according to a socio-economic classification comprising seven super-groups split into 52 sub-groups. The map is beautifully arranged allowing patterns in the vast amount of information to become clear at local, regional and national scales. In a single map, they have managed to effectively display detailed information about 60 million people recorded in 40 census variables in over one million places. The colour gives the appearance of a stain-glassed window inviting you to explore the information at different distances. Sans serif type is a good choice to tie in with the clean regular lines of the map itself and transparency allows large labels to be placed unobtrusively.

Butler[32] suggests that 'visualizing data is like photography. Instead of starting with a blank canvas, you manipulate the lens to present the data from a certain angle'. His social graph of 500 million facebook users cleverly demonstrates this philosophy. He asks 'what might the locality of friendship look like between users of facebook' and takes the links between facebook user's location and the location of their friends, plots a black-blue-white great circle arc between them and the result is a detailed map of the world. There are no other geographical datasets, yet the shapes of continents, locations of cities and some international boundaries emerge. The map is made entirely out of human relationships. The black background contrasts well with the almost fluorescent lines to create a fibre-optic appearance that lights up the globe. The long distance curves contrast well with the shorter, almost straight, lines of local connections to create an intriguing spider-web pattern. The facebook logo is so widely known that the map needs no other data or contextual information to enable us to make sense of the theme or patterns.

28. Shneiderman, B. (1992). 'Tree visualization with tree-maps: 2-D spacefilling approach', ACM Transactions on Graphics, 11, pp. 92–99.

29. Wood, J. and Dykes, J. (2008). 'Spatially ordered treemaps', IEEE Transactions on Visualization And Computer Graphics, 14, pp. 1348–1355.

30. Wood, J., Slingsby, A. and Dykes, J. (2010). 'Layout and colour transformations for visualising OAC data', in Proceedings of the GIS Research UK 18th Annual Conference GISRUK 2010, ed. by Haklay, M., Morley, J. and Rahemtulla, H., pp. 455–462, UCL, London.

31. Wood, J., Slingsby, A., Dykes, J. and Radburn, R. (2010). Examples of graphics developed by giCentre and used by Leicestershire County Council, http://www.gicentre.org/clg/ (accessed 20 January 2012).

32. Butler, P. (2010). Visualizing friendships, http://www.facebook.com/note.php?note_id5469716398919 (accessed 20 January 2012).

FIG. 23. *London's Kerning* by NB Studio, 2006

Maps based entirely on typography are abstract representations of a landscape and have been used effectively as fills for land use and through repetition for linear networks. Type functions literally as well as to locate mapped features. This example, by NB Studio, was one of the first to gain wide attention. Prepared for The London Design Festival in 2006[33] as a commentary on social space, the large format poster went on to win the design week awards in 2007. The map shows only names of locations, streets or places. Larger fonts reflect more important spaces with smaller fonts representing a less celebrated space. Smaller type is used as a replacement for roads and view at a distance, the structure of the city emerges as the form, orientation and positioning combine to create landmarks and shapes that can be easily identified. The map is a great example of the power of typography in map-making and also illustrates how effective a single colour can be. Maps do not always need to be in colour to be visually stunning or effective.

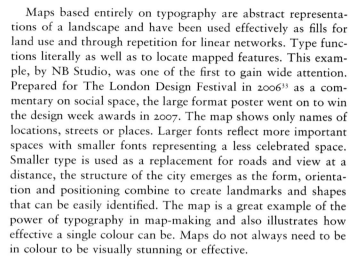

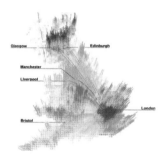

FIG. 24. *Redrawing the Map of Great Britain from a Network of Human Interactions* by Ratti et al., 2010

Regional boundaries defined for the purposes of administration inevitably split a country into arbitrary areas. Ratti et al.[34] questioned whether these boundaries respect the natural ways in which people interact by delineating space using an analysis of the network of over 12 billion individual telephone calls. What they achieved was a beautifully abstract map of Great Britain that shows people interact inside traditional boundaries. The map base is a grid of squares each comprising 3042 pixels. Each pixel is a node and its connection strength to every other pixel is shown by varying opacity. This creates a strong structural framework for the map with connecting lines showing the strongest 80% of links. Colours identify regions that emerge through analysis with the dark colour for London allowing lines connecting it with elsewhere to be clearly seen across lighter regions. The prudent use of labels to identify major cities is just enough to support interpretation and the leader lines tie them well to the map, clearing the map itself of labelling. The tilted viewing angle allows connections to be seen in a way that a plan view would not allow.

33. NB Studio. (2006). ISTD London's Kerning, http://www.nbstudio.co.uk/projects/istd-londons-kerning (accessed 18 January 2012).

34. Ratti, C., Sobolevsky, S., Calabrese, F., Andris, C., Reades, J, Martino, M., Claxton, R. and Strogatz, S. H. (2010). 'Redrawing the map of Great Britain from a network of human interactions', PLoS ONE, 5, p. e14248.

THEMATIC MAPS

FIG. 25. *Maps Descriptive of London Poverty* by Charles Booth, 1898–1889

Charles Booth, an English philanthropist and businessman, is renowned for his survey into life and labour in London at the end of the nineteenth century. Critical of the value of census returns as a way of identifying inequality, he set to work investigating poverty for which he was recognized by awards from the Royal Statistical Society and the Royal Society. The *Maps Descriptive of London Poverty* is an early example of social cartography.[35] Using Stanford's Library 6 inches to 1 mile Map of London and Suburbs, Booth coloured each street to indicate the income and social class of its inhabitants. This choropleth overlay on a light base is an early form of 'mashup' and used a rudimentary diverging colour scheme with black for the lowest class (Vicious, semi-criminal) through dark blues and into reds and yellows (upper-middle and upper classes, wealthy). Neighbouring colours were deliberately similar in hue so the map illustrated social transitions across space though strong gradients are easily seen when blacks and reds are in close proximity. The overlay is slightly transparent to allow the underlying basemap detail to be seen for interpretation.

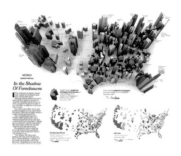

FIG. 26. *In the shadow of foreclosures* by the *New York Times*, 5 April 2008

Very few newspapers have a team dedicated to data visualization and information design. Many (e.g. Data Blog at *The Guardian*) produce data dumps that are the antithesis of design. *The New York Times*, however, creates consistently high quality maps and graphics to tell their stories. Small Labs Inc.[36] provide an excellent repository of over 300 superb examples of their work to date which are a catalogue of best practice in thematic map design. The map of foreclosures[37] displays multiple variables in a striking 3D graphic giving the map the look of buildings on a city landscape. The map labels do not dominate even though they add important contextual statistics. Subtlety is the key here. The fine san serif shape means it sits further in the background and doesn't obstruct the foreclosure shapes. The white US country background provides a neutral landscape for the buildings to emerge from. Simple, thin, solid black lines delineate the state lines. The map page is complemented by two traditional but expertly constructed choropleths and the overall page maintains a clean fresh appearance with excellent visual balance.

35. London School of Economics and Political Science. (2002). Charles Booth Online Archive http://booth.lse.ac.uk/ (accessed 19 January 2012).

36. Small Labs Inc. (2011). New York Times Infographics, http://www.smallmeans.com/new-york-times-infographics/ (accessed 19 January 2012).

37. New York Times. (2008). In the shadow of foreclosures, http://www.nytimes.com/imagepages/2008/04/05/business/20080406_METRICS.html (accessed 19 January 2012).

FIG. 27. *Detail of area around the Broad Street pump* by John Snow, 1854

John Snow was an English physician, the father of modern epidemiology and the inventor of anaesthesia among many successes.[38] His famous map, published later as part of his essay on the mode and distribution of deaths from cholera in Soho, London in 1854,[39] is a classic not only in cartography but also analytically. Snow used the map as an exploratory tool to establish that cholera was a water-borne disease. He was the first to propose cholera's mode of communication in this way and the map helped solve a critical problem of the time. Cartographically, Snow's map is often cited as the first to use separate thematic layers to determine a spatial relationship between variables. The beauty of the map is in its brilliant simplicity, showing only the detail required to make the link between deaths and water distribution. The mapping of deaths using a single symbol identified individuals for impact with multiple deaths being seen as a density due to their clustering. This is a form of hot spot mapping, but without the commonly employed surface of colours representing generalized data we often see today.

SCHEMATIC MAPS/DIAGRAMS

FIG. 28. *London Underground Pocket Railways map* by Harry Beck, 1933

The London Underground map was originally created by engineering draughtsman Harry Beck in 1931 and first published in 1933. Beck drew the diagram in his spare time and although London Underground was originally skeptical of his radical proposal, it became immediately popular and imitated across the globe for countless other metro and mass transit systems.[40] Beck realized that the physical locations of the stations was largely unimportant because the railway ran mostly underground so a schematic diagram was a more effective solution by which to navigate. The simplified map consisted of stations, colour-coded straight line segments which run vertically, horizontally or at 45° diagonals. Ordinary stations (marked as ticks) were differentiated from interchanges (diamonds, later to be replaced by circles). The map exaggerates the detailed central area and contracts the external areas. The map shows no relationship to above ground geography other than the River Thames. The same approach is still in use today by Transport for London,[41] though the map has gone through countless revisions and design changes the core characteristics remain.

38. Vinten-Johansen, P. (2006). The John Snow Archive and Research Companion, http://johnsnow. matrix.msu.edu (accessed 19 January 2012).

39. Snow, J. (1855). On the Mode of Communication of Cholera. London, Churchill.

40. Garland, K. (1994). Mr Beck's UndergroundMap, Capital Transport Publishing, London.

41. Transport for London. (2012). Maps, http://www.tfl.gov.uk/gettingaround/1106.aspx (accessed 19 January 2012).

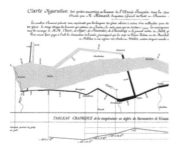

FIG. 29. *Carte figurative des pertes successives en hommes de l'Armée Française dans la campagne de Russie 1812– 1813* by Charles Minard, 1869

The Minard map was published in 1869 on the subject of Napoleon's disastrous Russian campaign of 1812. French civil engineer Charles Joseph Minard's maps and graphics are often cited among the best in information design. The schematic flow approach of this statistical graphic, which is also a map, displays six separate variables in a single twodimensional image: geography, time, temperature, the course and direction of the army's movement, and the number of troops remaining. The map has been described as one of the most complete statistical maps of all time.[42] It is superbly clear in its representation using only two colours with black emphasizing the death toll on the retreat. The horrific human cost is the story of the map as Napoleon entered Russia with 442,000 men and returned with barely 10,000 which included some 6000 that had rejoined from the north, symbolized as 1 mm of line width to 10,000 men. The geography of the battles is marked and the temperature is shown across the bottom supports interpretation. Napoleon underestimated the vastness of Russia and the inhospitable winters. Minard exquisitely mapped the total disaster.

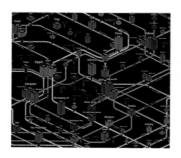

FIG. 30. *Web trend map 4* by iA Inc., 2009

This fascinating schematic maps the most influential internet domains and people onto the Tokyo Metro map. Each domain is assigned to individual stations on the Tokyo Metro map in ways that complement the characters of each.[43] Complementary websites are grouped to a line that suits it and the map produces inter-linkages among companies in multiple ways inspiring some intriguing interplay. For instance, Twitter is assigned the station with the biggest 'buzz' and Google is mapped at Shinjuku which is the world's busiest station. Web domains can be evaluated based on their position (proximity to a main line or hub representing importance), height (success measured in traffic, revenue and media attention) and width (stability as a business entity). The axonometric gridlines define street level with all subways positioned below the gridded surface. The Trend Setters are labelled using speech bubbles as if they are saying their name and other labelling works well to support differentiation of major trends. The colour palette is extraordinarily vivid and works well against pure black. Why map the internet in this way? As iA say themselves: because it works.

42. Tufte, E. (2001). The Visual Display of Quantitative Information, 2nd ed., Graphics Press, Cheshire.
43. Reichenstein, O. (2009). Web Trend Map 4, http://www.informationarchitects.jp/en/web-trend-map-4-final-beta/ (accessed 19 January 2012).

MAPS IN MASS MEDIA

FIG. 31. *Escape the map* by
Mercedes Benz, 2011

Escape the map is an online advertisement for Mercedes-Benz (http://www.escapethemap.co.uk/). It offers an immersive, interactive experience for the viewer who becomes a participant in the story. The work has drama, does not overtly force the brand and follows a 'choose your own adventure' type plot. It is sophisticated, memorable and unique. The map and map-related objects like the falling map pins and address (locator) are key metaphors in the story. The advert immerses you in the future, how we may use maps in cars and how important location is and will become not just for navigation but as a defining way of living. The heads up display on the electronic paper map works particularly well. The work clearly illustrates how popular maps have become in the mass media and there are numerous references to familiar online web map services, virtual globes and their techniques (such as facial blurring on Google Street View) as well as having a sub-plot that references social media such as Twitter. This familiarity with ubiquitous mapping and social media tools means that the advertisement hooks us into a familiar world to tell its futuristic story.

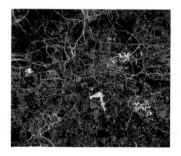

FIG. 32. *OpenStreetMap:
a year of edits* by Ito!, 2008

Ito! work with large transport datasets and have been pioneers in the creation of imaginative animations showing temporal activity in transport data (http://www.itoworld.com/static/data_visualisations.html). Their stunning visuals have been used for such diverse projects as displaying road traffic count data and the impact of the Icelandic ash cloud on air traffic in 2010. The first animation they produced displayed, in animated form, the entire year's worth of edits to the OpenStreetMap database. It paints a fascinating picture as the OpenStreetMap movement captures and publishes new features from single roads to entire countries. The simple temporal legend gives a sense of the speed at which data was acquired and reflects the impact and speed of OpenStreetMap itself in this era of democratized mapping and citizen science. The animation moves rapidly to reflect the equally rapid map changes to create a flickering movie. New edits light up the map across the black backdrop and then disappear as they are added to the core of the map suggesting database integration, an excellent way of showing the quantity and rapidity of change for such a large and important dataset.

FIG. 33. *The Wilderness Downtown* by Chris Milk, 2010

The Wilderness Downtown is an interactive music video/ movie featuring *We used to wait* by Arcade Fire and built as a way of experimenting with Google Chrome's interface (http:// thewildernessdowntown.com/). Constructed in collaboration with Google, it is without doubt one of the most innovative ways in which maps feature in a short film. Many of the components of the movie are based on the Google Maps suite; the viewer types in an address before launching the movie which then launches multiple inter-related Chrome windows to create a map-based journey through your neighbourhood that accompanies the music. The main hooded and mysterious character is seen running through the neighbourhood the viewer specified, across satellite images and amidst rotating Street Views. Animated components (flying birds, exploding trees) add to each window to use the maps in innovative ways as you become the eyes of the runner. The concept of allowing people to fly through their own neighbourhoods and stop and look at their own house as part of a music video is truly unique.

WEB MAPS

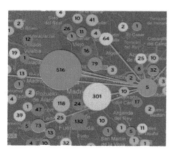

FIG. 34. *Madrizd Map of Knowledge* by Madrizd, 2010

The Madrizd Map of Knowledge (http://www.madrimasd.org/ mapa-conocimiento/) allows people to locate strategic information on the activities of companies and networks that form the Community of Madrid. The map acts as a visual network so shared interests, ideas, publications and projects can be seen. It is a striking design with highly saturated colours which become immediately engaging the moment you roll your mouse over the map symbology. The symbols move and reshape/size and interlinkages appear automatically. The application uses sound to emphasis different user interactions, augmenting visuals in interesting and useful ways. Users can switch between a wide range of basemap styles for different purposes. The satellite image has been reclassified to reflect the look and feel of the overlaying symbol and user interface. The floating symbols integrate well with the basemaps and morph at different scales to sizes that better reflect the position of the businesses as the scale increases. A navigable locator map shows you where you at all times and map controls are unobtrusive. The application is in Spanish, yet non-native speakers are able to use it and have a pleasant and engaging experience.

FIG. 35. *Google Maps* by Google, 2005

Lars and Jens Rasmussen's mapping company was acquired by Google in 2004 and the mapping landscape transformed on 8 February 2005 when Google Maps was released as a web-based product (maps.google.com). Google Maps (and numerous complimentary products) has revolutionized the way the people view, use and make maps and how they interact with their surroundings. The Google Maps API underpinned the democratization of online mapping to allow anyone to create geographically contextualized mashups. The design is recognizable and supports a strong, clear brand that is consistent at a local scale, globally. Integration of complementary functionality (e.g. routing, traffic information, overlay of social media and photographs, zooming, panning, querying and measuring) provides an application with a multitude of purposes that goes beyond a general reference map. The design is automatically modified depending on its use. For instance, secondary roads widen at particular scales when you overlay traffic information to show each direction of traffic flow. The appearance of 3D buildings and moving shadows at large scales (in some cities) represent the built environment like never seen before. Quite simply the map is, and continues to be revolutionary.

FIG. 36. *Immigration Explorer* by Matthew Bloch and Robert Gebeloff, the *New York Times*, 10 March 2009

Examples of good web maps are rare but this map by the *New York Times* is an exception.[44] Beginning with a clear story, user controls allow effortless mining of layers of information across an unobtrusive colour palette with sensibly deployed pop-ups. The temporal dimension can be explored and the switching between choropleth and proportional symbol maps perfectly marries the map type to the data. Users can select variables and modify their depiction such as changing the relative size of symbols to make the best use of screen size. Using terms like 'bubbles', rather than 'proportional symbols', simplifies terminology so it makes sense for the average reader. A simple, neutral basemap supports the overlay detail and zooming is enabled but not beyond levels not supported by the data. Symbology is transparent, allowing overlapping symbols and basemap detail to remain visible. Fine white outlines around only those symbols that overlap, subtle haloes and abbreviated labels at small scales which switch to full names at large scale are examples of a high attention to detail. This is a well-crafted web map that perfectly blends form with function using the medium appropriately.

44. Bloch, M. and Gebeloff, R. (2009). Immigration Explorer, New York Times, http://www.nytimes.com/interactive/2009/03/10/us/20090310-immigration-explorer.html (accessed 26 January 2012).

TOPOGRAPHIC MAPS

FIG. 37. *1 : 100 000 Topographic Map of Switzerland* (extract Beromunster sheet 65-3950-32) by Swisstopo, 2002

Swiss cartography is renowned for its accuracy, quality and artistry and no collection of the best topographic maps could ignore them. Swisstopo, the official name for the Swiss Federal Office of Topography (http://www.swisstopo.ch), is responsible for the production of topographic maps at a range of scales and while 1 : 25 000 is their most detailed map, this extract of the 1 : 100 000 series represents a range of excellent design principles. The use of colour in particular to vary label meaning, show quantities, represent or imitate reality and to decorate visually enlivens the map.[45] The Swiss style is well structured, maintains uniformity, uses white space effectively, contains beautiful typography and unrivalled depiction of relief on topographic reference maps. The typeface sets a classic tone using primarily 'antique' looking serifs that includes a unique combination of thick and thin strokes. Hill shading is in the classic Swiss style based on the work of Eduard Imhof.[46] The maps are rich in content and deliver complex information in a succinct, well-organized manner. Swisstopo topographic maps are truly works of art.

FIG. 38. *1 : 24 000 Quadrangle series* (extract from Stow, VT 44072D6) by USGS, 1968

The United States Geological Survey largest (in terms of scale and quantity) and best-known map series is the 7.5-minute or 1 : 24 000 quadrangle series. The scale is unique in national mapping being related to the measurement of 1 inch to 2000 feet. Each of nearly 57,000 maps is bounded by two lines of latitude and longitude covering 64 square miles in southern latitudes but, due to convergence of meridians, only 49 square miles in northern latitudes. The specification has been applied to many other geographies that the US mapped during military operations which demonstrates a high level of flexibility and versatility in the design. As a brand, the series is instantly recognizable and successful. The content serves both civilian and military purposes and supports varied usage. Marginalia is well structured and complex information delivered in a succinct, well organized manner. The series was officially completed in 1992 and while The National Map (http://nationalmap.gov/ustopo/) represents a new generation of digital products, the impact of the originals persists with new maps arranged in the 7.5-minute quadrangle format as well as retaining the same look and feel.

45. Tufte, 1990.
46. Imhof, E. (1982). Cartographic Relief Representation, Walter de Gruyter, Berlin, republished by Esri Press, 2007.

FIG. 39. *MasterMap®* by
Ordnance Survey, 2001

Not so much a map as a digital product that records every single fixed feature of Great Britain in a contiguous database, MasterMap® represents the most detailed, consistent and up-to-date geographical database of any country at a scale of 1 : 1250. Four separate layers contain topographic, transport, address and imagery data to form the full product. Every feature is assigned a Topographical Identifier that gives it a unique reference as well as attribute information to classify it and support mapping tasks. Continuous review means that the database is as current as the latest ground survey data capture and the product is versatile and flexible enough to suit a myriad of mapping purposes at different scales. The schema is robust and currently the database contains over 460 million individual features with extensive metadata. As a product, the release of MasterMap was, and remains, innovative and its scale and level of detail are unsurpassed. The uniqueness of its design is in the construction of a database that supports the mapping needs of a diverse user base.

FIG. 39. *MasterMap®* by Ordnance Survey, 2001

SUMMARY

We have presented our rationale for reasserting the relevance of design in map-making[47] and as an alternative to criticizing map-makers who are not classically trained in cartography for the plethora of errors, gaffes and sub-par approaches, we, instead, have sought to encourage them to look at examples of good design for inspiration. Of course, encouraging good design is not limited to amateur map-makers for even professionals should continually explore examples of design excellence and learn something new. By studying examples of quality, we encourage map-makers to explore ways of reverse engineering techniques to learn how to apply them in ways that support their own work.

The 39 maps presented here are by no means definitive or exhaustive but represent the broad consensus of 20 cartographers acknowledged in their field. They illuminate the concepts discussed in Field and Demaj[48] and show how they can be expertly applied. Of course, the list can never be complete and in compiling the examples presented, we have had to reject many which might have equal merit. And what of our own favourites? The work of Beck, Minard, Berann, Bollmann and National Geographic would rise to the top of the cartographic pile at the time of writing though frankly, tomorrow, others might be preferred. As with music or art, where we may express a particular favourite song or painting at one time or another, so it is the same with maps and that has been a fascinating aspect of this exercise. Debates between the authors and the cartographic experts we surveyed have been interesting and stimulating and served to prove that excellence in cartographic design remains at least in part, in the eye of the beholder.

47. Field and Demaj, 2012.
48. Field and Demaj, 2012.

DISCLAIMER

The thoughts and ideas expressed in this paper do not necessarily represent the positions, strategies or opinions of Esri.

ACKNOWLEDGEMENTS

The authors wish to thank the cartographic experts who took part in our survey and offered their top 10 well-designed maps. It was a hard job to whittle down the list from over 100 examples cited so apologies if we have missed out one of your favourites!

Small extracts of the maps presented in this paper have been used to display excellence in design. We have sought to contact all copyright holders to gain permission and we thank them for allowing us to reproduce the extracts here. We are also grateful to the International Cartographic Association Commission on Map Design for hosting online extracts of the maps presented here and for providing a mechanism for readers to discuss and debate the relative merits of the maps included, and to propose alternatives. Please visit mapdesign.icaci.org for more details.

Reflections of 'Reasserting Design Relevance in Cartography'

WILLIAM CARTWRIGHT AM

RMIT University

This essay reflects upon the two-part paper's significance in advancing the debate on 'good design', in a cartographic context. The papers are both informing and insightful. As well, I was pleased to see that they referred to a paper by Karssen from 1980 that covered 'Beauty in Maps'. I used this paper as a reference when preparing lectures in my early days as an academic. I think that Karssen's ideas are still applicable today.

Here I reflect on Field and Demaj's papers under four general headings:

- The Wall;
- The Drawing;
- The two 'Fs'; and
- The Conflict.

THE WALL

This section of the essay does not refer to the double album of 1978 by Pink Floyd (...*All in all it was just a brick in the wall. All in all it was all just bricks in the wall* ...) or the physical division between West and East Berlin, but the wall in my office. On it is not an output from *Google* or *Bing* (or any other computer-generated map product), but framed maps of the London Underground (Beck and pre-Beck), *The Tate Gallery by Tube* poster by David Booth, a world map incorporated into a 1950s insurance company calendar (with all of the British Commonwealth in bright red), a poster from the 1989 British Library exhibition *What Use is a map?* and a reproduction of a traditional Japanese landscape drawing. I like them (obviously); I like this eclectic collection of map-related artefacts that give me pleasure each time I look at them. And I like them for their aesthetics and their 'look-and-feel'.

I think that I'm not the only cartographer that has an eclectic collection of visual displays at home or work. We appreciate the impact of the art, the (hidden) science behind the image and the (acknowledged) technology behind the reproduction.

We take this all in one visual 'swoop' of the artefact. We just know that it 'works', and it works from our first viewing, and subsequent appreciation of the object.

THE DRAWING

We like to draw maps. The act of drawing links our mental map to the physical map. Drawing is the glue that holds-together our cognitive and constructed images of geography (Cartwright, 2012).

Goode and Darling (1904, p. 2), have said that we draw maps because:

(1) *geography deals primarily with space-relation, and all record comes sooner or later to a distribution in a map;*

(2) *'the place for the atlas is in the head', i.e., we need definite mental pictures of the map, at least of the fundamentals of space-relation;*

(3) *the study of the essentials, and the exercise of the muscles in reproducing the form, train the attention and fix a clear mental image.*

Our map drawings 'work' as an art form, as a technological outcome and as a 'communicatible' artefact. And, they must provide a 'balanced' and multi-faceted view of geography. In undertaking the drawing, we are continually considering the interplay between the cognitive and the physical map; how to make the physical map do our bidding and how best to modify our cognitive map to accord to the realities of production, reproduction and communication.

THE TWO 'FS'

The design of any efficient product depends upon addressing both Form and Function. Addressing just one of these design principles is fraught with danger. Form refers to the appearance and arrangement of visual and functional components. Form can be considered from two perspectives: Function — good visual design improves function; and Aesthetics — contributes to the visual appeal of a product. For maps, Form can refer to the optimization of colour, image quality, graphics, typography, visual organization and the visual design of information structures. For electronic maps, navigation and interaction tools that maximize the function, its aesthetic appeal needs to be added.

Function refers to the way in which a product operates and the way in which a product serves its purpose. Function is provided via appropriate information design and layout. For electronic maps, navigation design, interaction design and the development of useful interaction tools (for example, search tools and map interaction tools) must be included.

Form and Function are inseparable. One cannot work without the other in a successful design. For example, atlas design provides Function by enabling users to logically move throughout a publication. But also the Form of the overall look of each page and section — placement, colour, graphics and typography — determines whether the underlying structure of a complex publication are noticed, whether they are legible and whether their functionality is recognized.

In electronic map product design, Form should follow Function, meaning that design must support function. For example, the focus on content and interaction is addressed first then decisions are made relating to colour, fonts and graphics. But, for paper products, especially those that 'lean' towards the artistic, rather than scientific approaches to design, Function usually follows Form.

The actual design process is so complex and so closely linked with taste and aesthetics that choosing just one is impossible.

THE CONFLICT

Field and Demaj's papers identified a number of conflicts that map designers have to face, if they are to properly undertake the design of a map. These conflicts have perhaps always been what cartographers must work-through if the best design possible is sought.

I saw these conflicts identified by the authors:

- design versus production;
- design versus aesthetics;
- convention versus innovation;
- design 'satisfaction' versus usability;
- subjectivity versus objectivity;
- focused versus generalist;
- machine versus human (see also Cartwright, 2014 for my take on this topic); and
- science versus art versus technology.

In 1980, *The Cartographic Journal* published a paper by Aart J. Karssen entitled 'The Artistic Elements in Map Design'. Field and Demaj obviously see this as a key treatment on the topic of map design with which I strongly concur. Karssen saw:

- beauty in maps;
- beauty and generalization;
- beauty and symbolization; and
- beauty and colour.

He proposed that 'beauty' could be addressed in five map production stages. And, depending upon whether or not the subjective elements of map design were considered, the result of a map design process could result in just a 'map', or a 'beautiful map'. Karssen's diagram illustrating his five map production stages is reproduced in Figure 1.

Karssen said that subjective and objective elements could be designed for 'beauty'. He proposed the testing of both the subjective and objective elements of a design. Subjective elements he called 'proportions' — harmony, composition and clarity (aesthetic ratio). Objective elements that could be tested were seen to be location, accuracy of data/data sets and the appropriateness/accuracy of thematic information.

Field and Demaj have succinctly identified the conflicts that face contemporary cartographers when undertaking map design and brought them to the fore for a contemporary audience. As noted, they identify the need to balance the three elements of Art, Science and Technology. This is not a mean feat to accomplish and echoes the problem again highlighted by Karssen over 30 years ago when he noted:

> *Today, however, it is the managers, scientists, and technologists who take the leading role in cartography and those with artistic talents have a low evaluation. Indeed, in the future there will be much greater emphasis on the mechanisation and automation of the technical aspects and the cartographer will be more involved with design. In this case the cartographer must ensure that technology does not push the quality of map representation into second place and thus undermine the beauty of maps.* (Karssen, 1980, p. 127)

When undertaking map design, we strive to produce designs that are clear and aesthetically pleasing. The designs need to be able to be reproduced, communicated, transmitted, visualized and our intentions when undertaking the design understood.

Field and Demaj's papers provide a timely reminder of the importance of design, backed with authoritative examples and explanations. When looking at those maps

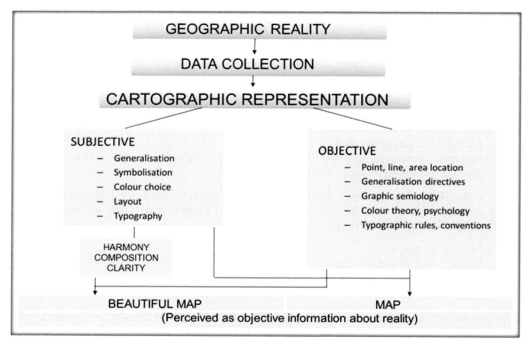

Fig. 1.   Karssen's five phases of map design, re-drawn from Karssen (1980, p. 127)

and also the map-related objects on our walls, we need to continually question how best to provide cartographic artefacts that have aesthetics, which are visually pleasing and 'work' as a conveyor of information about geography. No mean feat!

REFERENCES

Cartwright, W. E. (2012). 'What makes a good drawing', in **What Makes a Good Drawing**, ed. by Farthing, S., University of the Arts, London.

Cartwright, W. E. (2014). 'Man vs. machine: the application of map drawing machines to cartography and compromised views of geography', in **The Drawn Word**, ed. by Farthing, S. and McKenzie, J., Studio International, New York.

Goode, J. P. and Darling, F. W. (1904). 'Geography: the function of map-drawing', **The School Review**, 12, pp. 67–69.

Karssen, A. J. (1980). 'The artistic elements in map design', **The Cartographic Journal**, 17, pp. 124–127.